Statistical Models in S

sink ("file.out")

    intervening output sent to the file

sink()

Also available from Wadsworth & Brooks/Cole

R. Becker, J. Chambers, A. Wilks, *The New S Language: A Programming Environment for Data Analysis and Graphics*

P. Bickel, K. Doksum, J. Hodges, Jr., *A Festschrift for Erich L. Lehmann*

G. Box, *The Collected Works of George E. P. Box, Volumes I and II*, G. Tiao, editor-in-chief

L. Breiman, J. Friedman, R. Olshen, C. Stone, *Classification and Regression Trees*

G. Casella, R. Berger, *Statistical Inference*

J. Chambers, W. S. Cleveland, B. Kleiner, P. Tukey, *Graphical Methods for Data Analysis*

J. Chambers, T. Hastie, *Statistical Models in S*

W. S. Cleveland, M. McGill, *Dynamic Graphics for Statistics*

K. Dehnad, *Quality Control, Robust Design, and the Taguchi Method*

R. Durrett, *Lecture Notes on Particle Systems and Percolation*

R. Durrett, *Probability: Theory and Examples*

F. Graybill, *Matrices with Applications in Statistics, Second Edition*

L. Le Cam, R. Olshen, *Proceedings of the Berkeley Conference in Honor of Jerzy Neyman and Jack Kiefer, Volumes I and II*

E. Lehmann, *Testing Statistical Hypotheses, Second Edition*

E. Lehmann, *Theory of Point Estimation*

P. Lewis, E. Orav, *Simulation Methodology for Statisticians, Operations Analysts, and Engineers*

H. J. Newton, *TIMESLAB*

J. Rawlings, *Applied Regression Analysis*

J. Rice, *Mathematical Statistics and Data Analysis*

J. Romano, A. Siegel, *Counterexamples in Probability and Statistics*

J. Tanur, F. Mosteller, W. Kruskal, E. Lehmann, R. Link, R. Pieters, G. Rising, *Statistics: A Guide to the Unknown, Third Edition*

J. Tukey, *The Collected Works of J. W. Tukey*, W. S. Cleveland, editor-in-chief
  Volume I: *Time Series: 1949-1964*, edited by D. Brillinger
  Volume II: *Time Series: 1965-1984*, edited by D. Brillinger
  Volume III: *Philosophy and Principles of Data Analysis: 1949-1964*, edited by L. Jones
  Volume IV: *Philosophy and Principles of Data Analysis: 1965-1986*, edited by L. Jones
  Volume V: *Graphics 1965-1985*, edited by W. S. Cleveland
  Volume VI: *More Mathematical,1938-1984*, edited by C. Mallows

Statistical Models in S

Edited by
John M. Chambers
Trevor J. Hastie
AT&T Bell Laboratories

Wadsworth & Brooks/Cole Advanced Books & Software
Pacific Grove, California

Wadsworth & Brooks/Cole Advanced Books & Software
A Division of Wadsworth, Inc.

Printed in the United States of America
10 9 8 7 6 5 4 3 2

Library of Congress Cataloging-in-Publication Data
Statistical models in S / edited by John M. Chambers, Trevor J. Hastie.
      p.   cm.
   Includes bibliographical references and index.
   ISBN 0-534-16764-0
     1. Mathematical statistics—Data processing.   2. Linear models (Statistics)
3. S (Computer program language)   I. Chambers, John M., 1941–
II. Hastie, Trevor J., 1953–
QA276.4.S65   1991
519.5'0285'5133—dc20   91-17646
                                           CIP

ISBN 0-534-16764-0 (casebound)
ISBN 0-534-16765-9 (paperbound)

This book was typeset by the authors using a PostScript-based phototypesetter (Linotronic 200P). Figures were generated in PostScript by S and directly incorporated into the typeset document. The text was formatted using the LaTeX document preparation system (Leslie Lamport, Addison-Wesley, 1986).

UNIX is a registered trademark of AT&T in the USA and other countries.
PostScript is a trademark of Adobe Systems Incorporated.
The automobile frequency-of-repair data is copyrighted 1990 by Consumers Union of United States Inc., Yonkers, NY 10703. Reprinted by permission from CONSUMER REPORTS, April 1990.

Sponsoring Editor: *John Kimmel*
Editorial Assistant: *Nancy Miaoulis*
Production Editor: *Kay Mikel*
Manuscript Editor: *Carol Dondrea*
Cover Design: *Vernon T. Boes*
Cover Printing: *Phoenix Color Corporation*
Printing and Binding: *Arcata Graphics/Fairfield*

Preface

Scientific models — simplified descriptions, frequently mathematical — are central to studying natural or human phenomena. Advances in computing over the last quarter-century have vastly increased the scope of models in statistics. Models can be applied to datasets that, in the past, were too large even to analyze, and whole classes of models have arisen using intensive, iterative calculations that would previously have been out of the question. Modern computing can make an even more important contribution by providing a flexible, natural way to express models and to compute with them. Conceptually simple, "standard" operations in fitting and analyzing models should be simple to state. Creating nonstandard computations for special applications or for research should require a modest effort based on natural extensions of the standard software.

This book presents software extending the S language to fit and analyze a variety of statistical models: linear models, analysis of variance, generalized linear models, additive models, local regression, tree-based models, and nonlinear models. Models of all types are organized around a few key concepts:

- *data frame* objects to hold the data;

- *formula* objects to represent the structure of models.

The unity such concepts provide over the whole range of models lets us reuse ideas and much of the software.

Fitted models are objects, created by expressions such as:

```
mymodel <- tree(Reliability ~ ., cars)
```

In this expression, `cars` is a dataset containing the variable `Reliability` and other variables. Calling the function `tree()` says to fit a tree-based model and the formula

```
Reliability ~ .
```

says to fit `Reliability` to all the other variables. The resulting object `mymodel` has all the information about the fit. Giving it to functions such as `plot()` or `summary()` produces descriptions, including various diagnostics. Giving it to `update()`, along with changes to the formula, data, or anything else about the fit produces a new fitted model. The goal is to let the data analyst think about the content of the model, not about the details of the computation.

Many users will want to go on to develop ideas of their own, by using and modifying the underlying software. Making such extensions easy was one of the main goals of our software design and of the book's organization. The functions provided should be a base on which to build to suit your own interests. The material covered here is far from the whole story. We hope to see many new ideas worked out: improvements in efficiency and generality of the existing functions; specialization of the software to applications areas; extensions to new statistical techniques; and different user interfaces building on this software. In writing the book and distributing the software, we hope many of you will become involved in these exciting projects.

Reading the Book

The book is designed to accommodate different interests and needs. Each chapter covers a topic from beginning to a fair degree of depth. If your interests center on one topic, you can read right through that chapter, referring back to other chapters occasionally if you need to. If your interests are more general, you will be better off reading the beginning of several chapters (the first section to get the general ideas, or the first two if you want to do some computing). Skip the later sections of the chapters at first; they are likely to seem a bit heavy.

The book begins with three chapters of general and introductory material, including a first chapter that informally shows off the style by presenting a sizable example. However you plan to read the rest of the book, we strongly recommend reading this chapter first, to make later motivation clear. If you aren't sure whether the book is for you or not, the first chapter should help there also. The heart of the book, Chapters 4 through 10, deals with the statistical models, from linear models to tree-based models. Finally, the material in the appendices gives computational details related to all the previous techniques. In particular, Appendix A presents the computational core of our approach, a new system of object-oriented programming in S.

The chapters on specific kinds of models are organized into four sections, treating the topic of the chapter in successively greater detail. The first section introduces the statistical concepts, the terminology, and the range of techniques we intend to cover in the chapter. The intent is to let readers acquainted with the statistical topic match their understanding to the terminology and context we will be treating in later sections. Reading just the first section of each such chapter will give an overview of the contents of the book. The second section of each chapter introduces the basic S software with examples. Reading the first two sections of a chapter should allow you to start applying the ideas to your own data.

The third and fourth sections of the chapters introduce more advanced use of the software and explain some of the computational and statistical ideas behind the software. One or both of these sections will be recommended if you plan to extend the software or to use it in nonstandard ways, but you should probably wait to read them until you are familiar with the basic ideas.

As ideas from previous chapters come up, some back-referencing may be needed. However, once you have a grasp of the basic ideas about models and data, the individual chapters should be largely self-contained.

For purposes of learning the statistical methods—for example, in a course—this book should be combined with one or more texts treating the kinds of model being discussed. Bibliographic notes at the end of each chapter suggest some possibilities. For purposes of learning about statistical *computing*, the later sections of the chapters introduce numerical and other computational techniques, again with references for further reading.

The Plots

Although graphics is not an explicit topic of this book, good plotting is essential to our approach. We believe that examining the data and the models graphically contributes more than any single technique to using the models well. Skimming through the book anywhere between Chapter 5 and Chapter 9 should suggest the importance of the plots. We emphasize simple graphics expressions; for example,

```
plot(object)
```

should produce something helpful, for all sorts of `object`s. The plots in the book can all be done in S; we show them in PostScript output, but the software is device-independent. Several of the chapters feature new graphical techniques, such as a conditioning plot to show gradual changes in patterns. There are also plots with mouse-based interactive control, including a flexible plotting toolbox for additive models and interactive plotting for tree-based models.

The New Software

The software for statistical models to be described in this book is part of the 1991 version of the S language. The 1991 version is a major revision that incorporates, in addition to the statistical models software itself:

- A mechanism for object-oriented programming, using classes and methods. This new programming style pervades all the modeling software. It makes possible a simpler approach for ordinary computation, with a few generic functions applicable to all the kinds of model. Extensions of the software are easier and cleaner through the use of classes and methods. Appendix A describes the use of classes and methods in S.

- Extensions to the treatment of S objects in databases. There are a number of these extensions, the most important for the modeling software being the ability to attach S objects as databases. This capability allows formulas in models to be interpreted in terms of the variables in a single data object. The database extensions are described in Section 3.3.2.

- A new facility for interactive help. This you should find useful right away. The character "?" invokes interactive help about a particular object or expression:

  ?lm would give you help about the lm() function;

  ?myfit for some object myfit gives help concerning that class of objects;

  ?plot(myfit) gives you information in advance about using the function plot() on myfit.

  Typing ? alone gives help on the on-line help facility itself.

- A new set of debugging tools. These are chiefly new versions of the functions trace() and browser().

- A "split-screen" graphics system that allows flexible arrangements of multiple plots on a single frame.

- A large number of extensions to numerical methods, graphics, functions for statistical distributions, and other areas.

Relevant new features will be described as they arise throughout the book.

Some basic familiarity with simple use of S will be needed for this book, but you should be able to learn what you need either as you read the book or by spending a little while learning S beforehand. S is a large, interactive language for data analysis, graphics, and scientific computing. Other than the material in the present book, S is described in *The New S Language*, (Becker, Chambers, and Wilks, 1988). We will refer to this book by the symbol **S**, usually followed by a page or section number. The first two chapters of **S** will be enough to get you started.

Detailed Documentation

Appendix B contains detailed documentation for a selected subset of the functions, methods and classes of objects. Online documentation is available for these, and for all the other functions discussed in the book, by using the "?" operator.

Obtaining the Software

The S software is licensed by AT&T. Information on ordering S can be obtained by calling 1-800-462-8146. S is available either in source form or in compiled (binary) form. There is one version only of the source, while the binary comes compiled for a particular computer. For most use, we recommend a binary version, with support. Several independent companies provide S in this form; call the phone number above for more information. If you want S in source form, you can order it directly from AT&T.

The software you get must be the 1991 version or later. The version date should be shown when you receive the software, or you can check it before running S, by typing

```
S VERSION
```

The response should be a date. If the command isn't there or the date is earlier than 1991, you won't be able to run the software in this book. Check with the supplier of the software about getting an updated version.

The statistical modeling software is available as a library of S functions, plus some C and FORTRAN code. Depending on the local installation, you may get the statistics material automatically or may need to use a special command. In S, type the expression

```
> library(help=statistics)
```

to get the local documentation about the library.

As you read through the book, we recommend pausing frequently to play with the software, either on your own data or on the examples in the text. The majority of the datasets used in the book have been collected in the S library data:

```
> library(data)
```

will make them available. The figures in the book were produced using a PostScript device driver in S; the same S commands will produce plots on your own graphics device, though device details may cause some of them to look different.

Acknowledgements

This book represents the results of research in both the computational and statistical aspects of modeling data. Ten authors have been involved in writing the book. All are in the statistics research departments at AT&T Bell Laboratories, with the exceptions of Douglas Bates of the University of Wisconsin, Madison, and Richard Heiberger of Temple University, Philadelphia. The project has been exciting and challenging.

The authors have greatly benefited from the experience and suggestions of the users of preliminary versions of this material. All of our colleagues in the statistics research departments at AT&T Bell Laboratories have been helpful and remarkably patient. The various beta test sites for S software, both inside and outside

AT&T, have provided essential assistance in uncovering and fixing bugs, as well as in suggesting general improvements.

Special thanks are due to Rick Becker and Allan Wilks for their detailed review of both the text and the underlying S functions. Comments from many other readers and users have helped greatly: special mention should be made of Pat Burns, Bill Dunlap, Abbe Herzig, Diane Lambert, David Lubinsky, Ritei Shibata, Terry Therneau, Rob Tibshirani, Scott Vander Wiel, and Alan Zaslavsky. In addition to the authors, several people made valuable contributions to the software: Marilyn Becker for the analysis of variance and the tree-based models; David James for the multifigure graphics; Mike Riley for the algorithms underlying the tree-based models; and Irma Terpening for the local regression models. Lorinda Cherry and Rich Dreschler provided valuable software and advice in the production of the book.

Thanks also to those who helped supply the data used in the examples. For the wave-soldering experiments, we are indebted to Don Rudy and his AT&T colleagues for the data, and to Anne Freeny and Diane Lambert for help in organizing the data and for the models used. Thanks to Consumers Union for permission to use the automobile data published in the April, 1990 issue of *Consumer Reports*. Thanks to James Watson of AT&T for providing the long-distance marketing data and to Colin Mallows for the table tennis data.

It has been a pleasure to work with the editorial staff at Wadsworth/Brooks Cole on the preparation of the book; special thanks to Carol Dondrea, John Kimmel, and Kay Mikel for their efforts.

JMC & TJH

Contents

Chapter 1

An Appetizer

John M. Chambers
Trevor J. Hastie

This book is about *data* and statistical *models* that try to explain data. It is an enormous topic, and we will discuss many aspects of it. Before getting down to details, however, we present an appetizer to give the flavor of the large meal to come. The rest of this chapter presents an example of models used in the analysis of some data. The data are "real," the analysis provided insight, and the results were relevant to the application. We think the story is interesting. Besides that, it should give you a feeling for the style of the book, for our approach to statistical models, and for how you can use the software we are presenting. Don't be concerned if details are not explained here; all should become clear later on.

1.1   A Manufacturing Experiment

In 1988 an experiment was designed and implemented at one of AT&T's factories to investigate alternatives in the "wave-soldering" procedure for mounting electronic components on printed circuit boards. The experiment varied a number of factors relevant to the engineering of wave-soldering. The response, measured by eye, is a count of the number of visible solder skips for a board soldered under a particular choice of levels for the experimental factors. The S object containing the design, `solder.balance`, consists of 720 measurements of the response `skips` in a balanced subset of all the experimental runs, with the corresponding values for five experimental factors. Here is a sample of 10 runs from the total of 720.

```
> sample.runs <- sample(seq(720),10)
> solder.balance[sample.runs,]
    Opening Solder Mask PadType Panel skips
162       S   Thin A1.5     D6     3     6
 75       S  Thick A1.5     L6     3     1
653       L   Thin   B3     L8     2     4
117       L   Thin A1.5     W9     3     0
 40       M  Thick A1.5     D6     1     0
569       L  Thick   B3     L9     2     0
229       M  Thick   A3     L7     1     0
788       S  Thick   B6     L4     2    30
655       L   Thin   B3     W9     1     0
129       M   Thin A1.5     L4     3     1
```

We can also summarize each of the factors and the response:

```
> summary(solder.balance)
 Opening       Solder        Mask         PadType    Panel          skips
 S:240     Thin :360    A1.5:180    L9    : 72    1:240    Min.   : 0.000
 M:240     Thick:360    A3  :180    W9    : 72    2:240    1st Qu.: 0.000
 L:240                  B3  :180    L8    : 72    3:240    Median : 2.000
                        B6  :180    L7    : 72             Mean   : 4.965
                                    D7    : 72             3rd Qu.: 6.000
                                    L6    : 72             Max.   :48.000
                                    (Other):288
```

The physical and statistical background to these experiments is fascinating, but a bit beyond our scope. The paper by Comizzoli, Landwehr, and Sinclair (1990) gives a readable, general discussion. Here is a brief description of the factors:

   Opening: amount of clearance around the mounting pad;

   Solder: amount of solder;

   Mask: type and thickness of the material used for the solder mask;

   PadType: the geometry and size of the mounting pad; and

   Panel: each board was divided into three panels, with three runs on a board.

Much useful information about the experiment can be seen without any formal modeling, particularly using plots. Figure 1.1, produced by the expression

   plot(solder.balance)

is a graphical summary of the relationship between the response and the factors, showing the mean value of the response at each level of each factor. It is immediately obvious that the factor Opening has a very strong effect on the response: for levels

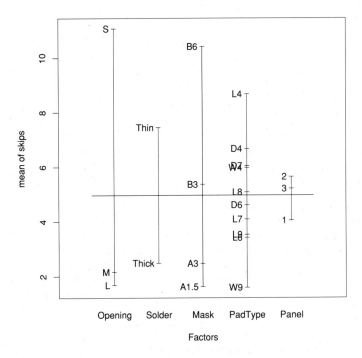

Figure 1.1: *A plot of the mean of* skips *at each of the levels of the factors in the solder experiment. The plot is produced by the expression* plot(solder.balance).

M and L, only about two skips were seen on average, while level S produced about six times as many. If you guessed that the levels stand for small, medium, and large openings, you were right, and the obvious conclusion that the chosen small opening was too small (produced too many skips) was an important result of the experiment.

A more detailed preliminary plot can be obtained using plot.factor(), which produces a separate boxplot for each factor:

```
plot.factor(skips ~ Opening + Mask)
```

We have selected two of the factors for this plot, shown in Figure 1.2, and they both exhibit the same behavior: the variance of the response increases with the mean. The response values are counts, and therefore are likely to exhibit such behavior, since counts are often well described by a Poisson distribution.

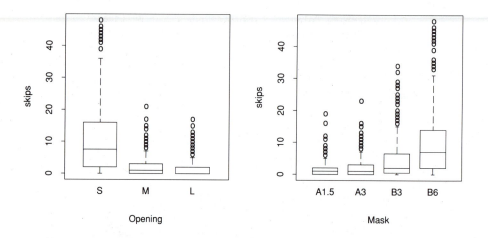

Figure 1.2: *A factor plot gives a separate boxplot of* skips *at each of the levels of the factors in the solder experiment. The left panel shows how the distribution of* skips *varies with the levels of* Opening, *and the right shows similarly how it varies with levels of* mask.

1.2   Models for the Experimental Results

Now let's start the process of modeling the data. We can, and will, represent the Poisson behavior mentioned above. To begin, however, we will use as a response sqrt(skips), since square roots often produce a good approximation to an additive model with normal errors when applied to counts. Since the data form a balanced design, the classical analysis of variance model is attractive. As a first attempt, we fit all the factors, main effects only. This model is described by the formula

        sqrt(skips) ~ .

where the "." saves us writing out all the factor names. We read "~" as "is modeled as"; it separates the response from the predictors. The fit is computed by

```
> fit1 <- aov(sqrt(skips) ~ . , data = solder.balance)
```

The object fit1 represents the fitted model. As with any S object, typing its name invokes a method for printing it:

```
> fit1
Call:
    aov(formula = sqrt(skips) ~ Opening + Solder + Mask + PadType +
```

```
          Panel, data = solder.balance)

Terms:
               Opening Solder   Mask PadType  Panel Residuals
  Sum of Squares 593.97 233.31 359.63  113.44 14.56   493.05
  Deg. of Freedom      2      1      3       9     2      702

Residual standard error: 0.83806
Estimated effects are balanced
```

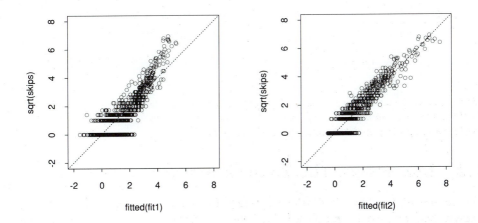

Figure 1.3: *The left panel shows the observed values for the square root of* skips, *plotted against the fitted values from the main-effects model. The dotted line represents a perfect fit. The fit seems poor in the upper range. The right panel is the same plot for the model having main effects and all second-order interactions. The fit appears acceptable.*

Once again, plots give more information. The expression

```
plot(fitted(fit1), sqrt(skips))
```

shown on the left in Figure 1.3, plots the observed skips against the fitted values, both on the square-root scale. The square-root transformation has apparently done a fair job of stabilizing the variance. However, the main-effects model consistently underestimates the large values of skips. With 702 degrees of freedom for residuals, we can afford to try a more extensive model. The formula

```
sqrt(skips) ~ .^2
```

describes a model that includes all main effects and all second-order interactions. We fit this model next:

```
fit2 <- aov(sqrt(skips) ~ .^2, solder.balance)
```

Instead of printing the fitted object, we produce a statistical summary of the model using standard statistical assumptions, in this case an analysis of variance table as shown in Table 1.1, with mean squares and F-statistic values.

```
> summary(fit2)
```

	Df	Sum of Sq	Mean Sq	F Value	Pr(F)
Opening	2	594	297	766	0.00
Solder	1	233	233	602	0.00
Mask	3	360	120	309	0.00
PadType	9	113	13	33	0.00
Panel	2	15	7	19	0.00
Opening:Solder	2	42	21	54	0.00
Opening:Mask	6	89	15	38	0.00
Opening:PadType	18	34	2	5	0.00
Opening:Panel	4	1	0	1	0.66
Solder:Mask	3	20	7	17	0.00
Solder:PadType	9	20	2	6	0.00
Solder:Panel	2	7	4	9	0.00
Mask:PadType	27	28	1	3	0.00
Mask:Panel	6	9	1	4	0.00
PadType:Panel	18	10	1	1	0.14
Residuals	607	235	0		

Table 1.1: *An analysis of variance table for the model* fit2, *including all main effects and second-order interactions. The columns give degrees of freedom, sums of squares, mean squares, F statistics, and their tail probabilities, nearly all zero here because of the very large number of observations.*

The function summary() is *generic*, in that it automatically behaves differently, according to the *class* of its argument. In this case fit2 has class "aov" and so a particular method for summarizing aov objects is automatically used. The earlier use of summary() produced a result appropriate for data.frame objects. The modeling software abounds with generic functions; besides summary(), others include plot(), predict(), print(), and update().

The fitted values are plotted in the right panel of Figure 1.3, and the improvement is clear. Of course, we really expect an improvement; including all the pairwise interactions costs us 95 degrees of freedom! We can see from the table that the F statistic column varies greatly for the second-order terms in the model. The three largest values, interestingly, are the interactions of three of the factors, `Opening`, `Solder`, and `Mask`. So an interesting intermediate model could be formed from just these interactions:

```
sqrt(skips) ~ . + (Opening + Solder + Mask)^2
```

This time we gave three factors, explicitly, for which we wanted interactions:

```
> fit3
Call:
     aov(formula = sqrt(skips) ~ Opening + Solder + Mask + PadType
         + Panel + (Opening + Solder + Mask)^2, data = solder.balance)

Terms:
                   Opening Solder   Mask PadType  Panel Opening:Solder
  Sum of Squares   593.97 233.31 359.63  113.44  14.56          41.62
  Deg. of Freedom       2      1      3       9      2              2

                   Opening:Mask Solder:Mask Residuals
  Sum of Squares          88.66       19.86    342.90
  Deg. of Freedom             6           3       691

Residual standard error: 0.70445
Estimated effects are balanced
```

The left panel of Figure 1.4 shows the observed/fitted values for this second-order submodel, which is comparable to the right plot of Figure 1.3. It uses far fewer degrees of freedom, achieves almost as good a fit, and also accounts for the departures missed by the main-effects model.

1.3 A Second Experiment

The results from the first experiment were valuable in the application, and subsequently a similar experiment was run at another AT&T factory. The results are recorded in the design object `solder2`. In part, the intention was to apply some of the lessons learned in the first experiment. The design was nearly the same as in the first experiment, and we can use `summary()` and `plot()` as before:

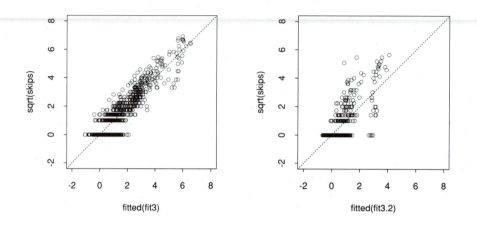

Figure 1.4: *The left panel is a plot of the square root of* skips *from the first AT&T solder experiment against the fitted values for the second-order submodel. The right panel is the same plot using the data from the second solder experiment.*

```
> summary(solder2)
 Opening       Solder        Mask          PadType      Panel          skips
 S:300      Thin :450    A1.5:180    L9      : 90     1:300    Min.    : 0.0
 M:300      Thick:450    A3   :270   W9      : 90     2:300    1st Qu.: 0.0
 L:300                   A6   : 90   L8      : 90     3:300    Median : 0.0
                         B3   :180   L7      : 90              Mean    : 1.2
                         B6   :180   D7      : 90              3rd Qu.: 0.0
                                     L6      : 90              Max.    :32.0
                                     (Other):360              NA's    :150
```

The summaries show some striking differences, especially that there are far fewer skips overall in this experiment. Only 17% of the runs from the second experiment had skips, compared to 66% from the first. Figure 1.5 shows a plot of the design, created by the expression plot(solder2). The plot suggests that in this case, factor Mask appears to have the largest effect. At first it may appear that the two experiments are almost unrelated—a little discouraging for the statistician, although the engineer is likely to be happy that the overall performance is substantially improved. As for modeling, if we start with the last model considered for the first experiment,

```
> fit3.2 <- update(fit3, data = solder2, na = na.omit)
```

and plot its fit in the right panel of Figure 1.4, the model does not appear to fit these data well at all. Notice the use of the na.action= argument in the call to

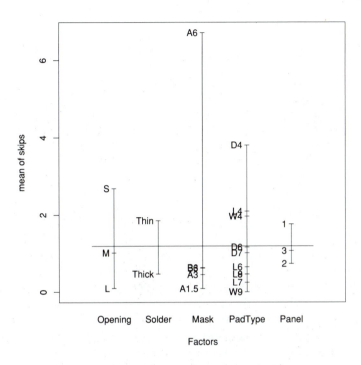

Figure 1.5: *A plot of the mean of* skips *at each of the levels of the factors in the second wave-soldering experiment. Compare with Figure 1.1.*

update(). There are 150 missing values for the variable skips in the solder2 data frame; using na.omit() causes all those cases with missing values for any of the variables specified in the model to be ignored in the model-fitting.

Interestingly, more careful analysis shows that the two experiments are not as unrelated as they initially appear to be. We must keep in mind that we can no longer use the square-root transformation with so many zero responses. More fundamentally, the statistical model should reflect more closely the way engineers would likely view the process. When (as one would certainly prefer) solder skips are a rare event, it is natural to imagine that the solder process has two states: a "perfect" state where no skips will be observed, and an "imperfect" state in which skips may or may not occur. From the view of the application, one is particularly interested in factors that relate to keeping the process in the perfect state.

When this more complicated but more plausible model is worked out in detail, it shows patterns in the second experiment that are largely consistent with those in the first. To see these results in detail, you will have to read on in the book. Some of the ideas, however, we can sketch here as a final appetizer.

Suppose we are only concerned with whether there are *any* skips, as measured by the logical variable

```
skips > 0
```

Although this variable is very different from the quantitative variable `sqrt(skips)` that we have studied so far, models for it can be handled in a very similar style. Specifically, a *generalized linear model* (GLM) using the binomial distribution is a natural way to treat such `TRUE, FALSE` or equivalently `1, 0` variables:

```
fit3.binary <- glm( skips > 0 ~ . + (Opening + Solder + Mask)^2,
    data = solder2, family = binomial)
```

What has changed here? The function `glm()` has replaced `aov()` to do the fitting, the response is now a logical expression, and a new argument

```
family = binomial
```

has been added. As you can imagine, `glm()` fits generalized linear models, and the new argument tells it that the fit should use the binomial family within the GLM models. Otherwise, the specification of model and data remain the same. Also, the object returned can be treated similarly to those we computed before using `aov()`, applying the various generic functions to summarize the model and study how well it works.

Another idea, somewhat complementary to using a binomial model, is to treat the response directly as counts, rather than using the square-root transformation. As we said early on in our discussion, the Poisson distribution is a natural model for counts, and usually works better than the transformation when the typical number of counts is small. The same generalized linear models allow us to model the mean of a Poisson distribution by the structural formula we used earlier. Let's apply this to the data from both experiments to compare the results:

```
> exp1.pois <- glm(skips ~ . + (Opening + Solder + Mask)^2,
    data = solder.balance, family = poisson )
> exp2.pois <- update(exp1.pois, data = solder2 )
```

We display the fits in Figure 1.6, using the square-root scale as before to compare these fits with those in Figure 1.4. The Poisson model appears to be an improvement over those in Figure 1.4, especially for the second experiment; the systematic bias for large counts is gone.

This is still not the end of the story. There are more zero values in the data from the second experiment than the Poisson model predicts. The binomial model

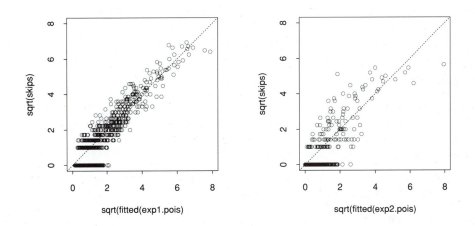

Figure 1.6: *The second-order model of Figure 1.4, treating the response* `skips` *as Poisson and using a log-linear link. The data are plotted on the square-root scale for comparison with Figure 1.4. The plot on the left corresponds to the first experiment, the right the second.*

handles this aspect, but we clearly can't just apply *both* binomial and Poisson models since they imply two incompatible explanations. An answer is to use a *mixture* of the two models, as the idea of perfect and imperfect states for the process suggests. Some runs are in the perfect state, and the binomial model lets us treat the probability of this; others are in the imperfect state, and for those the Poisson model can be applied. This model, called the Zero-Inflated Poisson, cannot be described as a single linear or generalized linear model. A full statistical discussion by the inventor of the technique is in the reference Lambert (1991). One version is described in Section 10.3, as an example of a general nonlinear model.

1.4   Summary

This has been a large plate of appetizers, and we will finish here. All the same, we have touched on only a few of the kinds of models that appear in the book, mainly the analysis of variance and generalized linear models. The book discusses models that fit smooth curves and surfaces, generalized additive models, models that fit tree structures by successive splitting, and models fit by arbitrary nonlinear regression or optimization. We also showed only a small sample of the diagnostic summaries and plots appearing in the rest of the book.

However, the general style to be followed throughout has been illustrated by the examples:

- The structural form of models is defined by simple, general formulas.

- Many kinds of data for use in model-fitting can be organized by data frames and related classes of objects.

- Different kinds of models can be fitted by similar calls, typically specifying the formula and data.

- The objects containing the fits can then be used by generic functions for printing, summaries, plotting, and other computations, including fitting updated models.

- The computations are designed to be very flexible, and users are encouraged to adapt our software to their own needs and interests.

In presenting our appetizer, we did not emphasize the last point heavily, but it is central to the philosophy behind this book. Even though a large number of functions and methods are presented, we intend these to be a starting point for the computations *you* want, rather than some rigid prescription of how to use statistical models.

Chapter 2

Statistical Models

John M. Chambers
Trevor J. Hastie

This is a book about statistical models — how to think about them, specify them, fit them, and analyze them. Statistical models are simplified descriptions of data, usually constructed from some mathematically or numerically defined relationships. Modern data analysis provides an extremely rich choice of modeling techniques; later chapters will introduce many of these, along with S functions and classes of S objects to implement them. All these techniques benefit from some general ideas about data and models that allow us to express what data should be used in the model and what relationships the model postulates among the data. You should read this chapter (at least the first two sections) for a general notion of how models are represented. You can do this either before you start to work with specific kinds of models or after you have experimented a little. Getting some hands-on experience first is probably a good idea—for example, by looking at the first two sections of Chapter 4 on linear models, or by experimenting with whatever kind of model interests you most.

The first two sections of this chapter introduce our way of representing models, and are likely to be all you need for direct use of the software in later chapters. When and if you come to modify our software to suit your own ideas, as we hope many users will do, then you should eventually read further into Sections 2.3 and 2.4.

Throughout the book, we will be expressing statistical models in three parts:

- a *formula* that defines the structural part of the model—that is, what data are being modeled and by what other data, in what form;

- *data* to which the modeling should be applied;

- the stochastic part of the model—that is, how to regard the discrepancy or residuals between the data and the fit.

This chapter and the next concentrate on the first two of these. They discuss how formulas are represented, what objects hold the data, and how the two are brought together. The rest of the book then brings together the three parts in the context of different kinds of models.

Formulas are S expressions that state the structural form of a model in terms of the variables involved. For example, the formula

```
Fuel ~ Power + Weight
```

reads "`Fuel` is modeled as `Power` plus `Weight`." More precisely, it tells us that the response, `Fuel`, is to be represented by an additive model in the two predictors, `Power` and `Weight`. There is no information about what method should be used to fit the model. Formulas of this general style are capable of representing a very wide range of structural model information; for example,

```
100/Mileage ~ poly(Weight, 3) + sqrt(Power)
```

says to fit the derived variable `100/Mileage` to a third-order polynomial in `Weight` plus the square-root of the `Power` variable. Transformations are used directly in the formula, and the basis for the polynomial regression in `Weight` is generated automatically from the formula. Here is a formula to fit separate B-spline regression curves within the two levels of `Power` obtained by cutting `Power` at its midrange:

```
Fuel ~ cut(Power, 2) / bs(Weight,df=5)
```

In the next example, nonparametric smooth curves will be used to model the transformed `Fuel` additively in `Weight` and `Power`, using 5 degrees of freedom for each term:

```
sqrt(Fuel-min(Fuel)) ~ s(Weight, df=5) + s(Power, df=5)
```

The details of these formulas will be explained later in the chapter.

The models above imply the presence of some data on `Fuel`, `Power` and `Weight`; in fact, reasonable models are inspired by data, since models without data are hard to think about. These data actually do exist, and form part of a large collection of data on automobiles described in Chapter 3 and used throughout the book; the present model relates fuel consumption to two vehicle characteristics. Part of the model-building process is collecting and organizing the relevant dataset, and looking at it in many different ways. Some of the useful views are simple, such as summaries and plots. The next chapter is about tools for organizing data into objects that are convenient both for studying the data directly and as input for more sophisticated procedures. For the moment we assume that such data organization has already taken place, and that all the variables referred to in formulas are available.

2.1 Thinking about Models

Models are objects that imitate the properties of other, "real" objects, but in a simpler or more convenient form. We make inferences from the models and apply them to the real objects, for which the same inferences would be impossible or inconvenient. The differences between model and reality, the *residuals*, often are the key to reaching for a deeper understanding and perhaps a better model.

2.1.1 Models and Data

A road map models part of the earth's surface, attempting to imitate the relative position of towns, roads, and other features. We use the map to make inferences about where real features are and how to get to them. Architects use both paper drawings and small-scale physical models to imitate the properties of a building. The appearance and some of the practical characteristics of the actual building can be inferred from the models. Chemists use "wire frame" models of molecules (by either constructing them or displaying them through computer graphics) to imitate theoretical properties of the molecules that, in turn, can be used to predict the behavior of the real objects.

A good model reproduces as accurately as possible the relevant properties of the real object, while being convenient to use. Good road maps draw roads in the correct geographical position, in a representation that suggests to the driver the important curves and intersections. Good road maps must also be easy to read. Any good model must facilitate both *accurate* and *convenient* inferences. A large diorama or physical model of a town could provide more information than a road map, and more accurate information, but since it can be used only by traveling to the site of the model, its practical value is limited. The *cost* of creating or using the model also limits us in some cases, as this example illustrates: building dioramas corresponding to every desirable road map is unlikely to be practical. Finally, a model may be attractive because of *aesthetic* features — because it is in some sense beautiful to its users. Aesthetic appeal may make a model attractive beyond its accuracy and convenience (although these often go along with aesthetic appeal).

Statistical models allow inferences to be made about an object, or activity, or process, by modeling some associated observable data. A model that represents gasoline mileage as a linear function of the weight and engine displacement of various automobiles,

```
Mileage ~ Weight + Disp.
```

is directly modeling some observed data on these three variables. Indirectly, though, it represents our attempt to understand better the physical process of fuel consumption. The accuracy of the model will be measured in terms of its ability to imitate the data, but the relevant accuracy is actually that of inferences made about the real object or process. In most applications the goal is also to use the model to understand or predict beyond the context of the current data. (For these reasons, useful statistical modeling cannot be separated from questions of the design of the experiment, survey, or other data-collection activity that produces the data.) The test data we have on fuel consumption do not cover all the automobiles of interest; perhaps we can use the model to predict mileage for other automobiles.

The convenience of statistical models depends, of course, on the application and on the kinds of inference the users need to make. Generally applied criteria include simplicity; for example, a model is simpler if it requires fewer parameters or explanatory variables. A model that used many variables in addition to weight and displacement would have to pay us back with substantially more accurate predictions, especially if the additional variables were harder to measure.

Less quantifiable but extremely important is that the model should correspond as well as possible to concepts or theories that the user has about the real object, such as physical theories that the user may expect to be applicable to some observed process. Instead of modeling mileage, we could model its inverse, say the fuel consumption in gallons per 100 miles driven:

```
100/Mileage ~ Weight + Disp.
```

This may or may not be a better fit to the data, but most people who have studied physics are likely to feel that fuel consumption is more natural than mileage as a variable to relate linearly to weight.

2.1.2   Creating Statistical Models

Statistical modeling is a multistage process that involves (often repeated use of) the following steps:

- obtaining data suitable for representing the process to be modeled;

- choosing a candidate model that, for the moment, will be used to describe some relation in the data;

- fitting the model, usually by estimating some parameters;

- summarizing the model to see what it says about the data;

- using diagnostics to see in what relevant ways the model *fails* to fit as well as it should.

The summaries and diagnostics can involve tables, verbal descriptions, and graphical displays. These may suggest that the model fails to predict all the relevant properties of the data, or we may want to consider a simpler model that may be nearly as good a fit. In either case, the model will be modified, and the fitting and analysis carried out again.

If we started out with a model for mileage as a linear function of weight and displacement, we would then want to look at some diagnostics to examine how well the model worked. The left panel of Figure 2.1 shows `Mileage` plotted against the values predicted by the model. The model is not doing very well for cars with high

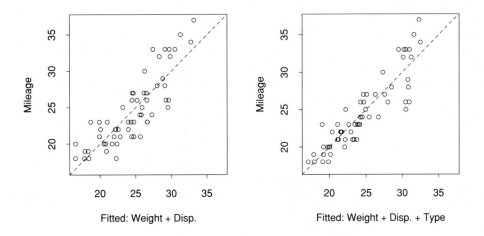

Figure 2.1: *Mileage for 60 automobiles plotted against the values predicted by a linear model in weight and displacement (the left panel) or weight, displacement and type of automobile (the right panel).*

mileage: they all fall above the line. The change to 100/`Mileage` helps some (there is a plot on page 104). If we add in a coefficient for each type of car (compact, large, sporty, van, etc.) the fit improves further. In practice, we would continue to study diagnostics and try alternative models, seeking a better understanding of the underlying process. This model is our most commonly used simple example, and will recur many times, to introduce various techniques.

Research in statistics has led to a wide range of possible models. Later chapters in this book deal with specific classes of models: traditional models such as linear regression; recent innovations, such as models involving nonparametric smooth curves or tree structures; important specializations such as models for designed experiments, and general computational techniques such as minimization, which can be used to fit models not belonging to any of the standard classes. This rich choice of possible models is of real benefit in analyzing data. Whenever we can specify a model that is close to our intuitive understanding or is able to respond to some observed failure of a standard model, chances are we will more easily discover what is really going on. A limited computational or statistical framework that requires us to distort or approximate the model we would like to fit makes such discovery more difficult. It can also hide from us some important information about the data. The methods presented in this book and the functions that implement the methods are designed to give the widest possible scope in creating and examining statistical models.

Of course, all this rich variety will only be helpful if we can use it easily enough. We must be able to carry out the steps in specifying the models without too much effort on our part. The fitting must be accurate and efficient enough to be used in practical problems. There must be appropriate summaries and diagnostics so that we can assess the adequacy of the models. In later chapters, each of these questions will be considered for the various classes of models.

Fortunately, many different classes of models share a substantial common structure. The steps we listed above apply to many models, and important summaries and diagnostics can be shared directly or, at worst, adapted straightforwardly from one class of models to another. The organization of the S computations for the various classes of models is designed to take advantage of this common structure.

This chapter describes a way to express the structural formula for the model. What about the data? For the moment we can assume the data are around in our global environment, and simply refer to variables by name. In Chapter 3 we describe *data frames*, a more systematic way of organizing and providing the data for a model. Depending on the class of models, formulas and data frames may be all we need to specify; for example, if we are using linear least-squares fitting, there is not much more to say in step 3. Other kinds of models may require some further specifications; generalized linear models, for example, require choosing link and variance functions. The choice of the kind of model and the provision of these additional specifications fix the stochastic part of the model to be fit.

2.2   Model Formulas in S

The modeling formula defines the structural form of the model, and is used by the model-fitting functions to carry out the actual fitting. Most readers will already be familiar with conventional modeling formulas, such as those used in textbooks or research papers to describe statistical models, as in (2.1) below. The formulas used in this book have evolved from mathematical formulas as a simpler and in some ways more flexible approach to be used when computing with models.

A formula in S is a symbolic expression. For example,

```
Fuel ~ Weight + Disp.
```

just stands for the structural part of a model. If you evaluate the formula, you will just get the formula. In particular, use of a formula such as the one above does not depend on the values of the named variables; indeed, the variables need not even exist! The expression to the left of the "∼" is the *response*, sometimes called the *dependent* variable. In this case the response is simply the name `Fuel`. The right side is the expression used to fit the response, made up in an additive model of *terms* separated by "+". The variables appearing in the terms are called the `predictors`. Experienced S users are by now probably very curious, so this comment is for their benefit: "∼"() is an S function that does nothing but save the formula as an unevaluated S expression, a *formula* object.

The formula above expresses most of the ingredients of a statistical model of the form

$$Fuel = \alpha + Weight\beta_1 + Disp.\beta_2 + \varepsilon \tag{2.1}$$

For most of the models in this book, the formula does not specifically refer to the parameters β_j in the linear model. These can be inferred and so we save typing them. In a sense, we also avoid mental clutter, in that the names of the parameters are not relevant to the model itself. When we come to general nonlinear models in Chapter 10, however, the formula will have to be completely explicit, since it is no longer additive.

The formula makes no reference to the errors ε either. These, of course, are the stochastic part of the model specification. When formulas are used in a call, say to the linear regression model-fitting function

```
lm(Fuel ~ Weight + Disp.)
```

we complete the rest of the modeling specification; `lm()` assumes the mean of `Fuel` is being modeled by the linear predictor, and uses least squares to compute the fit. Expressions such as the one above were encountered in Chapter 1; in fact all the model-fitting functions take a formula as their first argument, and in most cases the same formula can be used interchangeably among them (hopefully with different consequences!).

The formula above is equivalent to

```
Fuel ~ 1 + Weight + Disp.
```

where the 1 indicates that an intercept α is present in the model. Since we usually want an intercept, it is included by default; on the other hand, we can explicitly exclude an intercept by using -1 in the formula

```
Fuel ~ -1 + Weight + Disp.
```

In using formulas it is important to keep in mind that we are writing a shorthand for the complete model expression. In particular, there is no operation going on that adds `Weight` and `Disp.`; the operator "+" is being used in a special sense, to separate items in a list of terms to be included in the model. The formula expression is, in fact, used to generate such a list, from which the terms and the order in which they appear in the model will be inferred. This inference poses no problem for most models, but with complicated formulas, some care may be needed to understand the model implied. The remainder of this section gives enough information for most uses of model formulas; Section 2.3.1 provides a complete description.

2.2.1   Data of Different Types in Formulas

The terms in a formula are not restricted to names: they can be any S expression that, when evaluated, can be interpreted as a variable. For example, if we wanted to model the logarithm of `Fuel` rather than `Fuel` itself, we could simply use that transformation in the formula

```
log(Fuel) ~  Weight + Disp.
```

A variable may be a *factor*, rather than numeric. A factor is an object that represents values from some specified set of possible *levels*. For example, a factor `Sex` might represent one of two values, `"Male"` or `"Female"`. Readers familiar with S might wonder what happened to the *category*, which is also an object with levels. Factors have all the features of categories, with some added class distinctions; in particular there is a distinction between *factors* and *ordered factors*. Factors can be created in a number of ways, as will be discussed in Section 3.2. For the moment the distinction between factors and categories is not important, and we will simply refer to them as factors.

Factors enter the formula in the same way as numeric variables, but the interpretation of the corresponding term in the model is different. In a linear model, one fits a *set* of coefficients corresponding to a factor. Consider the model

```
Salary ~  Age + Sex
```

where `Salary` and `Age` are numeric vectors and `Sex` is a two-level factor. This is now shorthand for a model of the form

$$Salary_i = \mu + Age_i\beta + \begin{cases} \alpha_M & \text{if } Sex_i \text{ is } Male \\ \alpha_F & \text{if } Sex_i \text{ is } Female \end{cases} + \varepsilon_i \qquad (2.2)$$

where α_F and α_M are two parameters representing the two levels of `Sex`. The coding of factors proceeds from observing that this model is equivalent to one in which the factor is replaced by one "dummy" variable for each level—namely, a numeric variable taking value 1 wherever the factor takes on that level, and 0 for all other observations. In this case, for example, suppose `XMale` is a dummy variable set to 1 for all `Male` observations and `XFemale` is set to 1 for all `Female` observations. The original model is then equivalent to

```
Salary ~ Age + XMale +XFemale
```

Often in models such as this not all of the coefficients can be determined numerically; for example, in (2.2) we could replace μ by $\mu + \delta$, and then compensate by replacing α_F and α_M by $\alpha_F - \delta$ and $\alpha_M - \delta$. Numerically such indeterminacies can be detected by collinearities in the variables used to represent the terms (Xmale and Xfemale add to a vector of ones, which is also used to represent the constant μ), and will be handled automatically during the model-fitting. Occasionally, you may want to control the parametrization of a term explicitly; Section 2.3.2 will show how.

Other non-numeric variables enter into the models by being interpreted as factors. A logical variable is a factor with levels "TRUE" and "FALSE". A character vector is interpreted as a factor with levels equal to the set of distinct character strings. A category object in S will be treated as a factor in the modeling software. Section 3.2.1 deals with these issues in more detail.

A term in a formula can also refer to a matrix. Each of the variables represented by the columns of the matrix will appear linearly in the model with its own coefficient. However, the entire matrix is interpreted as a single term.

To sum up so far, the following S data types can appear as a term in a formula:

1. a numeric vector, implying a single coefficient;

2. a factor or ordered factor, implying one coefficient for each level;

3. a matrix, implying a coefficient for each column.

Transformations increase the flexibility greatly, since the final element in this list is

4. any S expression that evaluates to a variable corresponding to one of the three types above.

To appreciate this last item, consider these examples of valid expressions that can appear as terms within a formula:

- (Age > 40), which evaluates to a logical variable;

- cut(Age,3), which evaluates to a three-level category;

- poly(Age,3), which evaluates to a three-column matrix of orthogonal polynomials in Age.

The classical computational model for regression is an X matrix and a coefficient vector β. The rich syntax of our modeling language allows us instead to think of each of the terms as an entity, even though they eventually will be expanded into one or more columns of a model matrix X in most of the models discussed. But the formulas and the modeling language put no restrictions on the form of a term or on the interpretation given to the term by a particular model-fitting function. The contribution of a term to the fit can often be thought of as a function of the underlying predictor; factors produce step functions, and terms based on functions like `poly()` produce smooth functions. See Section 2.3.1 and Chapter 6. For other models, like local regression and tree-based models, the contribution of the terms is interpreted differently. In particular, the contribution of a term to a tree-based model is invariant under monotone transformations of the variable.

2.2.2   Interactions

Terms representing the *interaction* of two or more variables lead to further shorthand in formulas. We may suspect that the effect of a variable in a model will be different depending on the level of some factor variable. In this case we need to fit an additional term in the model.

As an example, we consider some factors that describe the solder experiment in Chapter 1(these data are described in more detail in Chapter 3 and used throughout the book). `Opening` and `Mask` are two factors in the experiment, having three and five levels respectively. To allow for interactions, we will fit a term for each of the individual factors and in addition a coefficient for each level of the interaction— that is, for each combination of levels for the two factors. This is expressed in the formula language as

```
Opening + Mask + Opening:Mask
```

which implies fitting coefficients for the 3 levels of `Opening`, the 5 levels of `Mask`, and the 15 levels of their combination. The idea behind this separation into *main effects* and *interaction effects* is that for simplicity, we would prefer the interactions to be absent; by fitting them separately, we can examine the additional contribution of the interaction terms. (Once again, not all these coefficients can be determined independently.)

Rather than writing out the three terms, we allow a special use of the "`*`" operator in formulas to imply the inclusion of the two terms that are operands of the "`*`" *and* of their interaction. Thus

```
Opening * Mask
```

is equivalent to the previous expression.

When one of the variables is numeric, the interaction notation is still recognized, but it reduces to fitting coefficients for the factor variable and separate coefficients for the numeric variable within each level of the factor (see Section 2.3.1 for details).

Interactions may be defined between more than two variables; for example,

```
Opening * Mask * Samt
```

is interpreted to produce terms for each of the individual variables, for each of the two-way combinations, and for the three-way combination (that is, a coefficient for each of the $3 \times 5 \times 2$ levels of the factor defined jointly by all three variables). Another form of interaction is known as *nesting*, which we discuss in Section 2.3.1.

The full repertoire of special operators in formulas is discussed in Section 2.3.1. The same section discusses how formulas are interpreted, which may be relevant if your application is very specialized. Try, however, to build up formulas in as simple and unambiguous a way as possible.

2.2.3   Combining Data and Formula

The data and the formula for a model come together when we actually fit a particular model—e.g., when we estimate coefficients. The model-fitting functions will generate an appropriate internal form for the data in preparation for the fitting. For linear models and most of their extensions, this form is the model matrix or X matrix, in which one or more columns correspond to each of the terms in the model. Experienced modelers may have imagined the construction of this model matrix while reading the previous section, a tedious task traditionally regarded as part of the "art" of regression. The function `model.matrix()` does just this; in its simplest form it takes a single `formula` argument (with or without a response) and produces a matrix. Try it on a simple formula and see what happens! While it might be comforting for you to read Section 2.3–4 to see how we construct the ingredients of this matrix, such detailed knowledge is not necessary for standard use of the techniques we present in later chapters.

Nonstandard situations that may make model matrices of more interest include the handling of very large problems, where the size of the model matrix may force the use of special techniques, and various kinds of updating, subsampling, and iterative computations using some of the observations in the data. In these computations, practical considerations may require working directly with the model matrix.

The columns of model matrices contain coded versions of the factors and interactions in the model. The particular choice of coding will be of concern only if you want to interpret particular coefficients; Section 2.3.2 discusses how to control the coding. Section 2.4.3 contains further discussion of model matrix objects. That section is intended for those who need or want to know how the computations actually take place. In particular, to develop a *new* approach to fitting models, not covered by any of the chapters of the book, you would need to understand something about the steps that go into creating a model matrix.

2.3   More on Models

The third section of each chapter in this book expands on the S functions and objects provided in the chapter. Here we discuss the options and extended versions available that add new capabilities to the basic ideas. The material in section 3 should usually be looked at *after* you have tried out the essentials presented in section 2 on a few examples. Experience shows that, after trying out the ideas for a while, you will have a better feeling for how to make use of the functions, and will begin to think "This would be a bit better if only ...." Section 3 is intended to handle most of the "if-only." When the extra feature needed is not here, and there is no obvious way to create it by writing some function yourself, the next step is to look at section 4, which reveals how it all works. There you can learn what would be involved in modifying the basics. However, you should not take that step before thoroughly understanding what can be done more directly.

2.3.1   Formulas in Detail

In Section 2.2 we introduced model formulas and gave some examples of typical S expressions that can be used to give a compact description of the structural form of a model. Simple model-fitting situations can often be handled by the simple formulas shown there, but the full scope of model formulas allows much more detailed control. In this section, we give the full syntax available and explain how it is interpreted to generate the terms in the resulting model. Unless otherwise stated, we will always be talking about linear or additive models, in which the coefficients to be fitted do not have to appear explicitly in the formula. Formulas as discussed here follow generally the style introduced by Wilkinson and Rogers (1973), and subsequently adopted in many statistical programs such as GLIM and GENSTAT. While originally introduced in the context of the analysis of variance, the notation is in fact just a shorthand for expressing the set of terms defined separately and jointly by a number of variables, typically factors. Its application is therefore much more general; for example, it works for tree-based models (Chapter 9), where there is no direct link to linear models. Two additional extensions appear in our use of formulas:

- a "variable" can in fact be an arbitrary S expression, and

- the response in the model is included in the formula.

Of course the "any expression" in the first item had better evaluate to one of the permissible data types: numeric vector, factor (including categories and logicals), or matrices. The discussion here focuses on special operators for the predictors, and so, in the examples below, we will omit the response expressions.

  A model formula defines a list of terms to appear in a model. Each term identifies some S expression involving the data. This expression, in a linear model, generates one or more columns in a model matrix. These columns, each multiplied by the appropriate coefficient, are the contribution of this term to the fit. For other types of models, the contribution of the term may be computed in a slightly different way, but in any case the *expanded* definition of the model corresponds to this list of terms. A corresponding expanded formula has one expression for each term, separated by + operators. You will hardly ever write fully expanded formulas, but any formula you do write will first be expanded (and simplified) before being evaluated. In this section we proceed by first discussing the meaning of expanded formulas. Then we give the complete rules by which arbitrary formulas are expanded.

Interaction and Nesting

Expanding a formula reverses the process shown in Section 2.2 of choosing a short-hand for a formula. For example, the formula

    Opening * Mask

says that we want a model which fits coefficients for Opening, for Mask, and for the interaction of the two. When Opening and Mask are factors, this means a coefficient for each level of the factor; if either is a numeric variable or a numeric matrix, there will instead be one coefficient for the variable or for each column of the matrix. (We will discuss the meaning of interaction in this case later in this section.) In the more customary textbook notation,

- a factor:factor interaction represents a term of the form γ_{ij}, which is a set of IJ constants for each cell in the two-way table obtained by crossing the two factors (assuming the factors have I and J levels, respectively);

- a factor:numeric interaction represents a term of the form $\beta_j x$, or a varying slope model, in which the coefficient of the numeric variable x is different for each of the J levels of the factor;

- a numeric:numeric interaction represents a term βxz, where xz is simply the pointwise product of the variable x with the variable z. This is probably the least meaningful form of interaction, but of course the syntax allows far more meaningful terms to be created in cases such as this. For example, poly(x,z,2) will specify a bivariate quadratic surface in the two numeric variables x and z.

The formula Opening * Mask in expanded form is then

    1 + Opening + Mask + Opening:Mask

The formula above brings in factors in a *crossed* model; that is, the model says that the individual factors should be included and, in addition, that the contribution of one factor to the fit may change depending on the level of the other factors.

Nested terms in a model, on the other hand, arise when the levels of one factor are only meaningful within a particular level for some other factor or combination of factors. For example, suppose we have some geographic data in which the variable state defines the state for each location and the factor county indexes counties within each state. Clearly, county was generated by coding whatever county names appeared for each state, so that level 1 of county means something different in different states. In this context, a main effect for county is meaningless, and a typical model will fit a main effect for state and then look at the coefficients for county within each level of state. In expanded form this could be written

```
1 + state + county:state
```

However, to emphasize that the last term is thought of as a nested term, not an interaction, we allow (and encourage) writing the model as

```
1 + state + county %in% state
```

The formula written above in expanded form has the shorthand notation

```
state / county
```

meaning "state and then county within state." Notice that while factors joined by * can be permuted without changing the meaning (except for the order of the expanded terms), factors joined by / can never be meaningfully permuted: if county is nested in state, then state cannot be nested in county.

While the model implies a coefficient for each level of each term, in practice the coefficients have built-in dependencies. When a model matrix is created that represents a particular model, columns coding each term are included for all the coefficients that can be estimated; this is the condition for a *valid* coding of the model. One would like the individual coefficients to be meaningful in terms of the overall model and to avoid too many redundant coefficients that will have to be removed in the fitting. In Section 2.4.1, the general rules for coding will be outlined, but for practical purposes you need not worry about the coding unless you want to understand or control the specific choice of coefficients.

The coding of factors depends on the overall model. In the two-factor crossed model, main effects are included for both of the factors in the interaction term. All the possible coefficients for the interaction term can be estimated by representing each factor by contrasts among the levels of the individual factors. The contrasts will be chosen by default in a standard way, but the coding can be controlled, as shown in the next section. Unordered factors are coded as successive differences using the Helmert contrasts (Section 2.3.2), and ordered factors are coded to give a polynomial fit to a hypothetical underlying numeric variable. In the nested model, there is no main effect for county, so that the coding of county %in% state proceeds differently: state is coded by dummy variables and county by contrasts. This produces the computational equivalent of "county within `state`" and also guarantees that the model will fit as many coefficients as can be meaningfully defined. The details of coding affect only the meaning of the individual coefficients estimated. Any valid representation will give the same contribution to the overall fit in the model for each of the terms. If you don't care about the individual coefficients, leave the default coding in place; if you do care, look at page 32 to see how to change it.

When one of the variables in an interaction is numeric, the term will be computed formally the same way, but some extra remarks are needed on how crossed and nested interactions are interpreted. A numeric factor is always "coded" as itself. In interactions, the numeric factor will be multiplied by either the dummy variables or the contrasts for a factor. Consider a simple example using the automobile data: suppose `Weight` is numeric and `Foreign` is a logical variable, which will be turned into a factor with two levels corresponding to `FALSE` and `TRUE`. Both the crossed formula

    Foreign * Weight

and the nested one

    Foreign / Weight

make sense, but they mean something different. Consider the nested version. As before, this expands into the main effects for `Foreign` followed by `Weight` within each level of `Foreign`. In terms of the actual coefficients, one coefficient will be fitted for `Weight` using only data from level 1 of `Foreign` and one using data for level 2. There will only be one coefficient for `Foreign`, estimating the contrast between the two levels. This is equivalent to fitting a model to observations for which `Foreign` is `TRUE` as

$$\mu + \alpha_F + \beta_1 \times \texttt{Weight}$$

and another model to observations for which `Foreign` is `FALSE` as

$$\mu - \alpha_F + \beta_2 \times \texttt{Weight}$$

There are four coefficients: the intercept μ, the contrast α_F for `Foreign`, and coefficients β_i for `Weight` within each level of `Foreign`. This formula therefore corresponds to the concept of fitting "separate slopes" to the different levels of the factor.

The crossed formula fits main effects for both `Foreign` and `Weight`, and then fits the product of `Weight` with the coded contrasts of levels for the factor. In terms of specific coefficients this is

$$\mu + \alpha_F + \beta \times \texttt{Weight} + \gamma \times \texttt{Weight}$$

when `Foreign` is `TRUE` and

$$\mu - \alpha_F + \beta \times \texttt{Weight} - \gamma \times \texttt{Weight}$$

when `Foreign` is `FALSE`. Again there are four coefficients, but this time there is an overall slope β for `Weight` and a contrast γ estimating the interaction of `Foreign` and `Weight`. This is an appropriate way to code the model if we want to look at an overall fit to `Weight` and *then* to examine whether something substantial would be added to the model by allowing the regression to depend on the level of the factor. The distinction between the crossed and nested versions is not so strong here as when both predictors are factors, because a numeric factor is always just itself, but the treatment is entirely analogous. (Section 2.4.1, where we discuss how the coding works, will show *why* the computations can be the same.)

When *both* factors in an interaction are numeric, the formula expands as usual; now the pure interaction `x:z` amounts to fitting an ordinary product. In this case, however, you may really have wanted to use * or / in its ordinary S sense, in which case you ought to have protected the expression with the identity function `I()`, as we will show when we go into details about general formulas next.

Syntax of Formulas

We now give the full rules for writing model formulas. A model formula is created by separating a response term from the predictor terms by the operator $\sim$; the response can be absent. Expressions appearing in a model formula are interpreted as ordinary S expressions, except for the following operators:

    +   -   *   /   :   %in%   ^

The operator – is used to delete terms; for example,

    Padtype * Opening * Mask - Padtype:Opening:Mask

deletes the third-order interaction term that was implied by the expansion of the * operator, so that the formula expands to

    Padtype + Opening + Mask + Padtype:Opening + Padtype:Mask + Opening:Mask

As in this example, the – operator is useful for compactly dropping a few inter-actions, when we are prepared to assume these particular terms are negligible. A simple use of – is to exclude the intercept from a model:

    Yield ~ Mass - 1

In Chapter 1 we describe the `update()` function for changing fitted models, typ-ically by altering the formula. The "-" operator plays a special role there as well (illustrated again in the first example in the list below).

The use of ":" to denote interaction is a break from the traditional Wilkinson and Rogers syntax, where "." is used instead. A "." is a valid part of a name in S, as in `wind.speed`, so it could not serve as an interaction operator. A single "."

does have a special meaning though; it serves as the *default* left or right side of a formula wherever that makes sense. We made use of "." in some of the examples in Chapter 1. Other examples are

- `update(lmob, . ~ . - Age)` is used to update the fitted linear-model object `lmob` by modifying its formula and then refitting it. The "." on the left of ~ implies that the response is the same as in `lmob`, while the "." on the right of ~ gets replaced by whatever was on the right in the formula used to fit `lmob`.

- `lm(Mileage ~ ., data = car.test.frame)`; here the "." is interpreted relative to the data frame `car.test.frame`, which is a dataset to be used in fitting the linear model. Data frames are described in Chapter 3. The "." here means that all the variables in `car.test.frame`, except `Mileage`, are to be used additively, which is equivalent to the explicit formula

  `Mileage ~ Price + Country + Reliability + Type + Weight + Disp. + HP`

- `lm(skips ~ .^2, data = solder.balance)`; similar to the previous item, except all the main effects and second-order interactions of the variables in `solder.balance` are to be used.

The following table summarizes the special meanings of the operators in formulas:

Expression	Meaning
$T \sim F$	T is modeled as F
F_a + F_b	Include both F_a and F_b
F_a - F_b	Include all F_a except what is in F_b
F_a * F_b	F_a + F_b + F_a : F_b
F_a / F_b	F_a + F_b %in% (F_a)
F_a : F_b or	The factor jointly indexed by F_a and
F_b %in% F_a	F_b
$F^\wedge m$	All the terms in F crossed to order m

The expression T is a term (with no special operators included), but F, F_a, and F_b can be arbitrary formulas, not just single terms. The operators in the table are special in their semantics (that is, in the way that S interprets them) but they otherwise act as they would in ordinary expressions, with the same precedence and association they would normally have (see ⟦S⟧, Section 3.2.6). Parentheses can be used to change the grouping implied by precedence rules—for example, to force a combination of terms to act like one term. The formula

  `Panel / (Opening * Mask)`

says to fit all the terms in `Opening * Mask`, within each level of `Panel`.

Slightly more subtle is

```
(Panel + Mask)/ Opening
```

which expands to

```
1 + Panel + Mask + Opening  %in%    (Panel + Mask)
```

The parentheses around `Panel + Mask` here do not get expanded further, and the last term is equivalent to

```
Opening:Panel:Mask
```

The last item in the table is a shorthand for creating interactions. For example

```
(Opening + Mask + Panel)^2
```

expands to the same formula constructed from using "-":

```
Padtype + Opening + Mask + Padtype:Opening + Padtype:Mask + Opening:Mask
```

Composite Terms in Formulas

The special meaning of the operators applies only at the top level in the formula expressions, and only on the right of the "∼". If the operators appear as the arguments to other functions, they behave as they always do in S. As the user certainly intended, the term

```
atan( Length / Width )
```

fits a single coefficient to the ordinary value of the S expression, and does not treat / as a nesting operator. Similarly,

```
sqrt (x - min(x))
```

does not treat - specially. It is also possible to *force* the operators to be treated in an ordinary way, by using the identity function, `I()`. This function returns its argument and exists only to protect special operators. For example, to fit as a single term the product of `Length` and `Width`, use

```
I( Length * Width )
```

to prevent the operator `*` from getting its special interpretation.

As emphasized before, any variables in the formula (either the response or the factors in the terms) can be arbitrary S expressions, so long as they evaluate to objects having a valid data type, namely: numeric variables or matrices, factors including ordered factors, and non-numeric variables which will be converted into factors.

Matrices that appear in the formula are treated as a single factor. This is how special curves can most easily be generated, and functions are provided that generate suitable matrices for common kinds of curves. The following are some examples:

- The expression `poly(x, degree)` returns a matrix whose columns are an orthogonal basis for fitting a polynomial of degree `degree` in the numeric variable `x`. Similarly `poly(x,y, degree)` returns the matrix of bivariate polynomial terms of degree no more than `degree`, and so on.

- The expression `bs(x, df)` returns a matrix which is a B-spline basis for piecewise-cubic regression on `x`. The parameter `df` is the degrees of freedom, which determines the number of interior knots. These knots are automatically placed by the function; otherwise, the `knots` argument can be used to place them explicitly.

See Chapters 6 and 7 for a general discussion of composite terms such as these.

Functions can be used that produce factors and categories as well; for example, `ordered(x,breaks)` will return an ordered factor cutting the numeric variable `x` at the breakpoints `breaks`. This is similar to the S function `cut()`, which produces a category. Expressions that produce factors or categories can be used in conjunction with the special operators, so that

```
cut(Weight,5) * Country
```

creates a five-level category from the numeric variable `Weight` and then uses it in a crossed model with the factor `Country`. Similarly

```
( Age < 45 ) * Cholesterol
```

creates different linear trends in `Cholesterol` for people under and over 45. The function `codes()` produces numbers to represent the levels in a factor or ordered factor; so if `Opening` is an ordered factor with levels `Large`, `Medium`, and `Small`, then the expression

```
codes(Opening)
```

implies a term linear in the numbers 1, 2, and 3, coding the three levels.

As always in S, you can write any functions of your own to create other suitable variables. The expressions can be more complicated than function calls as well:

```
group * (if(all(x)>0)log(x) else log(x-min(x)+.01))
```

Exotic expressions like this are perfectly legal but hard to read and not good programming style. A better approach is to define a function, say

```
plus.log <- function(x)
   if(all(x)>0)log(x) else log(x-min(x)+.01)
```

and then write the formula as `group * plus.log(x)`.

2.3.2 Coding Factors by Contrasts

On page 21 we noted that factors entering a model normally produce more coefficients than can be estimated. This is true regardless of the data being used; for example, the sum of all the dummy variables for any factor is a vector of all ones. This is identical to the variable used implicitly to fit an intercept term. This is *functional* overparametrization, as opposed to *data-dependent* overparametrization in which the number of observations is not large enough to estimate coefficients or in which some of the variables turn out to be linearly related. The functional problem is removed in most cases before any model-fitting occurs, by replacing the dummy variables by a set of functionally independent linear combination of those variables, which is arranged to be independent also of the sum of the dummy variables. For a factor with k levels, $k-1$ such linear combinations are possible. We call a particular choice of these linear combinations a set of *contrasts*, using the terminology of the analysis of variance. Computationally, the contrasts are represented as a k by $k-1$ matrix.

Any choice of contrasts for factors alters the specific individual coefficients in a model but does not change the overall contribution of the term to the fit. All the model-fitting functions choose contrasts automatically, but users can also specify the contrasts desired, either in the formula for the model or in the factor variable itself. By default, contrasts are chosen as follows:

- unordered factors are coded by what are known as the *Helmert* contrasts, which effectively contrast the second level with the first, then the third with the average of the first and second, and so on;

- ordered factors are coded so that individual coefficients represent orthogonal polynomials if the levels of the factor were actually equally spaced numeric values.

If this choice of contrasts is adequate, no user action is needed.

The simplest way to alter the choice of contrasts is to use the function `C()` , with usage `C(factor, contrast)` in the formula. The first argument is a factor, the second a choice of contrast. It returns `factor` with the appropriate contrast matrix attached as an attribute. The choice can be made in three ways:

- By giving the name of a built-in choice for contrasts: `helmert`, `poly`, `sum`, or `treatment`. For example, `C(Opening, sum)` uses the function `contr.sum()` to generate the appropriate sized contrast matrix. We will explain the meaning of these choices below.

- By giving a function, which when called with either a factor or the number of levels of the factor as its argument, returns the k by $k-1$ matrix of constraints: `C(Opening, myfun)` calls `myfun(Opening)` to generate the contrast matrix (if `myfun` exists as a function).

- By giving the contrast matrix *or* some columns for the contrast matrix. `C(Opening, mymat)` uses the matrix `mymat` as the contrast matrix.

The function `C()` tests for each of these cases to determine how it has been called.

The four standard choices correspond to four functions to generate particular flavors of contrasts. The polynomial contrasts are the result of the function `contr.poly()`:

```
> contr.poly(4)
           L     Q           C
[1,] -0.6708204  0.5 -0.2236068
[2,] -0.2236068 -0.5  0.6708204
[3,]  0.2236068 -0.5 -0.6708204
[4,]  0.6708204  0.5  0.2236068
```

The coefficients produced by this transformation of the dummy variables correspond to linear, quadratic, and cubic terms in a hypothetical underlying numeric variable that takes on equally spaced values for the four levels of the factor. In general, `contr.poly` produces $k - 1$ orthogonal contrasts representing polynomials of degree 1 to $k - 1$.

Similarly, the function `contr.helmert()` returns the Helmert parametrization. The first linear combination is the difference between the second and first levels, the second is the difference between the third level and the average of the first and second, and the jth linear combination is the difference between the level $j + 1$ and the average of the first j—for example,

```
> contr.helmert(4)
  [,1] [,2] [,3]
1  -1   -1   -1
2   1   -1   -1
3   0    2   -1
4   0    0    3
```

These two are the default choices.

The **sum** choice and the corresponding function `contr.sum()` produce contrasts between each of the first $k - 1$ levels and level k.

```
> contr.sum(4)
  [,1] [,2] [,3]
1   1    0    0
2   0    1    0
3   0    0    1
4  -1   -1   -1
```

This corresponds to a parametrization got by applying the constraint that the sum of the coefficients be zero.

The `treatment` form of coding is commonly used in models for which the first level of *all* the factors is considered to be the standard or control case, and in which one is interested in differences between any of the nonstandard or treatment situations. As constraints on the coefficients, this is usually expressed as saying that any coefficient in which any of the factors appears at its first level is set to 0. The equivalent coding uses the dummy variables for levels 2 through k. The function `contr.treatment()` gives this coding:

```
> contr.treatment(4)
  2 3 4
1 0 0 0
2 1 0 0
3 0 1 0
4 0 0 1
```

This is a legitimate coding, in that it captures all the coefficients. However, it is not a set of contrasts, in that the columns do not sum to zero and so are not orthogonal to the vector of ones. For applications to linear models in designed experiments, the coefficients will not be statistically independent for balanced experiments. This complicates the interpretation of techniques such as the analysis of variance, so that the control-treatment coding should generally not be used in this context. For some other models, such as the GLM models in Chapter 6, the lack of orthogonality is less obviously a defect, since the assumptions of the models do not produce statistical independence of the estimated coefficients anyway. Probably for this reason, the control-treatment coding is popular among GLM modelers, since what it lacks in orthogonality it gains in simplicity.

Any of these can be selected in a formula to override the default. You can also implement any function you like, perhaps by modifying one of the four standard functions, to produce a different set of contrasts. The matrix must be of the right dimension and the columns must be linearly independent of each other and of the vector of all ones. If this fails, model-fitting with complete data will produce singular models. An easy way to test this condition is to bind a column of 1 to the matrix and pass the result to the `qr()` function. The value of this function has a component `rank` that is the computed numerical rank of the matrix. For a set of contrasts on k levels, the rank should be k—for example

```
> qr(cbind(1,contr.treatment(4)))$rank
[1] 4
```

A function to generate contrasts must also, by convention, take either the levels of the factor or the number of levels as its argument. See any of the four standard functions for code to copy.

The third way to specify contrasts is directly by numeric data. You can start from the value of one of the functions, but a more typical situation in practice is that you want to estimate one or more specific contrasts, but will take anything suitable for the remainder of the $k - 1$ columns. Suppose `quality` is a factor (unordered) with four levels:

```
> levels(quality)
[1] "tested-low"  "low"    "high"   "tested-high"
```

Suppose that we want the first contrast of `quality` to measure the difference between tested and nontested—that is, levels 1 and 4 versus levels 2 and 3—and we don't care about the other contrasts. Then we can give the factor in the formula as

```
C(quality, c(1, -1, -1, 1))
```

Two additional contrasts will be chosen to be orthogonal to the specified contrast. If we had wanted the second contrast to be between the two `low` and the two `high` levels, we would have supplied `C()` with the matrix

```
      [,1] [,2]
[1,]    1    1
[2,]   -1    1
[3,]   -1   -1
[4,]    1   -1
```

and one further column would be supplied.

One additional detail is sometimes needed. Sometimes the user is willing to assert that *only* some specified contrasts in the levels of a factor can be important; the others should be regarded as *known* to be zero and omitted from the model. This is risky, of course, but is done in some experiments where the number of runs is limited and the user has considerable prior knowledge about the response. The specification can be done by giving `C()` a third argument, the number of contrasts to fit. For example, suppose we are fitting polynomial contrasts to an ordered factor, `Reliability`, and assert that no more than quadratic effects are important. The corresponding expression in the model would be

```
C(Reliability, poly, 2)
```

Since the ith contrast generated by `contr.poly()` corresponds to an orthogonal polynomial of degree i, this term retains only linear and quadratic effects.

The function `C()` combines a factor and a specification for the contrasts wanted, and returns a factor with those contrasts explicitly assigned as an attribute. The companion function `contrasts()` extracts the contrasts from a factor, and returns them as a matrix. The contrasts may have been explicitly assigned as an attribute or may be the appropriate default, according to whether the factor is ordered or not. If you want to set the contrasts for a particular factor *whenever* it appears, the function `contrasts()` on the left of an assignment does this. In the example of one specific contrast,

```
> contrasts(quality) <-  c(1, -1, -1, 1)
> contrasts(quality)
              [,1] [,2] [,3]
  tested-low     1 -0.1 -0.7
         low    -1 -0.7  0.1
        high    -1  0.7 -0.1
 tested-high     1  0.1  0.7
```

two additional linear combinations have been added to give a full contrast specification. Now, `quality` will have this parametrization by default in any formula, with the opportunity still available to use the `C()` function to override. As with `C()`, the function `contrasts()` on the left of an assignment takes an optional additional argument, `how.many`, that says to assign fewer than the maximum number of contrasts to the factor.

You can also change the default choice of contrasts for *all* factors using the `options()` command in S, once you know a little more about how coding is done.

```
> options()$contrasts
          factor        ordered
   "contr.helmert" "contr.poly"
```

shows us what the defaults are. These options are the names of functions that provide contrasts for unordered and ordered contrasts, respectively. To reset the defaults, use:

```
options(contrasts=c("contr.treatment","contr.poly"))
```

Redefining one or both of the elements changes the default choice of contrasts. The effect of using `options()` to change the default contrast functions lasts as long as the S session; each time S is started up, the permanent default is assumed. If you really want to have your own private default coding every time you run S, you can invoke `options()` automatically via the `.First()` function ($\boxed{S}$, Section 3.4.9). Notice that explicit choices for individual factors can still be used to override the new default coding by assigning the contrasts as before.

Strictly speaking, the term *contrast* implies that all the linear combinations are contrasts of levels. In this case, the sum of the numbers in any column of the matrix should be zero. *Orthogonal* contrasts have the additional property that the inner products of any two columns of the contrast matrix is also zero. The Helmert and polynomial contrasts have both these properties. The contrast and orthogonal contrast properties are particularly important for linear models in designed experiments. Otherwise, the choice of contrasts can introduce artificial correlations between coefficient estimates, even if the design is balanced. Additional details on the implications of contrasts for fitted models appear in Section 5.3.1 in the context of analysis of variance, and in Section 6.3.2 in the context of GLMs.

2.4   Internal Organization of Models

In this section, as in the fourth section of later chapters, we will reveal how things work: the internal structure of the objects that have appeared earlier in the chapter and the computational techniques used in the functions. This section is *not* required reading if you only want to use the functions described so far. It will be useful, however, for those who want to extend the capabilities and/or to specialize them in a serious way for particular applications. Extensions and modification of the software we provide is not only allowed but is one of the goals of our approach to statistical software in this book. Rather than trying to provide a complete approach to the topics we cover (probably an impossible task anyway), we present functions and objects that form a kernel containing the essential computations. The functions include what we see as the natural approach to common, general use of the statistical methods. The classes of objects organize the key information involved, with the goal of making subsequent use of the information as easy and general as possible.

Users with special needs, and researchers who want to extend the statistical techniques themselves, will want to go beyond what we provide. Understanding the material in this section will likely help.

2.4.1   Rules for Coding Expanded Formulas

This section gives the rules underlying the coding of factors in the expanded formulas of Section 2.3.1. To produce a model matrix for use in linear models, factors and their interactions are represented by columns of numeric data, either dummy variables or contrasts. To be valid, the representation must estimate the full linear model. Since such models are generally overparametrized, there will be many different valid representations in this sense. The goal of a particular representation is to be meaningful and reasonably parsimonious. The actual coefficients estimated should mean something in the application of the model. For example, a coefficient value significantly different from zero should say something useful about the data. A parsimonious parametrization is desirable numerically, since the size of the model matrix can in some cases be much larger than necessary. The mathematical discussion that follows provides the basis for understanding how the representation can be chosen for various models.

Each term in an expanded formula can be written using only one special operator, ":". Suppose we have an expanded formula with p factors:

$$F_1, F_2, \cdots, F_p$$

and m terms:

$$T_1 + T_2 + \cdots + T_m$$

The F_j need not be simple variables, but can be essentially arbitrary S expressions. The expanded term T_i can always be written as an interaction of 0, 1, or more of the F_j; say,

$$F_{i_1} : F_{i_2} : \cdots : F_{i_{o_i}}$$

where $1 \leq i_j \leq p$. The value of o_i is the *order* of T_i—that is, the number of factors. We will assume that the expanded formula is sorted by the order of the terms, so that all terms of order 1 appear first, then all terms of order 2, and so on.[1] The intercept term is the only term of order 0, and is written as 1. If it is present, it comes first.

As discussed in Section 2.2, a factor corresponding to F_j can be represented in the model by a matrix whose columns are the dummy variables corresponding to each level of F_j. The interaction of F_{j_1} and F_{j_2} is represented by the matrix containing all possible products of pairs of columns from the matrices representing the main effects. A three-way interaction is represented by all products of columns of this matrix and columns of the matrix representing F_{j_3}, and so on. For details, see the function `column.prods()`, which carries out just this computation.

If all factors were represented by dummy variables, there would be nothing more to the interpretation of expanded formulas. However, both numerical and statistical arguments require more careful coding of factors. Usually, coefficients cannot be estimated for all the levels of the factor. For example, the sum of all the dummy variables for any main effect is the constant 1, and so is functionally equivalent to the intercept. Coding all levels by dummy variables would produce a model matrix with more columns than necessary (in some cases *many* more), and the model matrix would nearly always be singular, so that numerical solutions would not produce estimates for all the requested coefficients. These are the numerical reasons for choosing a good coding for the terms, but the statistical reason is more important—namely, to allow a *meaningful* choice of coefficients for the particular model. The functional dependencies among the dummy variables in the terms imply that only certain linear combinations of the coefficients for the dummy variables are estimable. The goal is to represent those linear combinations so that the individual computed coefficients are useful for the particular model. Section 2.3.2 showed how this coding could be controlled.

The following rule specifies which factors should be coded by dummy variables and which should be coded by contrasts in producing the columns of the model matrix:

> Suppose F_j is any factor included in term T_i. Let $T_{i(j)}$ denote the *margin* of T_i for factor F_j—that is, the term obtained by dropping F_j from T_i. We say that $T_{i(j)}$ has appeared in the formula if there is some term $T_{i'}$ for $i' < i$ such that $T_{i'}$ contains all the factors appearing in $T_{i(j)}$. The usual case is that $T_{i(j)}$ itself is one of the preceding terms. Then F_j is

---

[1] The only ordering that we actually need is that any term T_i appear in the formula *after* its margins. It does not make sense for a factor to appear in the formula after some interaction including that same factor.

coded by contrasts if $T_{i(j)}$ has appeared in the formula and by dummy variables if it has not.

In interpreting this rule, the empty term is taken to be the intercept.

The application of this rule corresponds to generating a matrix—the model matrix—with n rows and some number of columns, to represent the whole model. This matrix comes from binding together the columns of the matrices produced by the rule for each term. We can compare this matrix with the overspecified but valid coding we would get if we used dummy variables for all the factors. Our rule is valid if the dummy variables, say X^*, introduced for term T_i in this overspecified coding, can be represented as a linear combination of columns from the matrices produced by our rule for terms up to and including T_i, for all i.

Here is an informal proof that the rule is valid. Start with an inductive assumption: suppose that the rule is valid for terms of order less than the order of T_i; specifically, for any such term, assume that its dummy matrix can be written as a linear combination of the matrices given by our rule for that term and those of its margins that are in the formula. Suppose F_j is one of the factors for which the rule says we can use contrasts. Let X_j be the n by $k_j - 1$ matrix of contrast variables for F_j, and X_j^* the corresponding n by k_j matrix of dummy variables. We will need to refer to the lth columns of these matrices; let's call them $x_{j:l}$ and $x_{j:l}^*$. Any column of X^* can be written as the product of one column from each of the dummy matrices, X_J^*, for factors J in T_i, so in particular it can be written as:

$$\left(\prod_{J \in T_{i(j)}} x_{J:l_J}^* \right) x_{j:l}^*$$

Note that this is ordinary multiplication of the n-vectors, not matrix product. Now look at the two parts of the above expression separately, the left part in parentheses and the single vector on the right.

1. By the inductive assumption, the left part is a linear combination of our matrices for $T_{i(j)}$ and *its* margins.

2. From the definition of a valid coding of the individual factors, the right part is a linear combination of 1 and the $x_{j:r}$

If we were to expand these two linear combinations, the result would be a linear combination of column products from our coding for T_i and for its various margins. Therefore, the inductive assumption holds for T_i as well. By looking directly at the cases of the empty term and terms of order 1, the inductive assumption holds for these cases, and so is believable in general. This is not quite precise; in particular, extra arguing needs to be added for the (somewhat strange) case that $T_{i(j)}$ is not in the model, but is contained in some other preceding term of order equal that of T_i.

The argument above should nevertheless be sufficiently convincing for our purposes here.

The rule does not always produce a minimal coding; that is, in some cases there may be functional dependencies between columns of the matrix representing T_i and those representing earlier $T_{i'}$. In particular, this will be the case again when there is no $T_{i'}$ that is exactly equal to $T_{i(j)}$. However, models of that form are usually questionable; for most sensible model formulas, the rule above produces a minimal and meaningful coding of the terms.

Numeric variables appear in the computations as themselves, uncoded. Therefore, the rule does not do anything special for them, and it remains valid, in a trivial sense, whenever any of the F_j is numeric rather than categorical.

2.4.2   Formulas and Terms

Formula objects pass through an intermediate stage before being combined with the data. This stage produces objects of class `"terms"`, which contain the formula after it is processed to have, in a convenient form, all the information needed to create the model. Users of model-fitting functions will not see this intermediate stage, but those of you who want to modify model-fitting techniques or to create a new class of models may find it helpful to know what information the `terms` objects contain. A `terms` object is an object of mode `"expression"` with extra attributes. The elements of the expression are the individual terms in the expanded right side of the formula:

```
> form1 <- skips ~ Panel * Opening
> terms1 <- terms(form1)
> as.vector(terms1)
expression(skips, Panel, Opening, Panel:Opening)
```

Now let's consider the attributes:

```
> names(attributes(terms1))
[1] "formula"     "factors"     "order"       "variables"
[5] "term.labels" "intercept"   "response"    "class"
```

The meaning of the attributes is as follows:

`"formula"`: the actual formula used to construct terms, in this case the contents of `form1`:

```
> attr(terms1,"formula")
skips ~ Panel * Opening
```

There is a generic function `formula()` for extracting formulas from a variety of objects; in this case, `formula(terms1)` would extract the formula from the `terms` object.

"factors": a matrix with factors along the rows and terms along the columns. The *j*th column says what factors appear in the *j*th term and also whether they are coded as contrasts or dummy variables. Values in the column are 1 for contrasts, 2 for dummy variables, and 0 if the factor does not appear in the term:

```
> attr(terms1,"factors")
        Panel Opening Panel:Opening
  skips     0       0             0
  Panel     1       0             1
Opening     0       1             1
```

The coding is specified according to the general rule in the previous section.

"order": a vector giving the order of each term: 1 for main effects, 2 for second-order interactions, and so on:

```
> attr(terms1,"order")
[1] 1 1 2
```

"variables": an expression whose elements are the expressions for each of the variables, including the response (remember that these need not simply be names):

```
> attr(terms1,"variables")
expression(skips, Panel, Opening)
```

"term.labels": the character form of the terms, only included to save repeated deparsing later:

```
> attr(terms1,"term.labels")
[1] "Panel"        "Opening"      "Panel:Opening"
```

which can also be extracted using the labels() generic function.

"intercept": a logical variable that will be TRUE unless the term -1 appears in the formula. Notice that the intercept term does not appear in the expression vector itself nor in the term labels. The label "(Intercept)" is used to label coefficients, etc. corresponding to the intercept.

"response": which variable in the "variables" attribute is the response (0 if there is no response specified).

Since all the model-fitting functions include the terms object for a particular model in the object that represents the fitted model, you can use the information above to conveniently get at information about pieces of the model when designing new summary functions or modifying the model-fitting. Section 7.4 describes some additional arguments in the call to terms that add to its flexibility.

2.4.3 Terms and the Model Matrix

The process of putting together data and formula to construct a model matrix involves three basic steps:

1. Convert the formula into a `terms` object, in which all the interactions and nested terms have been expanded, and any simplifications resulting from subtractions, parentheses, dots, and powers have been applied.

2. Compute a *model frame* from the terms and the data, containing variables corresponding to the expressions needed to compute the terms defined in step 2.

3. Generate the model matrix itself from the model frame.

A model frame is a special type of data frame, described in the next chapter. For the moment, simply think of it as a list of the response variable and variables corresponding to all the terms in the formula. The function `model.frame()` uses the terms object, specifically its attribute `variables`, to determine which expressions will be used in generating the model matrix. It returns a special data frame containing those variables. Notice that there is no restriction on the expressions appearing in the formula for the terms. The names of the variables will not necessarily be syntactic names in S; if one of the terms is `log(Fuel)` then the corresponding variable will have name `"log(Fuel)"`. Two other, optional computations take place during the evaluation of the model frame. If a `subset` argument is supplied, the corresponding subset will be extracted from each computed variable before it is inserted into the model frame. Similarly, if a `na.action` function is supplied either as an argument or as an attribute to the data frame, this function will be applied to the model frame. See Section 3.3.3 for further details.

Once the model frame has been computed, it is used to generate a model matrix, with columns corresponding to each of the terms in the model formula. A model matrix is a numeric matrix of suitably coded dummy variables, contrasts, or numeric variables, plus some attributes related to the model.

We don't have to worry about the steps described above; the fitting functions such as `lm()` do the work for us, and will return both the model frame and model matrix if requested. On the other hand, we can create a model matrix from some data directly from the `model.matrix()` function. To illustrate the structure in model matrices, we will compute a model matrix from a market study data frame described in the next chapter. The model chosen will use the numeric variable `usage` along with a complete model (main effects and interaction) for two factors, `nonpub` and `education`:

```
> model1 <- model.matrix(~ usage + nonpub * education)
> print(model1[1:5,], abbreviate = T)
   (Intrc) usage nonpub education1 education2 education3 education4
1        1     9      1         1        -1        -1        -1
2        1     2      1         1        -1        -1        -1
7        1     3     -1        -1        -1        -1        -1
8        1    .1     -1         1        -1        -1        -1
10       1     2     -1        -1        -1        -1        -1

   education5 nnp:edc1 nnp:edc2 nnp:edc3 nnp:edc4 nnp:edc5
1         -1        1       -1       -1       -1       -1
2         -1        1       -1       -1       -1       -1
7         -1        1        1        1        1        1
8         -1       -1        1        1        1        1
10        -1        1        1        1        1        1
> dim(model1)
[1] 1000   13
```

Notice we used no response in the formula; had one been there it would have been ignored. The model matrix has as many columns as are required by the coding of the expanded formula. The `abbreviate=` argument to the printing method for `matrix` objects abbreviates the column labels. The unabbreviated column labels are

```
> dimnames(model1)[[2]]
 [1] "(Intercept)"        "usage"              "nonpub"
 [4] "education1"         "education2"         "education3"
 [7] "education4"         "education5"         "nonpub:education1"
[10] "nonpub:education2"  "nonpub:education3"  "nonpub:education4"
[13] "nonpub:education5"
```

Model matrices have additional attributes:

```
> names(attributes(model1))
[1] "dim"       "formula"    "class"     "order"    "term.labels"
[6] "assign"    "dimnames"
```

The `class` attribute has value `c("model.matrix", "matrix")`, which means that model matrices inherit from the more general class `"matrix"`. The `formula`, `order` and `term.labels` attributes are retained from the `terms` object. The `assign` attribute is a list, with length equal to the number of terms. The elements of `assign` define which columns of the matrix belong to the corresponding terms:

```
> attr(model1,"assign")
$"(Intercept)":
[1] 1
```

```
$usage:
[1] 2

$nonpub:
[1] 3

$education:
 1 2 3 4 5
 4 5 6 7 8

$"nonpub:education":
 1  2  3  4  5
 9 10 11 12 13
```

The model.matrix function produces the matrix of predictors for linear and generalized linear regression and anova model-fitting routines; other model-fitting functions, such as those that build trees, require other constructions. These are discussed further in the relevant chapters, as well as in Section 3.3.3. There we also expand on the sequence of steps needed to create model frames and model matrices, to allow facilities for weights, missing data, and subsets.

Bibliographic Notes

The formula language described in this chapter was inspired by the the Wilkinson and Rogers (1973) formula language used in the package GLIM. Several of the enhancements introduced here, such as poly(), for example, were mentioned in the Wilkinson and Rogers paper, but not fully implemented in GLIM.

Chapter 3

Data for Models

John M. Chambers

This chapter describes the general structure for data that will be used throughout the book. In particular, it introduces the *data frame*, a class of objects to represent the data typically encountered in fitting models.

Section 3.1 presents some datasets that recur as examples throughout the book. S functions to create, manipulate, modify, and study data frames are described in Section 3.2. Section 3.3 discusses the computations on data frames and related classes of objects at a detailed level, suitable if you want to modify functions dealing with these objects.

As with Chapter 2, the ideas in this chapter underlie all the computations for various models in the following chapters. To get a general view of our approach to data, you should read some of this chapter before going on to specific models. Sections 3.1 and 3.2 should be plenty. Your data analysis will benefit from studying graphical and other summaries of the data *before* any commitment to a particular model. This chapter describes a number of such summaries and also shows how to apply S functions generally to the data in data frames. Therefore, we recommend reading through the first two sections of the chapter before fitting particular models.

3.1 Examples of Data Frames

The statistical models discussed in this book nearly always think of the underlying observational data as being organized by *variables*—statistical abstractions for different things that can be observed. Values on these variables can be recorded for a number of *observations*—points in time or space, for example. A *data frame* is an object that represents a sequence of observations on some chosen variables. In later sections of the chapter we will describe functions to create, modify, and use these objects. Here, we are only concerned with the concepts involved, not the specific functions.

Data frames clearly have the flavor of matrices; in particular, in thinking and in computing both, the variables can be treated as columns and the observations as rows of a matrix-like structure. Data frames can also act as *frames* in S; that is, the variables can be thought of as separate objects, named by the corresponding variable name in the data frame. In particular, model formulas as discussed in Chapter 2 use variable names in just this way. Data frames are more general than matrices in the sense that matrices in S assume all the elements to be of the same mode—all numeric, all logical, all character string, etc. In contrast, the variables in a data frame can be anything at all, so long as each variable is indexed by the same observations. The variables can even be matrices; in fact, matrix variables in data frames are very useful in model-fitting.

The essential concept to keep in mind throughout the book is that data frames support matrix-like computations, with variables as columns and observations as rows, and that, in addition, they allow computations in which the variables act like separate objects, referred to by name.

Time now for some examples. Statistical software for models, as for any other data analysis, can only be fully appreciated when it is seen working on substantial applications. In this book, we present both examples that are as realistic as practicable as well as the usual small examples that illustrate specific points conveniently. In the next three subsections, three fairly substantial sets of data are introduced. Later in the chapter, we will present some details of the procedures that get the data into the data frame as well as techniques for computing with the data. These datasets will be made available in S if you execute the expression

```
library(data)
```

which attaches a library of datasets for use in computations. Attaching this library will let you experiment with the examples in this book.

3.1.1   Example: Automobile Data

Our first example is a somewhat recreational one, acknowledging the lasting passion for the automobile, in the United States and elsewhere. Suppose we are interested in understanding the properties of different automobiles, such as their fuel consumption or their reliability. Are expensive cars more reliable, or less fuel-efficient? What about differences due to country of manufacture?

A wealth of data related to these questions is published by the magazine *Consumer Reports*, with one monthly issue per year devoted to automobiles, in addition to detailed test reports throughout the year, all providing testimony to the automobile's central role in our society. The annual issue contains several tables of data, of which we will use three: overall summaries, dimensions, and specifications. The reader of *Consumer Reports* with an interest in data analysis could record some of this information in the computer and organize it suitably for analysis. We will now illustrate how this might be done. Let's consider the three tables individually, showing objects in S corresponding to each. In Section 3.2.2 we show how S objects are generated from such data.

These tables are naturally represented as data frames, with rows corresponding to automobile models. The summary section of the issue, for instance, records five variables for price, country of origin, overall reliability, fuel mileage per gallon, and type of car. The S object `cu.summary` is a data frame containing the data. There are 117 models; let's look at a sample of 10. We will generate a sample from the row names of the data frame, sorting them alphabetically to make it easy to refer back to the printed table. Data frames always have row names, in this case the model names as they appeared in *Consumer Reports*. These names can be used to extract the corresponding rows of the data frame:

```
> summary10 <- sample(row.names(cu.summary), 10)
> summary10 <- sort(summary10)
> cu.summary[summary10,]
                    Price  Country Rel. Mileage     Type
  Acura Integra 4  11950    Japan   5      NA     Small
     Audi 100 5  26900  Germany   NA      NA    Medium
      BMW 325i 6  24650  Germany    4      NA   Compact
 Chevrolet Lumina 4  12140     USA   NA      NA    Medium
    Ford Festiva 4   6319    Korea    4      37     Small
    Mazda 929 V6  23300    Japan    5      21    Medium
 Mazda MX-5 Miata  13800    Japan   NA      NA    Sporty
  Nissan 300ZX V6  27900    Japan   NA      NA    Sporty
 Oldsmobile Calais 4   9995     USA    2      23   Compact
 Toyota Cressida 6  21498    Japan    3      23    Medium
```

Other data frames, `cu.dimensions` and `cu.specs`, represent the dimensions and specifications data. The set of rows (automobile models) changes from one table to another, so it will be natural to start with separate data frames and consider merging them when the data analysis demands it. In Section 3.2 we illustrate a variety of computations on data frames.

3.1.2 Example: A Manufacturing Experiment

Designed experiments to improve quality are important tools in modern manufacturing. In such experiments, a number of variables are chosen as factors to vary in the experiment. Several levels (typically two or three) are chosen for each factor.

The levels may be values of some numeric variable, such as the quantity of some ingredient or the diameter of an opening on a machine. They may also simply be alternatives, such as the choice of an ingredient in a recipe. The experiment is planned as a number of runs, on each of which all the factors are supposed to be set at predetermined levels. After each run, one or more response variables is recorded. An observer may examine the output of the run and record statistics, or some test of the output may be carried out and recorded.

A designed experiment will be represented in our work by a `design` object—a special class of objects that inherit all the properties of data frames. The variables in the data include design factors as above, plus some number of observed responses, plus perhaps other uncontrolled variables. The values of the factors are, in principle, prechosen according to the design selected by the statisticians and engineers to best achieve the goals of the study given any prior information available about the process and the constraints imposed by cost and other practical considerations. The factors may be ordered or unordered, according to whether we believe they represent some underlying, perhaps unobservable, numeric quantity. The entire design will usually be selected from some statistical definition (as we discuss in Chapter 5). Using this design, the experiment is run. To the design object are then added the responses and other observed variables to produce a data frame describing the experiment.

You have already seen designs if you have read Chapter 1. There we discussed experiments to study alternatives in the wave-soldering procedure for mounting electronic components on boards. Two experiments performed on the wave-soldering process were shown there. Let's return to that example and consider it more closely. Factors in both of the experiments were:

- opening: the amount of clearance around the mounting pad allowed in the solder mask;

- solder amount;

- mask: a composite factor coding the type and thickness of the material used for the solder mask;

- pad type: the geometry and size of the pad on which the component was to be soldered;

- panel: each board was divided into three panels at the beginning, middle, and end of the board (allowing three different experimental units to be defined in each board).

Such experiments are difficult and important exercises in pursuing quality in manufacturing, so our general goal of providing precise and flexible computational facilities is needed to get the most information from the experiment. As we saw in Chapter 1, these experiments also provided some interesting challenges in data analysis. An article by Comizzoli, Landwehr, and Sinclair (1990) discusses the background to this and similar experiments. It is an interesting example of the interaction between materials science and statistics, and well worth reading.

The design object `solder` contains all the data from the experiment at the first site:

```
> sampleruns <- sample( row.names(solder), 10)
> solder[ sampleruns,]
        Opening Solder Mask PadType Panel skips
865           M   Thin   B6      W9     1     3
844           M   Thin   B6      D4     1     5
 40           M  Thick A1.5      D6     1     0
689           M   Thin   B3      L9     2     0
636           L   Thin   B3      D4     3     5
836           L   Thin   B6      W9     2     0
757           M  Thick   B6      L4     1     5
493           M   Thin   A6      L6     1     2
440           S  Thick   A3      L7     2     7
834           L   Thin   B6      L8     3     9
```

The variable `skips` is the response: a count of the number of visible soldering skips on the particular run. The other variables are the factors described in the experiment; for example, `Solder` is solder amount, and `Mask` is a combined factor for solder mask type and thickness (combined to economize on the size of the design). The row names are just the run numbers in the experiment.

The dataset `solder.balance` shown in Chapter 1 was extracted from `solder`. The design object `solder2` contains the data from the second wave-soldering experiment.

3.1.3 Example: A Marketing Study

For the third example, we look at data used in a marketing study at AT&T. A survey of 1000 households was carried out to characterize customers' choice of a long-distance company.[1] Although relatively recent (1986), the study may already be regarded as historic, in the sense that it took place during the aftermath of the divestiture of the local telephone companies in the United States from the Bell System. As part of the divestiture process, each telephone subscriber was asked to pick a primary long-distance telephone company.

Data in the study we will examine were obtained from three sources: a telephone interview survey of selected households; telephone data based on service and billing databases; and demographic data, taken from a separate marketing database. The study hoped to develop a model to predict the household's tendency to pick AT&T. The model would help the marketing group target their efforts more effectively.

Variables from the survey included in our version of the data are:

---

[1]This example was kindly provided for us by James W. Watson.

- the income level of the household;

- the number of times the household had moved in the last five years;

- the age, education level, and employment category of the respondent.

Variables from the telephone records are:

- whether the household picked AT&T;

- the average monthly telephone usage of the household;

- whether the household had an unpublished telephone number;

- whether the household participated in some special AT&T plans or calling card service (before the choice of long-distance company).

Demographic variables included are:

- measures of affluence, such as mean income per household, average number of cars per household, and median housing value;

- household size (average number of people per household);

- the racial patterns, in particular the percent black population;

- the employment profile, in terms of percent professional, percent college graduate, and percent management;

- a clustering, presumably based on more extensive demographic data, that assigns a cluster number to each census block group.

Substantially more than this already large amount of data could have been included; many demographic variables were dropped when they appeared to add little to the predictions.

This example illustrates many of the features that make analysis of business data interesting but challenging. The data are highly "ragged": although we will treat all the variables as observations on the 1000 respondents to the survey, there are many missing values. The demographic data are unavailable for about 30% of the observations, and in any case are not observed on the individual households but rather on demographic entities determined by their addresses. The telephone data were derived (no doubt with considerable difficulty) from databases not originally intended to support data analysis at all. Nearly all the data, but particularly the survey data, can be expected to be based at least partly on subjective opinion, open to questions about meaning, and prone to errors.

The ten variables in the survey data are in data frame `market.survey` and the complete data, including nine additional demographic variables, are in `market.frame`. Both data frames have 1000 rows. Instead of sampling the data as we did in previous examples, let's make some graphical summaries of entire variables:

```
attach(market.frame)
plot(age)
plot(nonpub,usage)
```

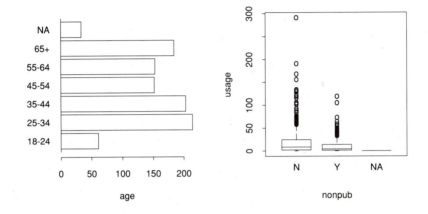

Figure 3.1: *Two graphical summaries of the marketing data,* market.frame. *The left panel shows the number of observations for each level of the ordered factor* age. *The right panel plots the distribution of* usage *(monthly telephone usage) separately for the two levels of* nonpub.

The style of attaching a data frame and exploring the variables through a variety of interactive graphics and other computations is a rewarding one, and we recommend it as a preliminary before doing any formal modeling.

3.2 Computations on Data Frames

We proceed now to show how computations on data frames can be carried out in S. As you use the modeling software described in this book, you will find yourself carrying out the following general computations involving data frames:

- setting up some data as a data frame;

- extracting, inserting, or modifying variables and/or rows of the data frame;

- plotting and summarizing the variables;

- creating a fitted model with a model formula involving the variables in the data frame;

- examining the fitted model using the data; for example, by plotting residuals in various ways against variables in the data frame.

Section 3.2.2 presents functions to create data frames. Section 3.2.3 provides methods to use and modify data frames. Some general summary functions for data frames are described in Section 3.2.4. Additional plots and summaries using any of the features of S can be carried out by attaching the data frame, as described in Section 3.2.3.

Throughout, the appropriate computations are more intuitively obvious if we keep in mind the two ways to think of data frames. First, a data frame can be regarded as a matrix with the variables being the columns and the observations the rows, but with the variables not restricted to any particular mode or class of object. Second, a data frame can be used as a frame within which the variable names define separate objects.

3.2.1   Variables in Data Frames; Factors

Variables in data frames can be anything that is indexed by the set of rows. However, variables that can be used for statistical models in this book are of three forms:

1. numeric vectors;

2. factors and ordered factors;

3. numeric matrices.

In each case, the variable is implicitly indexed by the rows of the data frame—the observations. That is, numeric and factor variables must have length equal to the number of observations, and matrix variables must have the same number of rows as there are observations.

Numeric vectors and matrices should be familiar to any S user. We use the term "numeric variable" throughout this book to distinguish this type of variable from a factor. This replaces the terminology "quantitative variable" or "continuous variable" for such variables. Factors are not described in $\boxed{\text{S}}$, and are introduced in Section 2.2.1. A factor is an object that represents repeated values from some set. The set is the `levels` attribute of the factor and is represented by a character vector. For example, suppose `sex` is a factor of length 10 with levels `"Male"` and `"Female"`.

```
> sex
 [1] Male    Male    Male    Female Female Male    Female Male    Male
[10] Male
```

When operating on factors, think of them as containing character strings, but with only strings from the `levels` allowed. Factors can contain missing values, while character vectors can not. On the other hand, factors are *not* numeric objects. Most basic functions behave as if they were operating on character data:

```
> sex == "Male"
 [1] T T T F F T F T T T
> sex+1
Error in call to "+": Non-numeric first operand
```

This is the key difference from categories, in which the numeric coding used in the implementation is not transparent to the user. Occasionally, one needs to replace a category by some corresponding numeric codes; in this case, use the function `codes()`:

```
> codes(sex)
 [1] 2 2 2 1 1 2 1 2 2 2
```

Some care is taken to make factors be well defined by the `levels` as a set; for example, neither the comparison nor the call to `codes()` depends on the ordering of the elements of `levels(sex)`. That is, whether the first element in `levels(sex)` is `"Male"` or `"Female"` should make no difference to any calculation involving `sex`. The levels of a factor form a *set*.

Users who became familiar with categories as described in $\boxed{\mathbf{S}}$ can think of factors as reimplementing categories as a class. Compared to categories, factors behave more strictly according to the model that says they are repeated values from some set of possible values.

An extension of factors is the class of `ordered` factors. Conceptually, these differ from factors in that the levels are taken to be ordered; now, the levels are not a set but a vector with elements in increasing order. A factor or a vector can be turned into an ordered factor by the function `ordered()`:

```
> osex <- ordered(sex, c("Male", "Female"))
> osex
 [1] Male   Male   Male   Female Female Male   Female Male   Male
[10] Male

Male < Female
```

The second argument is the levels in their desired order. Ordered factors can be created with any number of levels. For modeling, an ordered factor with only two levels behaves identically to an unordered factor. With three or more levels, the choice of contrasts in a linear model will be different for an ordered factor (see Section 2.3.2). Models that use different fitting methods from those in linear models may treat ordered factors differently; see, for example, Section 9.4 for tree-based models.

The function `ordered()` can also be used on the left of an assignment, with the same effect. For example,

```
> osex <- sex
> ordered(osex) <- c("Male", "Female")
```

would have the same effect as the previous computation to create `osex`.

Variables in data frames can contain missing values. The computations described in this chapter to create and manipulate data frames allow missing values and do nothing to remove them. The model-fitting methods, on the other hand, mostly do not allow missing values, so that some computations need to be done to the data before fitting the model. By default, most of the fitting functions will generate an error if the missing values are still there when the model frame is computed (see Section 3.3.3).

The functions in Section 3.2.2 to create data frames convert character or logical vectors to factors, since the modeling software interprets non-numeric variables as factors. If necessary, one can prevent this conversion; indeed, as we said, variables *can* be anything at all, so long as indexing them by the rows makes sense. All that is needed to ensure that a variable to be created in a data frame does not get converted is to make the expression that defines it an argument to the `I()` function. Thus, while the character vector `state.abb` would be converted to a factor, the expression `I(state.abb)` would stay a character vector. Section 3.3.1 goes into the implications of nonstandard variables in data frames. However, the three forms listed on page 52 are the most commonly useful, particularly since most statistical models are incapable of handling more general variables directly.

3.2.2   Creating Data Frames

New data frames can be created in a number of ways: by reading in data from an external file; by binding together objects of various kinds; by replacements or additions to existing data frames; and by other specialized methods. We begin by discussing how to read in an external table-like file, using the function `read.table()`. Starting on page 59, we discuss the function `data.frame()`, which combines other objects into a data frame. On page 62 we introduce a more specialized function, `expand.grid()`, to create a data frame over a regular grid of values. In addition to these specific functions, a list object can be coerced to be a data frame; see page 63.

Reading Tables of Data

Data frames are naturally analogous to printed tables, with the columns of the table as variables, the column labels as variable names, and the row labels as row names. Data from such a table should start as a text file outside of S and be made into a data frame by calling the function `read.table()`.

Consider the automobile data on page 46. The original data (from the viewpoint of a reader of *Consumer Reports*) consist of a number of tables along with verbal summaries that implicitly define additional tables. Specifically, `cu.summary` is a data frame containing information selected from the summary descriptions in the magazine. This table is a good example of data entry starting from a non-computerized source. To enter such data, you should begin by deciding what data you want, and by then having the data recorded as a text file. Often, you may have received data in this form, or close to it. Tables produced from other software (for example, spreadsheet software) often resemble printed tables, perhaps after cleaning up special control information.

The function `read.table()` is designed to read files that look like printed two-way tables. Let's begin with a simple example, a file that is essentially like the printout on page 47:

```
                   Price Country Reliability Mileage    Type
  Acura.Integra.4  11950   Japan           5      NA   Small
      Audi.100.5  26900 Germany          NA      NA  Medium
      BMW.325i.6  24650 Germany           4      NA Compact
Chevrolet.Lumina.4 12140     USA          NA      NA  Medium
   Ford.Festiva.4   6319   Korea           4      37   Small
    Mazda.929.V6  23300   Japan           5      21  Medium
 Mazda.MX.5.Miata 13800   Japan          NA      NA  Sporty
 Nissan.300ZX.V6  27900   Japan          NA      NA  Sporty
Oldsmobile.Calais.4 9995     USA           2      23 Compact
Toyota.Cressida.6 21498   Japan           3      23  Medium
```

If this is on a file named `"auto1"` it can be read in and turned into a data frame by a simple call to `read.table()`.

```
> somedata <- read.table("auto1")
> dim(somedata)
[1] 10  5
> dimnames(somedata)
[[1]]:
 [1] "Acura.Integra.4"    "Audi.100.5"         "BMW.325i.6"
 [4] "Chevrolet.Lumina.4" "Ford.Festiva.4"     "Mazda.929.V6"
 [7] "Mazda.MX.5.Miata"   "Nissan.300ZX.V6"    "Oldsmobile.Calais.4"
[10] "Toyota.Cressida.6"

[[2]]:
[1] "Price"       "Country"       "Reliability" "Mileage"       "Type"
```

Let's look at this example in a little more detail to see what is going on. The file contained fields separated by "white space", one or more blanks or tabs. The first line of the file contained five fields, meant to be the names for the variables; the remaining lines had six fields, the first being the row label and the rest the data for this observation. Some of the fields are numeric; others are character strings. The character strings will be turned into factor variables, as we can see by the following:

```
> sapply(somedata, data.class)
     Price   Country Reliability  Mileage      Type
 "numeric"  "factor"  "numeric"  "numeric"  "factor"
```

(The function `data.class()` returns the class of an object, but for objects without a class it tries a little harder to figure out what the object might be.) Notice that the numeric fields could have `NA` values in the input.

This is a good general form for a file to input with `read.table()`. You don't need to have either the variable names or the row labels; if they are missing, you can construct them later on. For example, suppose file `"auto2"` contained:

```
11950   Japan      5      NA    Small
26900  Germany     NA     NA   Medium
24650  Germany     4      NA  Compact
12140    USA       NA     NA   Medium
 6319   Korea      4      37    Small
23300   Japan      5      21   Medium
13800   Japan      NA     NA   Sporty
27900   Japan      NA     NA   Sporty
 9995    USA       2      23  Compact
21498   Japan      3      23   Medium
```

The expression `read.table("auto2")` turns this into a data frame.

The following options control the behavior of `read.table()`:

- Character strings by default cannot contain internal white space, hence the "." characters used in the row labels. You can allow white space by quoting the strings, by using an explicit default field separator character, or by organizing the data in fixed-format fields. In the example, we could quote all the model names used as row labels:

  ```
  "Acura Integra 4" 11950    Japan         5       NA    Small
  ```

  The field separator is the argument `sep=` to `read.table()`; to use a non-blank as a separator, say `":"`, we would need to replace all nonsignificant blanks by this character:

  ```
  Acura Integra 4:11950:Japan:5:NA:Small
  ```

  Notice that we edited out all extra blanks; when an explicit separator is used, blanks are significant in character fields.

- An optional argument `row.names=` to `read.table()` allows the call to specify where the row names come from, either giving explicit labels or specifying one of the fields to be used as row names instead of a variable. By default, the first field will be used for row names, if it is a non-numeric field with no duplicates. If all else fails, the row names are just the row numbers. They are *never* null and must be unique.

- Names for the variables can be supplied as the argument `col.names=`. By default, variable names will be read from the header line, if any; otherwise, the defaults are `"V1"`, etc. Explicit `col.names` override the default, but `read.table()` will still try to read over a header line so that unwanted column labels can be ignored. Names can also be changed after creating the object by using the function `names()` to assign new variable names to the data frame.

- The header line can be explicitly forced in by the argument `header=T`. This is needed if there is a column label for every column, as would be the case if the first line of the file `"auto1"` were

```
Model Price Country Reliability Mileage    Type
```

since in this case the first line has the same number of fields as the remaining lines, not one fewer.

Sometimes input data come in "fixed-format" fields; that is, it is known that each field starts in the same column on each line. In this case, the appearance of a particular separator character won't indicate a new field. The printed version of the automobile data on page 47 would make such a file. With a little tedious counting, we can find the starting columns of the fields. Providing this to `read.table()` reads in the data:

```
> columns <- c(1, 21, 27, 35, 47, 55)
> somedata <- read.table("auto",sep=columns, header = T)
```

With the explicit columns used to put in separators, the first line now looks just like the remaining lines, so `read.table()` can't tell automatically that there is a header line; instead, we used the explicit `header=` argument.

If you are not too sure about the column positions, it may be wise to create a new file with separators included, using the function `make.fields()`. Type `?make.fields` to see the detailed documentation for this function.

If you have been involved in many data-acquisition projects, your reaction to any list of possibilities such as the above may well be that serious projects tend to fall in the "none-of-the-above" category. If so, ad-hoc work will be needed, inside or outside S, to convert the data into a suitable form. Work outside S can involve any tools operating on text files; text editors (especially if they are reasonably programmable) and special languages like *awk* are helpful. Within S, two functions that may be useful are `count.fields()` and `scan()`. These functions count the number of fields in each line of a file and read data from a file. As an example of using them, consider the following sort of data. On each line of a file, we have recorded two variables, say X and Y, and then an arbitrary number of values of a third variable Z, corresponding to events measured under conditions given by X and Y. This is a common form of data; for example, imagine that the data record locations in the sky and the observed magnitude of all the stars within a certain radius of that location. The problem with such data is that there are a variable number of observations associated with each row. Putting aside the analysis for the moment, what is a convenient way to read the data in?

Here is one technique, broken into three steps. First, we get the number of fields on each line. Second, we read in all the data, ignoring lines. Third, the number of fields per line is used to extract X, Y, and Z. The following might be the data, on a file "sky.data":

```
10 30  5.9 10.9  8.2
20 40 13.8
30 50 10.7  8.8
40 60  9.8 11.0
50 30 13.4 11.9
60 40 9.2
70 50 9.4
80 60 13.4  7.2
```

Here are the first two steps:

```
> f <- count.fields("sky.data")
> f
[1] 5 3 4 4 4 3 3 4
> data <- scan("sky.data")
```

Now the problem is to split the data into the three variables. Doing this with a loop over the lines is fairly simple, and I will leave that as an exercise. The following technique is more typical of the advanced S user. It's not needed for understanding data frames, but you might find it interesting and helpful. We compute the positions of the X and Y values in the vector data, extract these values, remove them from data, and finally form Z by splitting what is left according to the original line.

```
> nline <- length(f)
> Xpos <- c(1, 1 + cumsum(f[-nline]))
> X <- data[Xpos]
> Y <- data[Xpos+1]
> data <- data[-c(Xpos, Xpos+1)]
> Z <- split(data, rep(1:nline,f-2))
```

The S function cumsum() gives the cumulative sum of its argument. Convince yourself that the expression above gives the indices in the data of all the X values. The call to split() splits the data according to the values of its second argument: all the values corresponding to a 1 into the first element, and so on. The call to rep()

returns as many 1s as there are Z fields on the first line, followed by as many 2s as there are Z fields on the second line, etc.

Having formed the three variables, one can combine them into a data frame using the `data.frame()` function to be described in the next section:

```
> sky.data <- data.frame(X, Y, Z=I(Z))
```

The function `I()` is used to keep Z as a single variable of mode `"list"`. The standard use of the model-fitting functions cannot treat such variables directly, so some additional calculations will be needed. For example, linear models to fit Z as a function of X and Y could use

```
> Zmean <- sapply(Z,mean)
> W <- sapply(Z,length)
```

to get a response `Zmean` and weights `W`, assuming all the individual observations were independent. Two possibilities for more direct use of this sort of data would be as a response in a general linear model in Chapter 6 (with a special family definition) or in a nonlinear model as in Chapter 10.

Combining Variables into a Data Frame

If the variables to be included in a data frame already exist in one or more S objects, the function `data.frame()` is the usual way to combine those objects into a data frame. For example,

```
> state <- data.frame(state.abb, state.center, state.x77)
```

takes data in the three arguments and creates a data frame combining all of them.

The arguments to `data.frame()` are an arbitrary number of objects, each of which will contribute one or more variables in the data frame returned. These arguments can be more general than items 1–3 on page 52, including the following kinds of objects:

1. Numeric vectors, factors and ordered factors: these each contribute a single variable.

2. Character or logical vectors: these are converted into factors. The levels will be the set of distinct values in the vector. The factor will not be ordered; the function `ordered()` will convert the data into an ordered factor.

3. Matrices: each column creates a separate variable in the data frame. Column names of the matrix, if any, are used for variable names.

4. Lists: like matrices, these contribute one variable for each component of the list, according to the rules we are outlining here, applied recursively. (For example, a character vector that is a component of a matrix is turned into a factor.)

5. Data frames: the variables in the data frame become variables in the result, essentially without processing.

If any argument to `data.frame()` is of the form `I(x)`, then `x` will be inserted as is into the data frame as a single variable. In particular, logical vectors, character vectors, matrices, lists, and data frames will *not* be converted as described above if protected by `I()`. Needless to say, if you put something in that makes no sense, later calculations with the data frame are likely to run into trouble. Basically, though, you can put anything you like into a data frame and compute away with it, as we illustrated in creating `sky.data` above.

As `data.frame()` proceeds through its arguments, it attempts to compute variable names and row names. For row names, the function takes the first reasonable candidate that comes along: a names attribute for a vector argument, row labels for a matrix, or row names for a data frame. This can all be overridden by supplying the argument `row.names=`. The value of this argument can be one of the variables in the resulting data frame, specified either by number or by name, in which case that variable becomes the row names attribute and is deleted as a variable. Alternatively, the row names argument to `data.frame()` can supply a specific vector of names. Wherever they come from, row names must be unique.

Variable names can be specified by naming the actual argument to `data.frame()`. For matrix, data frame, or list arguments, the appropriate column labels, variable names, or names attribute will be used (with an actual argument name pasted on if supplied). A rule is enforced that variable names in a data frame must be syntactically S names, made up of letters, numbers and ".", and not starting with a number. Variable names must also be unique. The function `make.names()` is used to ensure both these conditions; see its documentation for details of the algorithm used. You can get any variable names you want, simply by assigning the `names()` attribute of the data frame after creating it. Remember, though, that when data frames are attached or used with formulas in models, life will be much simpler if the variable names are really names that can appear in S expressions.

To illustrate, let's generate a data frame in which the rows correspond to the states of the United States. The standard S database contains several objects with related data ([S] , page 658): `state.name` and `state.abb` are character vectors with the name and its official abbreviation, `state.center` is a list whose components are the x and y co-ordinates for plotting on maps, `state.region` and `state.division` are factors for regions and divisions, and `state.x77` is a matrix with some demographic data. These are just the sort of data for which a data frame is a convenient structure, with the states corresponding to rows, and the variables containing different kinds of information about the states.

```
> state <- data.frame(state.center, state.x77,
+    row.names = state.abb)
> state[1:5,]
```

	x	y	Population	Income	Illiteracy	Life.Exp
AL	-86.7509	32.5901	3615	3624	2.1	69.05
AK	-127.2500	49.2500	365	6315	1.5	69.31
AZ	-111.6250	34.2192	2212	4530	1.8	70.55
AR	-92.2992	34.7336	2110	3378	1.9	70.66
CA	-119.7730	36.5341	21198	5114	1.1	71.71

	Murder	HS.Grad	Frost	Area
AL	15.1	41.3	20	50708
AK	11.3	66.7	152	566432
AZ	7.8	58.1	15	113417
AR	10.1	39.9	65	51945
CA	10.3	62.6	20	156361

The first argument was a list. Its two components, x and y, became separate variables. The second argument was a 50 by 7 matrix, providing seven more variables. The column labels of state.x77 were:

```
> dimnames(state.x77)[[2]]
[1] "Population"     "Income" "Illiteracy"     "Life Exp"
[5]     "Murder"   "HS Grad"       "Frost"         "Area"
```

Notice that these were automatically converted to legal names (by changing blanks to dots). The optional row.names= argument forced the row names to be the abbreviations in state.abb.

Now we will put in some additional information, and use argument names to data.frame() to control the variable names.

```
> state <- data.frame(state.center, state.x77,
+    name=state.name, region=state.region, row.names = state.abb)
> state[1:5,]
```

	x	y	Population	Income	Illiteracy	Life.Exp
AL	-86.7509	32.5901	3615	3624	2.1	69.05
AK	-127.2500	49.2500	365	6315	1.5	69.31
AZ	-111.6250	34.2192	2212	4530	1.8	70.55
AR	-92.2992	34.7336	2110	3378	1.9	70.66
CA	-119.7730	36.5341	21198	5114	1.1	71.71

	Murder	HS.Grad	Frost	Area	name	region
AL	15.1	41.3	20	50708	Alabama	South
AK	11.3	66.7	152	566432	Alaska	West
AZ	7.8	58.1	15	113417	Arizona	West
AR	10.1	39.9	65	51945	Arkansas	South
CA	10.3	62.6	20	156361	California	West

The last two arguments would have produced variables state.name and state.region; we supplied shorter names. If we had named either the second or third arguments,

that name would have been pasted together with the individual component or column names. Using `center=state.center`, for example, would give:

```
     center.x center.y Population Income Illiteracy Life.Exp
AL   -86.7509  32.5901       3615   3624        2.1    69.05
AK  -127.2500  49.2500        365   6315        1.5    69.31
AZ  -111.6250  34.2192       2212   4530        1.8    70.55
AR   -92.2992  34.7336       2110   3378        1.9    70.66
CA  -119.7730  36.5341      21198   5114        1.1    71.71
```

The state names were converted to a factor by the general rule given above. Occasionally, you may want to retain them as a character vector, if they should never be
used as a factor in modeling. The function `I()` can again be used for this purpose:

```
> state <- data.frame(name = I(state.name), state.center, state.x77)
```

by using the `I()` function as mentioned before.

Data Frames from Regular Grids

For some applications, one wants to generate pseudo-observations that form a grid
over some specified variables. When such data are to be given to a function that
expects a data frame as its argument, the grid needs to be used to generate the
corresponding data frame. The `predict()` methods, in particular, expect new data
to be a data frame. They return the values that the fitted model would predict to
correspond to the observations in this data frame. Suppose `Weight` and `Disp.` from
`cu.specs` were used as predictors in some model. To look at predictions from the
model over the range of the two variables, we might ask for all the pairs of values
from the two vectors:

```
> pretty(Weight)
[1] 1500 2000 2500 3000 3500 4000
> pretty(Disp.)
[1]   50 100 150 200 250 300 350
```

There are $6 \times 7 = 42$ pairs of values. Since prediction and other similar computations
usually want their input as a data frame, the function `expand.grid()` takes marginal
specifications of the values and creates a data frame with all the combinations:

```
> WDgrid <- expand.grid(Weight = pretty(Weight), Disp. = pretty(Disp.))
> dim(WDgrid)
[1] 42  2
> WDgrid[1:5,]
  Weight Disp.
1   1500    50
2   2000    50
```

```
3    2500    50
4    3000    50
5    3500    50
```

The call to `expand.grid()` should provide the names for the variables in the new frame, along with the marginal values. To make it easy to alter the specifications of the grid, the entire set of arguments can be replaced by a single list argument:

```
spec <- list(Weight = pretty(Weight), Disp. = pretty(Disp.))
WDgrid <- expand.grid(spec)
```

The grid variables can also be factors; the analogue to a range of values for a numeric variable is the `levels` for a factor. If the argument to `expand.grid()` is such a character vector (or any character vector), the corresponding variable in the result will be a factor with each level appearing the same number of times:

```
> WDgrid2 <- expand.grid(Type = levels(Type), Weight = range(Weight))
> WDgrid2
      Type Weight
1  Compact   1845
2    Large   1845
3   Medium   1845
4    Small   1845
5   Sporty   1845
6      Van   1845
7  Compact   3855
8    Large   3855
9   Medium   3855
10   Small   3855
11  Sporty   3855
12     Van   3855
```

Keep in mind that the total size of the grid can grow very quickly, since the number of rows is the product of the length of all the arguments!

The rows of the data frame produced by `expand.grid()` are ordered in "standard" order for a multiway array defined by the arguments to `expand.grid()`. So, for example, any vector whose elements correspond to the rows of `Wgrid2` could be made into a 6 by 2 matrix. This is, in fact, what some of the prediction methods arrange to do (see, for example, Section 8.2.4). The same operation can be done in general by giving the corresponding vector, along with the data frame, to the function `make.grid()`. Another example of `expand.grid()` is given on page 81, in illustrating its use with the `coplot()` function.

Coercing to a Data Frame

Since a data frame has characteristics both of a matrix and of a list, it is reasonable that either of these structures could be turned into a data frame. So they can, by calling `as.data.frame(object)`. This is the preferred way to create a data frame if the variables are naturally built up in an ordinary matrix or in a list. Any matrix can be turned into a data frame, but the operation is only likely to be useful if you plan subsequently to add new variables of a different class to the matrix or if the matrix contains character data that you want to treat as factors. If the data are all numeric, the data frame has no more information than the matrix.

Lists to be turned into data frames should have all elements of the same length and should have non-null, unique names.

3.2.3   Using and Modifying Data Frames

Two approaches to working with data frames are useful, either using matrix-like computations or attaching the data frame and using the variables as separate objects. The matrix view is most often useful when working with more than one data frame at a time and when extracting or replacing rows within a single data frame. To introduce new variables, to revise individual variables, or to study the variables interactively, the best approach is to attach the data. Some examples are given starting on page 67.

Data Frames as Matrices

Data frames can be treated as matrices in calls to most of the basic functions treating arrays: subsets and elements, `dim()`, `dimnames()`, and functions based on those. If `x` is a data frame, then

```
x[i,]; x[,j]; x[i,j]
dim(x); dimnames(x)
nrow(x); ncol(x)
```

produce results corresponding intuitively to their behavior on matrices. For example, `x[i,]` produces a new data frame by using `i` to index the rows of `x`. The indexing by `i` can use numeric, logical, or character values. Similarly, `x[,j]` indexes on the columns (the variables) and `x[i,j]` on both. When a single column is selected, the result is by default the variable, not a data frame containing one variable. This action can be suppressed by including the argument `drop=F`, following the rules applied to arrays ($\boxed{S}$, page 128). For example, if `stats` was some statistic computed for each of the variables in `market.frame`, the expression

```
market.frame[, stats > cutoff , drop=F]
```

ensures that the extracted object is still a data frame, even if it has only one column. A single row by default remains a data frame—there is no generally useful object corresponding to rows of a data frame. If you really want to, however, you can cause the single row to be dropped to a list by including the argument `drop=T`.

Variable names and row names may be abbreviated in selecting subsets, so long as the character strings given match the beginning of one unique name. In the `state` data, `state[,"P"]` would select the `Population` variable, but `State[,"I"]` would fail because it would partially match both `Income` and `Illiteracy`. Names given in replacement expressions must match exactly in order to replace rows or variables. These are just the standard S rules for partial matching, extended to data frames. Row and column replacements can specify positions not currently in the data. The result will be to extend the data frame as necessary. The example on page 66 uses this to combine data frames.

The `dim()` attribute is just what one would expect: a numeric vector of length 2 containing the number of observations and the number of variables. The list returned by `dimnames()` contains the row names and the variable names as its two elements. Notice that by the construction of data frames, both of these should contain a set of unique names, in contrast to a matrix in which `dimnames()` or either of its elements could be empty. These two attributes can also be replaced or modified by putting the expression on the left of an assignment.

Computations that want to use data frames as ordinary matrices can convert them. The standard coercing function, `as.matrix()`, has a method for coercing data frames. The technique used by

```
as.matrix(x)
```

is to find the ordinary S mode required to represent the data. The most typical case, if any of the variables in `x` is not numeric, is that the matrix will be of mode `"character"`. It is also possible for the resulting matrix to have dimension different from that of `x`, if any of the variables in `x` was itself a matrix (see the discussion in Section 3.2.2).

A second kind of conversion to an ordinary matrix is provided by the function `data.matrix()`. This function tries its best to interpret the variables in the frame as *numeric* data. In particular, it converts any factors or ordered factors to numbers representing the levels, by calling the function `codes()` for each of them. In order not to lose information, the factor levels are kept in a list, as attribute `"column.levels"` of the resulting matrix. Like `as.matrix()`, `data.matrix()` will expand any matrix variables in the data frame.

Another relation between data frames and matrices arises during the fitting of models. Most of the models discussed in this book proceed by creating a numeric matrix that describes all the terms included in the model. These objects belong to the `"model.matrix"` class, and can be generated by calling `model.matrix()`. These matrices encode appropriately all terms in a model, to produce a numeric matrix suitable for fitting. We describe them in Section 2.4.3, in discussing how variables of various kinds can be coded numerically. In particular, factors and ordered factors are converted to numeric variables derived from the "dummy" variables that code the presence of each level of each factor.

The three functions each produce different results, if any variable in the data frame is a factor or an ordered factor. Which result is right depends on the circumstances. Models in Chapters 4–7 use `model.matrix()`, models in Chapter 8 use a special function `loess.matrix()`, the tree-based models in (Chapter 9) use `data.matrix()`, and printing methods tend to use `as.matrix()` or a similar calculation.

An Example: Combining Data

To illustrate data frames as matrices, we will combine some of the data in the three data frames of automobile data we discuss in Section 3.1.1 to form a single data frame. The goal is to match corresponding model names in the three data frames, and create a single data frame. Since the model names in the three original data frames are not quite consistent, we will use the S function `pmatch()` and some hand-editing to match rows. This sort of preliminary data cleaning is typical of most data analyses. All the details would take too much space to show, but we can convey the style of the computations. You can skip to the frames discussion on page 67 if you are not interested in the example.

We first try to match row names in the three data frames as well as possible. The practical problem is that the three data frames use slightly different ways to refer to the automobile models. Of the three original frames, `cu.dimensions` seems to have the cleanest set of row names; sorting them and editing a little by hand produces our 111 target row names, saved in `common.names`. The row names in the other data frames tend to have extra characters after some of the model names.

The function `pmatch()` finds row names in the data frames that contain exactly the common names, or a unique match with extra characters after the common names:

```
match.summary <- pmatch(common.names, row.names(cu.summary))
match.specs <- pmatch(common.names, row.names(cu.specs))
```

These two vectors contain row numbers in `cu.summary` and `cu.specs` that we can match to names in `common.names`. We will use these matches to bind together rows from all three data frames. There will be `NA`'s in the vectors where names didn't match uniquely. The next step is to try to match those names by hand. Editing of the matching vectors and perhaps of `common.names` will occur.

Once we are satisfied with the matches, the new data can be set up by

```
> car.all <- cu.dimensions[match.dims, ]
> car.all[, names(cu.summary)]  <- cu.summary[match.summary, ]
> car.all[, names(cu.specs)] <- cu.specs[match.specs, ]
> row.names(car.all) <- common.names
```

Notice that we appended the new columns to `car.all` using the names for the columns in the original frames. Although we started with the rows of `cu.dimensions`, we end up with a `match.dims` as well, after editing and sorting. It was essential to check before the above calculation that the variable names were unique:

```
> any(duplicated(c(names(cu.summary), names(cu.dimensions),
+     names(cu.specs))))
[1] F
```

Once uniqueness is assured, the computations above create a data frame with all the data, carrying over the individual variable names, and using `common.names` to name the rows.

As another example, done similarly, we construct a data frame with just the automobiles for which Consumers Union test data are available. Since the `Mileage` variable in the `cu.summary` data frame is missing unless the model was tested, we begin by selecting rows from `cu.summary`:

```
> ok <- !is.na(car.all[,"Mileage"])
> car.test.frame <- cu.summary[ok,]
```

From a process of matching row names from `cu.specs`, we then added some variables from `cu.specs`. Both `car.test.frame` and `car.all` will appear in examples in this and later chapters.

Data Frames as Frames or Databases

A *frame* in S is a mapping of names to objects for the purpose of evaluating S expressions using the names. The S evaluator maintains a frame to hold the arguments in a call to an S function, plus any local assignments, during the evaluation of the call. A *database* is a directory, S object, or other permanent construction that is attached via the `attach()` function to define a similar mapping of names to objects. The files in the directory, the components of the object, or whatever mapping the database implies make objects available by name to subsequent S expressions.

A data frame can be attached as a database and can be used as a frame for evaluation. In either case, each of the variables in the data frame becomes available, by name, as a separate object. The model-fitting functions to be described in later chapters all take a data frame as an optional `data` argument.

```
fuel.fit <- lm( Fuel ~ Weight + Disp., fuel.frame)
```

The variables in the data frame `fuel.frame` include all the names appearing in the formula. These variables will be automatically made available by name during the computation of the fit. Section 3.3.3 will discuss the details of what happens when data frames are used in model-fitting.

To do interactive computations with the variables in a data frame, you should attach it; for example,

```
> attach(cu.specs, 1)
> h <- Weight/Disp.
```

This is the recommended way to revise a data frame, whether changing existing variables or creating new ones. Notice the second argument to `attach()`: the data frame is attached in position 1, as the working data. As a result, ordinary assignments, such as that for `h` above, take place in the attached database. A secondary benefit is that variables in the data frame will not be hidden by objects in the previous working database (typically the local `.Data` directory).

Any calculations used to create or modify objects in the attached database do *not* modify the original data frame. In the example, `h` is created in the working data, an object internal to the S evaluator. The call to `attach()` initialized that object as a copy of the data frame `cu.specs`. From then on, assignments only changed the internal object. You can save a copy of this object at any time. The easiest, and the recommended, way to save it is at the time the attached database is detached; for example,

```
detach(1, save = "new.specs")
```

will save the revised data frame as the object `new.specs`, in the database in second place on the search list before detaching. This is typically the local `.Data` directory.

This sequence of steps is the recommended way to revise data frames:

- Attach a data frame as in position 1.

- Carry out any computations to revise or create variables.

- Detach and save the database in position 1.

For most purposes, you do not need to know any further details, but we provide a few here anyway, in case they may be relevant.

While the data frame is attached, its class is ignored in assignments. The objects created will not generally be valid variables in the data frame. For example, here is a natural way to take logs of a variable known to have zero, but not negative, values:

```
> attach(market.frame, 1)
> uu <- .0001*max(usage)
> logU <- log(usage + uu)
```

The object `logU` is a new variable, but `uu` is just a single number. Although the class of the attached database has not been lost, the internal object is not itself a valid data frame at this point. To turn it into such an object, the evaluator looks for a method for the generic function `dbdetach()`. There is such a method for data frames; specifically, it deletes all objects in the database that are not the right size to become variables. Only objects that have length or number of rows equal to the `row.names` attribute of the database are retained. Another class of objects can be used if you *do* want to retain these objects of arbitrary length along with the variables—see Section 3.3.4.

Attaching a data frame is also a good way to examine the results of a fit. Each kind of model provides methods for overall summaries and diagnostics. After looking at these, you will often want to examine the fit further, plotting variables from the data frame in a variety of ways along with components of the fit. Attaching the data frame makes such computations simple.

As always in S, the data frame attached or included as an argument can be any expression that evaluates to a data frame, not just the name of an object:

```
attach( car.all[1:50,] )
```

attaches a data frame from the first 50 rows of `car.all`.

3.2.4 Summaries and Plots

Plots and numerical summaries play a critical role in statistical modeling. Numerical summaries provide an incisive, although quite limited, quantification of aspects of the data such as the variation of measurements of a single variable or the degree of correlation between measurements of two variables. Plots have two roles: exploratory analysis before embarking on a first model, and diagnostic checking of fitted models. For a thorough, interactive analysis, the best approach is to attach the data frame to the search list and use as wide a range of appropriate computations in S as possible. In this section, we show the behavior of some generic functions that can be applied to entire data frames:

- `summary()`: print summaries;

- `plot()`: plot variables;

- `pairs()`: plot a scatterplot matrix;

- `coplot()`: plot, conditioning on other variables.

These functions are generic; that is, they have suitable methods for a wide range of objects. In the examples of this section, we will use the generic functions either with complete data frames

```
summary(car.all)
```

or with formulas

```
plot(Mileage ~ Weight, car.test.frame)
```

used to select or transform variables from the data frame. Of course, we can use
other S functions to plot and summarize, by giving rows or columns of a data frame
as arguments. For example, one function, `scatter.smooth()`, which we will also
show here, makes a scatterplot and adds a smooth curve.

The summary and plot functions are designed to give you an overall look at
the data. They try to do something reasonable in choosing information to print or
plot about the data variables and their interrelationships. Use them for an initial
view, to be followed by detailed study of interesting details. Generally, graphical
summaries are better for seeing unexpected features of the data than are printed
summaries. When there are *many* variables, there may simply be too many plots or
too much printed output, even for initial study. In that case, one approach is to look
at smaller subsets of variables. For example, from the `car.test.frame` constructed
above, we select the `Weight`, `Disp.` and `Mileage` variables for some simple analysis
of fuel economy:

```
> fuel.frame <- car.test.frame[,c("W","D","M")]
```

We will produce some plots of this data frame, and will use it in later model-fitting
examples.

Plots and summaries can be produced by using the generic `plot()` and `summary()`
functions with various classes of objects: data frames, individual variables, and
formulas. Giving a formula to a summary or plot implies that the response, if any,
and all the predictors in the formula should be displayed. As some examples will
show, this is a flexible and convenient way to produce summaries and plots.

Summaries

Calling `summary()` produces a printed summary of the variables:

```
> summary(car.test.frame)
        Price            Country     Reliability       Mileage
 Min.   : 5866   USA       :26   1    : 7       Min.   :18.00
 1st Qu.: 9870   Japan     :19   2    : 7       1st Qu.:21.00
 Median :12220   Japan/USA : 7   3    :12       Median :23.00
 Mean   :12620   Korea     : 3   4    : 6       Mean   :24.58
 3rd Qu.:14940   Germany   : 2   5    :17       3rd Qu.:27.00
 Max.   :24760   Sweden    : 1   NA's :11       Max.   :37.00
                 (Other)   : 2

       Type         Weight           Disp.            HP
 Compact:15   Min.   :1845   Min.   : 73.0   Min.   : 63.0
 Large  : 3   1st Qu.:2568   1st Qu.:113.5   1st Qu.:101.0
 Medium :13   Median :2885   Median :144.5   Median :111.5
 Small  :13   Mean   :2901   Mean   :152.1   Mean   :122.3
```

```
Sporty : 9    3rd Qu.:3243    3rd Qu.:180.0    3rd Qu.:143.5
Van    : 7    Max.   :3855    Max.   :305.0    Max.   :225.0
```

The summary for numerical variables, like `Price`, gives mean, median, smallest and largest values and first and third quartiles. For factors or ordered factors, like `Type` or `Reliability`, a table of counts is produced. For all forms, if there are missing values, the number of these will be printed.

Distribution Plots

A `plot()` method for the variables of a data frame generates plots summarizing the distribution of the variables. For numeric variables, quantile plots are shown; that is, if the data are in `x`, `sort(x)` is graphed against `ppoints(x)`. For factors, `plot()` graphs the counts for each level.

The method can be invoked by giving `plot()` a data frame. Where only some of the variables in the data frame are to be plotted, you can select a subset of the columns. This is illustrated in Figure 3.2:

```
plot(~ Country + HP, car.test.frame)
```

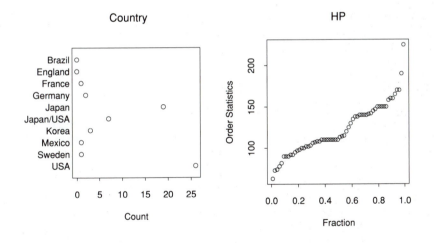

Figure 3.2: *Two distribution plots of data from* `car.test.frame`.

A flexible alternative is to use the model formulas to specify which variables to plot. A plotting method for formulas will produce scatter plots of the response (the left side of the "~" operator) against each of the terms (the expressions separated by "+" on the right side of the operator). If the left side is omitted, however, distribution

plots will be produced of each of the terms. The plots in Figure 3.2 could have been
made by:

```
> plot(~ Country + HP, car.test.frame)
```

This form is more flexible, in that any transformation or combination of variables
could have been specified in the formula. For a third option, the data frame could
have been attached, after which the formula can be given without a second argument:

```
> attach(car.test.frame)
> plot(~ Country + HP)
```

See Section 2.3.1 for further discussion of formulas.

Scatterplots, Scatterplot Matrices, and Smoothing

Let us focus now on three variables from `car.test.frame`: `Mileage`, `Disp.`, and
`Weight`. Our purpose is to study the dependence of fuel consumption on weight
and displacement. The scatterplot matrix is a useful graphical method for a first
look at the data to show inter-relationships among the variables:

```
> attach(car.test.frame)
> pairs(~ Mileage + Disp. + Weight)
```

The result is shown in Figure 3.3. To produce just the scatterplots of mileage
against weight and displacement, give a `formula` object to `plot()`:

```
plot(Mileage ~ Disp. + Weight)
```

Given a formula with a left side (a response), the plot method makes scatterplots
of the response against each of the terms on the right side. The result is shown
in Figure 3.4. The last two plots show that the variables are strongly associated,
and there is some suggestion of nonlinearities. In particular, the dependence of
`Mileage` on `Weight` appears to be somewhat curved. Let us add a smooth curve to
the scatterplot of `Mileage` against `Weight` to study the dependence more incisively:

```
> scatter.smooth(Mileage ~ Weight, span = 2/3)
```

The result is shown in the left panel of Figure 3.5. We have added the smooth
curve using a nonparametric regression procedure that is described in Chapter 8; the
argument `span` controls the amount of smoothness. The curve confirms the nonlinear
pattern. Since gallons per mile are as sensible for measuring gas consumption as
miles per gallon, it makes sense to attempt a straightening of the relationship by
an inverse transformation:

```
> scatter.smooth(100/Mileage ~ Weight, span = 2/3)
```

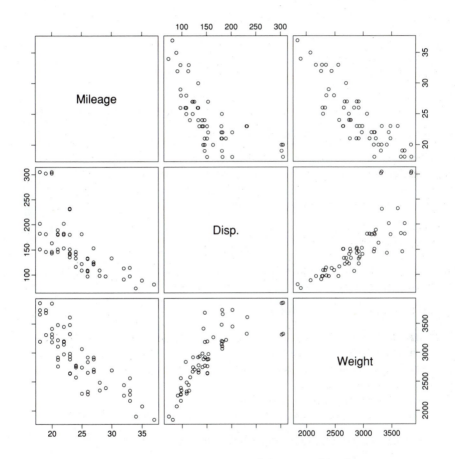

Figure 3.3: *Scatterplot matrix of measurements of three variables from* `car.test.frame`.

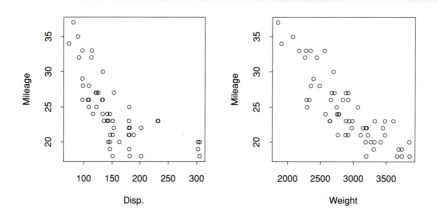

Figure 3.4: *Response against predictors.*

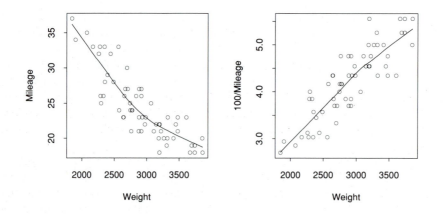

Figure 3.5: *Scatterplots of mileage and the inverse of mileage against weight with smooth curves added.*

The result is shown in the right panel of Figure 3.5. Now the dependence is nearly linear.

Since we have been having so much fun with cars, we will study the variables in the data frame `ethanol`, which are from an experiment with a single-cylinder automobile test engine (Brinkman, 1981). The data are graphed in Figures 3.6 and 3.7:

```
> plot(ethanol); pairs(ethanol)
```

The dependent variable, NO_x, is the concentration of nitric oxide, NO, plus the concentration of nitrogen dioxide, NO_2, in the engine exhaust. Concentration is normalized by the amount of work done by the engine, and the units are μg of NO_x per joule. One predictor is the compression ratio, C, of the engine. A second predictor is the equivalence ratio, E, at which the engine was run; E is a measure of the richness of the air and fuel mixture. There were 88 runs of the experiment, and as the name of the data frame suggests, ethanol was the fuel.

Coplots

A *conditioning plot*, or *coplot*, is a graphical method for seeing how a response depends on a predictor given other predictors. The technique is described in the second edition of the book *The Elements of Graphing Data*, (Cleveland, to appear). The function `coplot()` implements this graphical method for one or two given predictors. Coplots are used extensively in plotting with the local regression models in Chapter 8; look at the examples and discussion in that chapter for further motivation and details. The graphical technique is useful generally, however, so we introduce it here.

Figure 3.8 is a coplot of the ethanol data. The *dependence panels* are the 3×3 array of square panels and the *given panel* is at the top. On each dependence panel, NO_x is graphed against C for those observations whose values of E lie in an interval; thus, on the panel, we are seeing how NO_x depends on C for E held fixed to the interval. The intervals are shown on the given panel. As we move from left to right through the intervals in the given panel, we move from left to right and then bottom to top through the dependence panels. Figure 3.9 is a coplot of NO_x against E given C. Since C takes on five values, we have simply conditioned on each of these five values.

Figures 3.8 and 3.9 show us much about the ethanol data. For low values of E, NO_x increases with C, and for medium and high values of E, NO_x is constant as a function of C. Thus there is an interaction between C and E. Second, over the range of values of E and C in the dataset, NO_x undergoes more rapid change as a function of E for C held fixed than as a function of C for E held fixed. Finally, the plots show that the amount of scatter about the underlying pattern is small compared with the effect due to E and is moderate compared with the effect due to C.

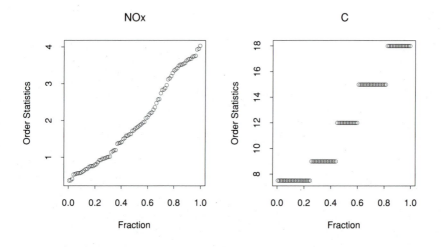

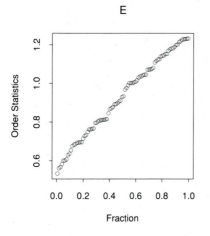

Figure 3.6: *Distribution plots for variables in* `ethanol`.

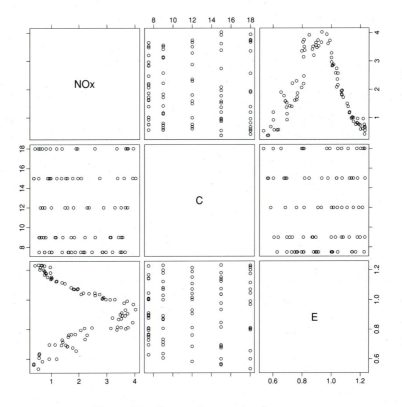

Figure 3.7: *Scatterplot matrix of* `ethanol`.

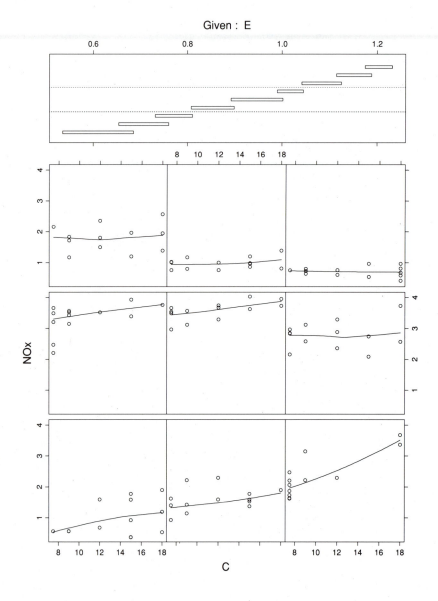

Figure 3.8: *Coplot of* NO_x *against* C *given* E.

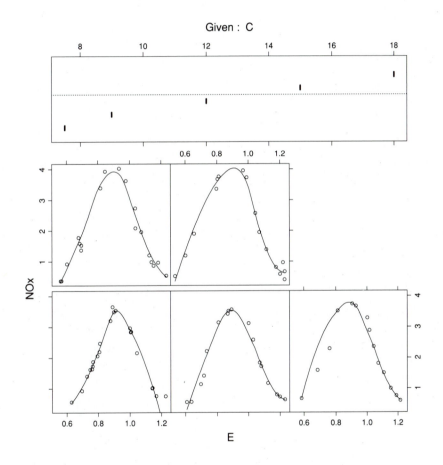

Figure 3.9: *Coplot of* NO_x *against* E *given* C.

Let us now see how we produced the two coplots, starting with Figure 3.8. First, the conditioning intervals were computed by the function `co.intervals()`.

```
> attach(ethanol)
> E.intervals <- co.intervals(E, number = 9, overlap = 1/4)
```

The result is a 9×2 matrix that gives the left endpoints of the intervals in the left column and the right endpoints in the right column:

```
> E.intervals
       [,1]  [,2]
[1,] 0.535 0.686
[2,] 0.655 0.761
[3,] 0.733 0.811
[4,] 0.808 0.899
[5,] 0.892 1.002
[6,] 0.990 1.045
[7,] 1.042 1.125
[8,] 1.115 1.189
[9,] 1.175 1.232
```

The intervals have the following properties: they contain approximately the same number of values of E, and the fraction of values shared by two successive intervals is approximately equal to `overlap`. Now we call `coplot()`:

```
> coplot(NOx ~ C | E, given.values = E.intervals, panel =
+    function(x,y) panel.smooth(x, y, degree = 1, span = 1))
```

The first argument is a `formula` that specifies the response, the predictor to plot against, and the given predictor; in our example we are graphing `NOx` against `C` given `E`. This is a special kind of formula, which uses the operator `"|"` to separate the predictor(s) from the conditioning variable. This operator is interpreted in formulas as "given", following its typical use in mathematics. With this interpretation, we read the formula as

"Model `NOx` by `C`, given `E`"

In the plotting functions, we extend "modeled" in a natural way to imply the informal process of looking at one variable as a function of another.

The argument `given.values` to `coplot()` specifies the conditioning values. For a numeric given predictor the values can be a two-column matrix as in the example, or can be a vector, in which case each element is both the left and right endpoint of an interval, so the intervals have length 0. We can also condition on the levels of a factor; in this case the argument is a character vector.

The argument `panel` takes a function of `x` and `y` that determines the method of plotting on each dependence panel; `x` refers to the abscissas of points on a panel and `y` refers to the ordinates. The default function is `points()`. In the above expression we have used the function `panel.smooth` to smooth the scatterplots. Notice a paradigm that is frequently useful in S: giving an in-line function definition to specify both a function and some optional arguments. In the example, we want `panel.smooth()` to be called with two specific values for optional arguments. This defines a new function, but instead of assigning it somewhere, we just include the definition in-line:

```
function(x, y) panel.smooth(x, y, degree = 1, span = 1)
```

The value of this expression is an S function, just what we need as the `panel` argument.

As with `scatter.smooth()`, the argument `span` to `panel.smooth()` controls the amount of smoothness. Also, the argument `degree` is specified. It controls the degree of the locally-fitted polynomials that are the basis of the smoothing method, and can take the values 1 or 2. Figure 3.9 was produced by the following expressions:

```
> C.points <- sort(unique(C))
> coplot(NOx ~ E|C, given.values = C.points, column.row = c(3,2),
+        panel = function(x,y) panel.smooth(x, y, degree = 2, span = 2/3))
```

In this case, `given.values` is a vector. The argument `column.row` has been used to specify the dependence panels to be arranged in an array with three columns and two rows.

Coplots of Fitted Functions

A coplot can also be used to display a surface fitted to a response as a function of two or three predictors. The points at which to do the plot will in this case be chosen by us, typically over a regular grid of values. The function `expand.grid()` will be used to generate a data frame corresponding to the grid. Let us look at one example. The data in data frame `air` are from an environmental study to determine the dependence of the air pollutant ozone on solar radiation, wind speed, and temperature:

```
> summary(air)
      ozone           radiation         temperature           wind
 Min.   :1.000   Min.   :  7.0    Min.   :57.00    Min.   : 2.300
 1st Qu.:2.621   1st Qu.:112.8    1st Qu.:71.00    1st Qu.: 7.400
 Median :3.141   Median :207.0    Median :79.00    Median : 9.700
 Mean   :3.248   Mean   :184.8    Mean   :77.79    Mean   : 9.939
 3rd Qu.:3.968   3rd Qu.:255.8    3rd Qu.:84.75    3rd Qu.:11.500
 Max.   :5.518   Max.   :334.0    Max.   :97.00    Max.   :20.700
```

Figure 3.10 is a scatterplot matrix of the data:

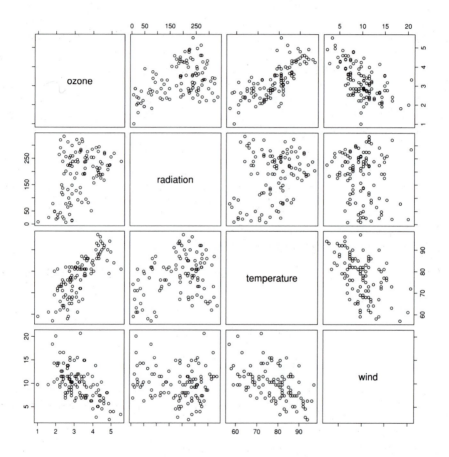

Figure 3.10: *Scatterplot matrix of* air.

```
> pairs(air)
```

In an analysis described in Chapter 8, a surface was fitted to ozone as a function of the three predictors. Suppose we want to make a coplot of the fitted surface against radiation, given wind and temperature. Denote the surface by $s(r, t, w)$. We want to condition on values of t and w and graph against r; that is, we graph $s(r, t^*, w^*)$ against r for various values of t^* and w^*. To do this, the surface was evaluated on a $50 \times 5 \times 5$ three-dimensional grid in the space of the predictors—50 values of radiation, 5 values of temperature, and 5 values of wind speed. Specifically, suppose object `air.marginal` is a list with three components: `radiation`, containing 50 equally spaced values of radiation from 125 to 250; `temperature`, containing 5 values of temperature from 70 to 85; and `wind`, containing 5 values of wind from 7.5 to 11.5 (a call to `seq()` will generate each of these). For plotting, we need a data frame with all the 1250 combinations of these values. The function `expand.grid()`, defined on page 62, turns the marginal values into a data frame:

```
> air.grid <- expand.grid(air.marginal)
```

`air.grid` is a data frame with 3 columns and $1250 = 50 \times 5 \times 5$ rows; the row values are the coordinates of the grid points:

```
> names(air.grid)
[1] "radiation" "temperature" "wind"
```

The surface values corresponding to this grid are an array, `air.fit`, with one dimension for each predictor:

```
> dim(air.fit)
[1] 50  5  5
> names(dimnames(air.fit))
[1] "radiation"   "temperature" "wind"
```

(How we construct these values is discussed in Chapter 8.) Now we add to the data frame `air.grid` to form a new data frame that has both the coordinates and the surface values:

```
> air.grid[, "fit"]  <- as.vector(air.fit)
```

The `as.vector()` is just to be certain that the fitted values enter as a vector; the definition of `expand.grid()` ensures that the rows of `air.grid` come in the same order as the values of the three-way array for `air.fit`. Finally, we can make the coplot:

```
> coplot(fit ~ radiation | temperature * wind, data = air.grid,
+    given.values = air.marginal[c("temperature", "wind")], type = "l")
```

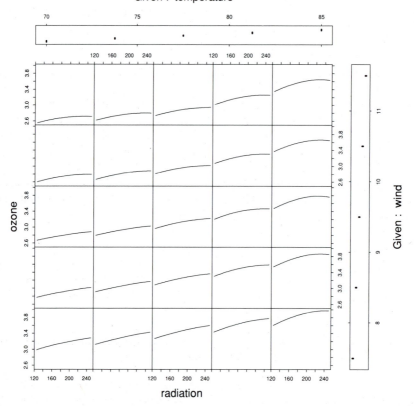

Figure 3.11: *Coplot of surface fitted to* `air`.

The result is shown in Figure 3.11. The dependence panels — that is, the graphs of the surface against radiation given wind speed and temperature — are the 5×5 array of square panels. The given panels, one for each conditioning predictor, are to the right and top. As we move up a column of dependence panels, the given values of wind speed increase, and as we move from left to right across a row of dependence panels, the given values of temperature increase. For example, the curve on the panel second from the bottom and third in from the left shows the surface given the second marginal grid value of temperature and the third marginal grid value of wind speed. Further study of the surface would construct two more coplots — ozone against temperature given radiation and wind, and ozone against wind given radiation and temperature.

3.3 Advanced Computations on Data

The discussion in the previous section should be adequate for using data frames in basic fitting and analyzing of models, as well as in other data analysis. We discuss next some topics that may be needed for more advanced use of models, and for adapting the functions to special applications. Section 3.3.1 explains in more detail how computations with data frames work. Section 3.3.2 discusses some new facilities for databases in S as they may be useful for model-fitting. Section 3.3.3 describes the intermediate model frame and model matrix objects. Finally, Section 3.3.4 introduces the parametrized data frame.

3.3.1 Methods for Data Frames

First, here is a look at how data frames work (the "private" view, in the terminology of Appendix A). The various methods for data frames use only a small number of internal quantities and assertions:

- The attribute `row.names` must be defined and must have length n, equal to the number of rows in the data frame.

- If `x` is a data frame, then `unclass(x)` should be a list with a non-null names attribute with unique names. The elements of this list are the variables (columns) in the data frame.

- The variables of the data frame must be interpretable as either a vector of length n or as a matrix with n rows.

Computations using a data frame as a frame require only that the names of the variables be unique, since any variable with a duplicate name will be inaccessible when the frame is attached. Variable names that are not syntactically names are inconvenient, since the only way to refer to a variable named `"Rev per mile"`, for example, is by

```
get("Rev per mile")
```

For this reason, `read.table()` and `data.frame()` convert all names to make them syntactically legal. The function `make.names()` can be used directly if you need to convert other strings to legal names.

Computations using a data frame as a matrix require that the variables behave either as vectors or as matrices defined over the n observations. They do not, however, otherwise constrain what is in the variables. For example, a vector of length n and mode `"list"` or `"character"` will work fine. Functions `"["()`, `"[["()`, `dim()`, and `dimnames()` all have methods that treat data frames as if they were matrices. Assignment methods also exist for each of these. The methods for `dim()` and `dimnames()` are fairly obvious, just using the `row.names` attribute to define the row dimensions and the row labels, and the length and names attribute of the unclassed list object for columns and column labels.

The method for `"["` replicates the behavior of the default subsetting for matrices ($\boxed{S}$, pages 127-129). Specifically, it treats expressions of the form

```
x[i, j]
```

where either `i` or `j` may be omitted, but not the comma (the expression `x[i]` is ambiguous since it has different matrix-like and frame-like interpretations). It also obeys the optional `drop=` argument. Either subscript can be numeric (positive or negative), logical, or character. Character subscripts are matched against the row names for `i` and the names for `j`. Matching is partial (via `pmatch()`) on extraction and strict on replacement, as with matrices. The semantics of subscripting can be summarized briefly as follows. The method creates a data matrix with one variable (column) for each variable extracted according to `j`. It then loops over the selected variables, selecting from each of those according to `i`. If the selected variable is a matrix, the rows implied by `i` are extracted; otherwise, the vector subset. It's worth noting that the method is rather deliberately simple at this stage: if `xj` is some selected variable, then the extraction is done directly by

```
xj[i]
xj[i, ]
```

in the two cases. The key point here is a reliance on methods for the variables; for example, a method `"[.factor"()` will ensure that extracted factors retain their factor-ness. If new classes of objects, including matrix-like classes, are to be used as variables in data frames, it is essential that methods be defined for them to make `"["()` work right.

Computations that apply `as.matrix()` to a data frame will also work, using a method for data frames. One difference from ordinary coercing to a matrix is important: if any of the variables in the data frame is itself a matrix or matrix-like, then the number of columns of the resulting matrix will not generally be the same as the number of variables. The method works as follows. It loops over the variables in the data frame, determining a common mode to represent the result. If all variables are numeric, so is the result. If some are non-numeric but none is recursive (list-like), then the result is a character matrix. Otherwise, the result is a matrix of mode `"list"`. In particular, factors are converted into character vectors, each element containing the corresponding level of the factor. If the factor originated as character data, this will be the obvious result, but remember that logical data are also converted to a factor and will not be automatically converted back to logical (not that this is likely to be a major problem in practical situations). The result of the method is a member of the class `"matrix"`, primarily to ensure it inherits a printing method capable of treating mode `"list"`.

As noted several times in earlier discussion, variables in a data frame can be considerably more general than the original description in Section 3.2.1. Once you try to use the data frame in any of the model-fitting in later chapters, however, you cannot expect variables other than those listed on page 52 to be interpreted sensibly when included in model formulas. Numeric vectors and numeric matrices are used as is. Factors and ordered factors are converted into coded numeric form, as is described in Section 2.4.3. Anything else will almost certainly cause an error. Other, future work on models could well make sense of more general variables, but the models discussed in this book do not.

3.3.2   Data Frames as Databases or Evaluation Frames

The version of S used in this book has a number of new features added since the publication of $\boxed{S}$. Appendix A discusses one key addition, the use of object-oriented programming. Another addition is an extension and formalization of S databases—that is, of the things that can be attached and detached in order to access objects by name in S expressions. In $\boxed{S}$, databases were always directories in the file system, and the objects were files, accessed by the name of the file. The current version allows databases to be list-like S objects, as well as compiled databases and user-defined classes of database objects. In addition, the object names can be arbitrary (nonempty) character strings, regardless of file system limitations.

Of these extensions, the ability to attach S objects as databases is used extensively in the book. The use of arbitrary object names is also important, implicitly, in permitting unrestricted naming of methods and other functions. The other extensions are not used. The whole topic of S databases in the current form is much too large to cover here, but in this section we outline some of the functions that deal with relevant aspects of databases.

First, a definition of the *search list*. At any time during an S session, this special object, internal to the S evaluator, defines the databases currently accessible by name in S expressions (see **S**, Section 5.4). In **S**, this object was in fact a character vector, although called a list. In the current version of S, it actually is a list. The elements of the list are the attached databases, which can be character strings (e.g., if they specify a directory), lists or other list-like S objects, or objects of special database classes. We refer to the object in element i as the database in position i of the search list.

When a name is used in an S expression, the evaluator matches that name to an object by looking successively in:

- the local frame of the currently evaluating function;

- frame 1, associated with the top-level expression;

- a session database, called database 0;

- each of the databases on the search list, in order.

This is essentially equivalent to **S**, page 118. What is new is the way the search is carried out, as well as the generality of the search list.

Each database in the search list can be examined by appropriate S expressions. Of course, the fundamental functions for this purpose are `get()`, `exists()`, `assign()`, and `remove()`. These get, test, and change the contents of the attached database. They behave consistently independent of whether the database was a directory, an object, or something else. In addition, `objects()` returns the names of all the objects known by name in the database, possibly restricted to those matching a regular expression in the sense of the *grep* command.

A few additional functions give useful information about the currently attached databases:

- `database.object(i)` returns the object that was attached as the database in position i.

- `database.type(i)` returns a character string giving the type of the database in position i. The possible types are `"directory"`, `"object"`, `"compiled"`, and `"user"`. Only the first two types are used in this book.

- `database.attr(which,i)` returns the value of the attribute `which` for the database object in position i.

The third function is particularly useful for the applications in this book. Data frames and classes inheriting from them have useful attributes in addition to the variables, `"row.names"` being an example. When the data frames are attached to the search list, it may be important to have access to these attributes. The function `database.attr()` handles this. For example, if we have attached the data frame `cu.specs`, as on page 68, we can obtain the row names by

```
database.attr("row.names")
```

Notice that we omitted the second argument giving the position of the database. In this case the function will search through the databases in order, returning the first non-null instance of the requested attribute, or `NULL`. Direct use of the function will be safer if an explicit position is included, but functions that suspect a data frame to have been attached but do not know the position, will ask `database.attr()` to search. This is the case, for example, with the `model.frame()` function. There is also an assignment function for database attributes, which can be used to modify attributes in the attached database while leaving the original data frame alone.

While the S evaluator is evaluating an expression, it creates temporary *evaluation frames*. Calls to S functions are evaluated in such frames, which are list-like objects whose components correspond to the arguments and to locally assigned objects. Each evaluation frame disappears when the evaluation of the function call is complete. An S function can create such a frame explicitly from a list or a data frame; for example,

```
> n <- new.frame(fuel.frame)
```

creates an evaluation frame and returns the index of the new frame in the vector of evaluation frames. The new frame initially contains copies of all the variables in the data frame `fuel.frame`. The returned index can be used as an argument to the function `eval()`:

```
eval(expr, n)
```

causes `expr` to be evaluated in the frame created by the call to `new.frame()`.

This is not something to be used directly in user-written expressions, but it provides an efficient mechanism for evaluating several expressions in the context of a particular data frame or other list-like object. In particular, it is used by the model-fitting functions for that purpose. Frames created by `new.frame()` disappear when the function calling `new.frame()` returns. A second useful mechanism, however, overrides this:

```
move.frame(n, to)
```

causes frame `n` to be handed over to frame `to`. This is an efficient mechanism for moving around intermediate results. See the documentation for `new.frame()` and `move.frame()` for further details.

Some of the other functionality described above for databases is available as well for temporary frames created by the evaluator:

- `sys.frame(i)` returns the ith frame in the evaluator.

- `frame.attr(which,i)` returns the value of the attribute `which` for the ith frame in the evaluator.

The concept of `database.type()` does not have an analogue for frames, since all frames are conceptually lists. There are a number of other functions that return quantities associated with the evaluator, but that are not meaningful for databases; for example, `sys.call()` and `sys.function()` give the call and the function definition.

3.3.3   Model Frames and Model Matrices

All the models in Chapters 4 through 7 prepare for fitting the model by essentially the same steps. Beginning with a formula, an optional data frame, and possibly other arguments such as `subset` or `weights`, they compute two intermediate forms of data:

1. a *model frame*, a data frame containing just the data needed to fit the model;

2. a *model matrix*, response variable, and possibly other information needed to carry out the fit.

The development of these objects from the formula, particularly emphasizing the `terms` object, is discussed in Section 2.4.3. We now continue that discussion, emphasizing the properties of the two intermediate data objects. Although the models discussed in Chapters 8 and 9 do not use model matrices, they do use model frames. From the model frame, they construct a different form of matrix suited to their own fitting methods, as described in those chapters.

The function `model.frame()` computes the model frame corresponding to a formula, using the `terms` object computed from such a formula. The variables in the model frame are the expressions for the terms of order 1 (the variables that appear as terms or as one of the arguments to `"*"`, `"/"`, or `":"` in an interaction). The response is also included. Optional weights and other special variables will be included in the model frame too, but under special names: `"(weights)"` for the weights, and so on. Enclosing the names in parentheses is intended to reduce the chance that they accidentally conflict with an actual variable name.

Let's develop an example to examine the contents of the model frame and model matrix. Suppose we decide to fit the model formula

```
100/Mileage ~ Weight + bs(Disp.) + Type
```

to the `car.test.frame` data. This will fit the transform of `Mileage` to a linear term in `Weight` plus a B-spline fit in `Disp.` plus a term in the factor `Type`. At the moment, however, it is not the fitted model that is of interest, but the objects common to all fitted models, specifically the model frame. Suppose we include the optional argument `weights=HP`, where `HP` is the horsepower variable in the same frame (don't ask for this to make statistical sense!). Finally, suppose the fit is only to be on the subset of the data for which `Weight<3500`. The actual call would be:

```
lm( 100/Mileage ~ Weight + bs(Disp.) + Type,
    car.test.frame, weights = HP, subset = Weight < 3500)
```

We're not interested in the fit, but in the model frame. Suppose we compute the model frame and store it in `mframe`. This can be done by giving `lm()` the argument

```
method = "model.frame"
```

Some representative rows of the model frame are as follows:

```
> mframe[sample.mframe, ]
```

	100/Mileage	Weight	bs(Disp.).1	bs(Disp.).2
Chevrolet Beretta 4	3.846154	2655	0.4264473738	0.1487607118
Chrysler New Yorker V6	4.545455	3450	0.3287921649	0.4117882453
Ford Mustang V8	5.263158	3310	0.0004951483	0.0377963202
Ford Probe	3.333333	2695	0.4264473738	0.1487607118
Hyundai Sonata 4	4.347826	2885	0.4413522746	0.1907077730
Mercury Tracer 4	3.846154	2285	0.2494567223	0.0287834680
Mitsubishi Wagon 4	5.000000	3415	0.4413522746	0.1907077730
Nissan 240SX 4	4.166667	2775	0.4433782744	0.2035636103
Toyota Corolla 4	3.448276	2390	0.2494567223	0.0287834680

	bs(Disp.).3	Type	(weights)
Chevrolet Beretta 4	0.0172977572	Compact	95
Chrysler New Yorker V6	0.1719115976	Medium	147
Ford Mustang V8	0.9617063693	Sporty	225
Ford Probe	0.0172977572	Sporty	110
Hyundai Sonata 4	0.0274681978	Medium	110
Mercury Tracer 4	0.0011070565	Small	82
Mitsubishi Wagon 4	0.0274681978	Van	107
Nissan 240SX 4	0.0311533408	Sporty	140
Toyota Corolla 4	0.0011070565	Small	102

In this model frame, there are four variables, one for each of the three terms in the model and one for the weights. The first term is an ordinary numeric variable, the second is a matrix returned by the B-spline function, and the third term is a factor. The matrix prints out as three columns, but the whole matrix is *one* variable in the model frame.

The model frame and model matrix are not usually intended for direct user interaction, but for use by other functions. Both objects have extra attributes that link them to the model. The model frame, in particular, contains as an attribute the `terms` object discussed in Section 2.4.3. The terms object contains information such as the identification of variable 1 in the model frame as the response.

A few of points of detail are worth noting. Since the model frame is not intended for human use, its variable names are left exactly as in the formula. In the example,

the second variable is itself a matrix, corresponding to the two orthogonal polynomials in `Weight`; this contributes two columns to the printout. The model frame
contains *only* the rows of the original data corresponding to the `subset` argument.

One further option affecting the model frame has not been shown. The `na.action`
argument to one of the model-fitting functions specifies a function that carries
out some action to detect missing values. The value returned by the `na.action`
function must be a data frame without missing values. Any missing values left
in the model frame cause an error in the model-fitting functions. Supplying the
argument as `na.omit` causes all rows containing any missing values to be dropped
from the analysis. Other strategies may be specified for particular kinds of models
or the user may supply a special function. Whatever is supplied as `na.action` must
be a function that takes a data frame as argument and returns another data frame
as result, with the same variables and containing no missing values.

The function `model.matrix()` takes a `terms` object and a model frame as arguments. It returns the matrix of linear predictors used in fitting linear models and
in models that derive from linear models, such as those of Chapters 6–7. The construction of the model matrix from the terms was discussed in Section 2.4.3. Here
we add a few points about the object itself.

The model matrix is a valid numeric matrix. It contains, in addition, attributes
that define its role in the model. The most important is `"assign"`. This is a list with
as many elements as there are terms. The elements of the list say which columns
of the matrix estimate coefficients to the corresponding term. For example, if `mm` is
a model matrix, the following computations would extract a matrix containing the
columns corresponding to the third term:

```
asgn <- attr(mm,"assign")
x3 <- mm[, asgn[[3]], drop=F]
```

This information is kept correct during the fitting of linear and related models, is
returned as an `"assign"` component of the fit, and is then used by the corresponding
summary methods. If `mm` is the model matrix corresponding to `mframe` above, the
assignment information is:

```
> attr(mm, "assign")
$"(Intercept)":
[1] 1

$Weight:
[1] 2

$"bs(Disp.)":
[1] 3 4 5
```

```
$Type:
[1]  6  7  8  9 10
```

Both the spline term, `bs(Disp.)`, and the factor `Type` contribute multiple columns to the model matrix.

The importance of the model matrix comes in large part because key numerical results (for example, the coefficients in fitting a linear model) are indexed the same way as the columns of the model matrix. Numerical fitting methods that work from the model matrix are expected to keep the `assign` attribute consistent with any changes they make. An important example is our standard fitting algorithm for linear models, `lm.fit.qr()`. If a linear model is over-determined, this method removes the aliased columns of the model matrix (by pivoting them to the right end of the matrix). At the same time, it modifies the `assign` attribute to show that these columns no longer contribute to the corresponding term. Whatever is done to the `assign` attribute of this nature should not change the number, order, or names of the list. These are used to match the terms for future calculations. Thus the `"assign"` component of a linear model can be used to select the coefficients corresponding to a particular term. The `assign` attribute is also kept correct by subsetting methods for model matrices.

3.3.4 Parametrized Data Frames

The matrix nature of data frames requires that each variable can be indexed by the rows of the data frame, either as a vector of n elements or a matrix with n rows. This restriction is reasonable enough in that it agrees with the intuitive view of data frames as corresponding observations on a number of variables. Occasionally, however, it would be convenient to have additional information accessible from the data frame that does not correspond to variables in this sense. For example, suppose we are analyzing the `solder` data and have decided to model the data in terms of a chosen power transformation of the `skips` variable. The chosen power might be kept as an object, say `power`. For clarity, it would be nice to write formulas in a style like

```
skips^power ~ .
```

In this way of thinking, we are regarding the chosen `power` as fixed (at least for the moment). It is then not a coefficient in the model, but neither can it be a variable in the data frame. We could just keep it separately, say in the working data, but this is also not very attractive, since it belongs with the `solder` data.

To handle such situations, we provide an extension to data frames called parametrized data frames, or `pframe`'s. The class of a `pframe` object is

```
c("pframe", "data.frame")
```

In addition to variables, `pframe` objects have parameters. These are accessed and set by special functions operating on the `pframe` object, but when the object has been attached for evaluation, they are accessible by name in the same way as variables. Parameters, however, are not assumed to be indexed by the rows of the data frame. The function `param()` can be used to extract or set a given parameter, while `parameters()` does the same for the complete list of all the parameters. These functions are analogous to `attr()` and `attributes()` for handling attributes (e.g., Ⓢ, pp. 143–146). For example,

```
param(solder, "power") <- .5
```

would set the parameter `"power"` and make expressions like `skips^power` work in any formula using `solder`.

Parametrized data frames can be attached in position 1 to create new or revised data, as we showed on page 68 for ordinary data frames. When they are detached, however, objects that are not variables will be retained as parameters rather than being dropped. Only a few special objects generated by the S evaluator (e.g., `.Last.value`) will be deleted. A convenient way to work with `pframe` objects is: first, assign any parameter so as to convert a data frame to a `pframe`; second, attach the object in position 1 and create or modify whatever parameters and/or variables you want; finally, detach and save the revised data frame.

Parametrized data frames come into their own in nonlinear models, where the coefficients must be explicitly included in the model formulas. In Chapter 10, parametrized data frames are used extensively to achieve effects such as holding some parameters constant at pre-specified values.

Chapter 4

Linear Models

John M. Chambers

This chapter presents S functions and objects for classical linear methods in statistics, in which a numerical response variable is predicted by linear combinations of other numeric or categorical variables. S functions and classes of objects described in this chapter fit linear models by least squares, and analyze the models by a variety of techniques. By modifying and extending the functions provided, you can specialize the modeling to your own applications or develop new statistical techniques for linear models.

The statistical, computational, and mathematical ground covered in this chapter deserve to be called "classic" in any sense. The statistical use of linear models goes back to Laplace and Gauss early in the nineteenth century and continues to underlie much of statistical modeling. The numerical techniques, also, are exceptionally reliable and well developed, representing some of the most successful results of numerical analysis. Many of the computational tools can be applied in other situations as well; in particular, many computations in later chapters will use linear least-squares computations as building blocks. The mathematical analysis of this topic, particularly the fundamental results of linear algebra and vector geometry, also serve as the basis for many other results.

If you expect to use linear models in your work, then you should read this chapter. The computational details described in Section 4.4 will likely be a useful reference for advanced work on other models as well, particularly those in Chapters 5–7. This chapter follows a style and organization of computations, around classes of objects and the functions that create and use them, that provides a perspective on linear models different from traditional treatments. The data analyst has greater freedom: summaries and displays are not restricted to a few predetermined printouts or tests. Instead, there is an essentially unlimited scope for getting the relevant analysis for particular problems. The theme is that one function, lm(), creates an object describing a fitted model, and other functions use this object to analyze or modify the fit. Optional arguments to lm() specialize the fitting. Fitted models can be updated in a simple and general way.

The functions and objects in this chapter provide a number of statistical techniques directly. Equally important in their design, however, is the goal of providing a powerful basis for nonstandard computations or for the implementation of new ideas. Even such a well-developed area of statistics as linear models provides many challenges and opportunities. New functions, new user interfaces, and new algorithms can all be built on the basis provided here.

4.1   Linear Models in Statistics

This chapter describes S functions and objects for linear models that use least-squares as a fitting criterion. Some of the discussion will also be relevant to other situations, either because the techniques do not depend specifically on the fitting criterion or because the least-squares computations form an essential basis for other techniques. The first section of the chapter outlines the statistical concepts we will use in discussing linear models. The goal is to set the background for the computational discussions that follow in the rest of the chapter. We don't aim to *teach* the statistics of linear models; for that, you should look at one of the many good books on the subject (some are mentioned on page 144), either in advance or while you are reading this chapter.

Linear regression models a numeric response variable, y, by a linear combination of predictor variables x_j, for $j = 1, \ldots, p$. Each of the variables was observed on the same n observations. The fitted values are the sum of coefficients β_j multiplying each of the x_j plus (usually) an intercept β_0. Using our "$\sim$" operator to mean "is modeled as", the linear model is:

$$y \sim \beta_0 + \beta_1 x_1 + \cdots + \beta_p x_p \tag{4.1}$$

Linear *least-squares* models estimate the coefficients to minimize the squared sum of residuals. If the response and predictors corresponding to the ith of n observations are $y_i, x_{i1}, x_{i2}, \ldots, x_{ip}$, then the fitting criterion chooses the β_j to minimize

$$\sum_{i=1}^{n} (y_i - (\beta_0 + \sum_{j=1}^{p} \beta_j x_{ij}))^2 \tag{4.2}$$

The standard statistical theory of linear models makes (4.1) more explicit by writing the model for the ith observation as:

$$y_i = \beta_0 + \sum_{j=1}^{p} \beta_j x_{ij} + \varepsilon_i \qquad (4.3)$$

and by making the following assumptions:

   i. the ε_i are independently and identically distributed;

   ii. the ε_i have mean zero and (finite) variance σ^2;

   iii. the ε_i are distributed according to the normal distribution.

Coefficients that minimize (4.2) define a fitted linear model, represented by what we will call a linear-model object. The object contains the estimated coefficients $\hat{\beta}_j$, the fitted values

$$\hat{y}_i = \hat{\beta}_o + \sum_j \hat{\beta}_j x_{ij}$$

and the residuals $y_i - \hat{y}_i$. The statistical assumptions define the distribution of various components of the fitted model. The residuals will be normally distributed with zero mean, and the estimated coefficient $\hat{\beta}_j$ will be normally distributed with mean β_j. The linear-model object also contains a set of effects, which we define formally in Section 4.4.1 on page 133. Under the statistical assumptions, they are independently and normally distributed, with variance σ^2. The elements of the squared effects are the contribution from fitting the corresponding coefficient to the standard analysis of variance breakdown. The effects are used extensively when linear models are applied in the context of the analysis of variance in Chapter 5.

The standard error σ can be estimated from the residuals. Estimated variances and covariances for the coefficients, fitted values, and residuals are known functions of the x_j multiplied by the estimate of σ.

These statistical characterizations lead to a variety of summaries and diagnostics, including plots. As with all models, both the structural form of the regression and its probabilistic characterization are at best simplifications that help us understand the data. Fortunately, linear models can use a particularly well-supplied box of tools to help assess and improve the model.

The model (4.1) is *additive*, representing y as the sum of p terms. In this view, $\hat{\beta}_j x_j$ is the contribution of the x_j term to the fit. Statistically, such a view needs to be taken cautiously since the contributions are not independent, but the view of (4.1) as an additive model is a useful one that carries over to generalizations such as those discussed in Chapters 6 and 7. The symbols β_j in (4.1) are not needed to convey the model. Just specifying the terms as x_j defines the structural form. Computer systems for additive models have evolved a notation in which the coefficients (including the intercept) are omitted. Our formula objects take that evolution further by including the response, adding some new notation, and allowing more general expressions for terms. For computational purposes we will write (4.1) as:

$$y \sim x_1 + \cdots + x_p \tag{4.4}$$

As we will see, this form is very convenient and extends well to related models in later chapters. It is important to remember, however, that some additional statistical specification is always involved before (4.4) fully defines a model that can be used in practice.

The variables in (4.4) were said to be numerical, but they can be more general than that. If x_j is a factor or an ordered factor, its presence in the formula stands for fitting a coefficient in the additive model for each level of the factor. Terms that are factors can therefore contribute more than one coefficient (that is, more than 1 degree of freedom) to the fit; still, it makes sense to regard the entire contribution as a single term. If x_j is a numeric matrix, the intention is to regard all the coefficients contributed by x_j as a single term in the model. As an example, choosing to fit a polynomial of degree d in one or more predictor variables usually means that the contributions of different basis vectors for the polynomials should be regarded as forming a single term. Similar comments apply to splines or other parametrized family of curves.

The response in a linear model cannot be a factor. Generalized linear models in Chapter 6 treat factors with two levels as response variables, and tree-based models in Chapter 9 will allow arbitrary factors. The response in a linear model *can* however be a numeric matrix, a generality that will not extend to some of the other models. The reason is that the solution to (4.2) is characterized essentially in terms of the matrix formed from the x_j. In particular the solution can be characterized by linear operators that are applied to y to generate coefficients, fitted values, residuals, and effects. If y is a matrix, the same linear operators apply, generating matrices instead of vectors.

The fitting functions in this chapter mostly use one particular numerical algorithm for linear least squares. The algorithm has been chosen for high accuracy and good reliability, and is based on the widely used LINPACK algorithm library. However, a recurring theme of this and some later chapters will be that it is the *objects* representing the fitted models that are key, not a particular algorithm that computes them. Modifications of our chosen algorithm or the use of an entirely different algorithm are perfectly acceptable, so long as the computations produce objects containing the necessary information in a form that the various summary and diagnostic methods can handle. Section 4.4.2 reviews some of the underlying theory behind linear least-squares models, to explain how the computations work and how you can alter them.

One extension of the standard model that will be discussed is that of *weighted* least squares, in which each of the squared residuals in equation (4.2) is given some specified weight, say $w_i \geq 0$. This generalization arises both directly (see page 111) and as a computational technique for implementing other models in terms of linear least squares. If it is known or suspected that the variance of the errors is *not* the same for all observations, then the standard fitting shown so far is not appropriate. Treatments for this situation include looking for a transformation of the data that makes the variance constant or using a model, such as one of those in Chapter 6, which incorporates the changing variance in the model itself. If, however, the variance is known (up to a constant), ordinary linear least-squares fitting can proceed by weighting the squared residuals proportionally to the inverse of the variance. For example, if each value of the response represents the average of observations on some known but varying number of replications for fixed values of the predictors, the variance will be inversely proportional to the number of replications, so that the number of observations can be used as weights in the fit. The statistical theory of linear models carries over essentially unchanged, by considering the model in terms of variables $w^{\frac{1}{2}}y$ and $w^{\frac{1}{2}}x_j$. For this to be valid, the weights (and the x_j as well) are assumed to be fixed and in particular not to involve y.

In addition to providing the basic fitting of the model (4.4), functions in this chapter address some other important aspects of using linear models:

- diagnostics, especially graphical, that look for aspects of the data that are *not* well explained by the model, often by looking at the residuals;

- examining the structural form of the model, to see the result of adding, dropping, or changing terms;

- summaries showing the inherent variability in the coefficients, fitted values, or predictions;

- summaries using the effects and other information to study the importance of individual terms in explaining the response.

Section 4.2.2 describes S functions to compute a number of analytical summaries and diagnostic plots. Variations on these and many other analytical results can be computed from the information in the objects, as described in Section 4.3.

4.2 S Functions and Objects

This section presents S functions for typical fitting and analysis of linear models. The function `lm()` returns an S object that we will call a *fitted linear least-squares model* object, or an `lm` object for short. Section 4.2.1 shows the fitting itself; Section 4.2.2, some summary functions; Section 4.2.3, some functions to compute predictions. The last sections, 4.2.4 and 4.2.5, present some useful options for doing the fitting and a powerful general technique for updating the fit.

4.2.1   Fitting the Model

The S function lm() creates a least-squares fit:

```
lm(formula, data)
```

where formula is the structural formula that specifies the model and data is the data frame in which the model is to be computed. Let's look at an example using data introduced in Chapter 3. On page 70, we created a data frame fuel.frame, including variables Fuel, Weight, and Disp.. A linear model that fits Fuel to Weight and Disp. is specified by the formula

```
Fuel ~ Weight + Disp.
```

which is read "Fuel is modeled as a linear combination of Weight and Disp.". The names occurring in the formula are interpreted in the frame fuel.frame. The fit,

```
fuel.fit <- lm(Fuel ~ Weight + Disp., fuel.frame)
```

does not produce printed tables or other summaries of the model. Instead, the lm object created represents the fit, and contains all the essential information, such as coefficients, residuals, fitted.values, and some other less obvious things. The lm object can be given to functions to produce summaries, or to functions with the names above (coefficients(), etc.) to get at the specific information in the model. Short forms of the commonly used extracting function names are provided to save typing: coef(), resid(), and fitted().

```
> coef(fuel.fit)
 (Intercept)      Weight         Disp.
   0.4789733 0.001241421 0.0008543589
```

Notice that the coefficients are printed with names, constructed automatically from the data and the formula. It is important to use the functions like coef(), rather than prying open the inner contents of the lm object, at least while you're getting used to the model-fitting. The functions can use *all* the information in the object to return a sensible result. Particularly with models that are derived from linear models, such as glm models, the raw components may be misleading.

The lm object itself can be printed, like any S object, by just giving its name:

```
> fuel.fit
Call:
lm(formula = Fuel ~ Weight + Disp., data = fuel.frame)

Coefficients:
 (Intercept)      Weight         Disp.
   0.4789733 0.001241421 0.0008543589

Degrees of freedom: 60 total; 57 residual
Residual standard error: 0.3900812
```

The style of the printing method for this and for other models is to show the simple information in the model that seems most likely to be relevant. The printing method tends to give you just what is in the fitted model, with little statistical embellishment. Another function, `summary()`, gives a more technical, statistical description of the model, using the statistical assumptions mentioned on page 97. We discuss this function on page 104.

Formulas are discussed extensively in Chapter 2 and will come up again repeatedly, but here are a few points to keep in mind. The individual terms in the formula can be any S expressions that evaluate to something that can be a predictor: numeric vectors, factors, ordered factors, or numeric matrices. Several coefficients in the model may correspond to one factor or matrix term. The special name "." may be used on the right of the "~" operator, to stand for all the variables in a data frame other than the response. Assuming that `Fuel`, `Weight`, and `Disp.` were the only variables in `fuel.frame`, the expression

```
fuel.fit <- lm(Fuel ~ ., fuel.frame)
```

would produce the same fit as before.

Terms in formulas are separated by "+"; therefore, if we want to have a *single* term equal to the sum of `Weight` and `Disp.`, the "+" sign must be protected. This is done by enclosing the term in the "identity" function, `I()`:

```
Fuel ~ I(Weight + Disp.)
```

The operators ":", "*", "^", "/", and "-" are also special on the right side of formulas. Terms that use these operators in their usual arithmetic sense should be protected by the `I()` function. The operators are special only as predictors, not in expressions for the response.

The terms for predictors can evaluate to numeric vectors, numeric matrices or factors. Logical or character vectors will be turned into factors. The response can be a numeric vector or a matrix. In the case of a matrix response, the coefficients, residuals, and effects will also be matrices, with the same number of columns as the response.

Models that include factors are discussed in great detail in Chapter 5, in the context of the analysis of variance. The model-fitting functions fit factors by, in principle, replacing them with the corresponding set of "dummy variables," variables that take the value 1 for observations with a particular level for the factor and 0 for all other observations. The details can be left for Chapter 5; for now, it's sufficient to know that factors can be included in models with the fitted values and residuals coming out correctly. With our `Fuel` example, it might be interesting to include the type of automobile as a predictor:

```
> fuel.fit2 <- lm(Fuel ~ Weight + Disp. + Type, fuel.frame)
> fuel.fit2
Call:
lm(formula = Fuel ~ Weight + Disp. + Type, data = fuel.frame)

Coefficients:
 (Intercept)    Weight     Disp.     Type1     Type2     Type3     Type4
      2.9606  4.164e-05 0.0075941  -0.14528  0.098084  -0.12409  -0.04358

    Type5
  0.19151

Degrees of freedom: 60 total; 52 residual
Residual standard error: 0.31392
```

By the way, using the function `update()` would have been simpler than rewriting the whole call to `lm()`:

```
fuel.fit2 <- update(fuel.fit, . ~ . + Type)
```

This very handy generic function allows nearly all common changes in a model to be created from the original model and the changed arguments; in addition, it uses the "." notation as shorthand for the previous left or right side of the formula. Section 4.2.5 will describe `update()`.

The factor `Type` has 6 levels,

```
> levels(Type)
[1] "Compact" "Large"   "Medium"  "Small"   "Sporty"  "Van"
```

which produce five coefficients, or contrasts. The individual coefficients are usually less important for factors than the overall contribution of the term to the fit. If you do want to know more about the choice of contrasts, see Sections 2.3.2 and 5.3.1. Another example with factors as terms will be shown on page 111.

It is worth emphasizing that a response or predictor in a formula is not restricted to being a name, but can be any S expression that evaluates to an object that can be used as that response or predictor. The variable `Fuel` was defined as `100/Mileage`, so we could have fit the same model by:

```
> fuel.fit2 <- lm(100/Mileage ~ Weight + Disp. + Type, car.test.frame)
```

The operator "/" is special in predictor terms, but not in the expression for the response, so there was no need to protect it. As with the response, the predictors on the right of the "~" can be anything that evaluates to numeric vector, matrix, or factor. Of course, the variables should all be defined on the same set of observations in order to be meaningful in the model.

Most analysis of linear models will benefit from organizing the data into a data frame, encouraging the kind of preliminary analysis discussed in Chapter 3. It is possible, however, to omit the `data` argument from the call to `lm()`, in which case the names in the formula will be evaluated in the usual way for arguments, meaning they typically will be the names of permanent S objects. As an example, suppose we want to reproduce the analysis of the "stack loss" data, a classic set of data in which the loss of ammonia in an industrial process (an indirect measure of the yield of the process) is fitted to three measures of conditions in the process. The response variable and the matrix of three predictors are supplied with S as the vector `stack.loss` and the matrix `stack.x` (**S** , page 657). The fit can be produced directly from these objects:

```
> stack.fit <- lm(stack.loss ~ stack.x)
> coef(stack.fit)
  (Intercept) stack.x.Air Flow stack.x.Water Temp stack.x.Acid Conc.
     -39.92         0.71564          1.2953            -0.15212
```

Notice that the formula referred only to the whole matrix of predictors, but the coefficients are labeled using the `dimnames` from the matrix. Nevertheless, expressing the formula in terms of `stack.x` means that the whole matrix should be regarded as a single term. More natural and more flexible would be to form a data frame and use the "." notation:

```
> stack <- data.frame(loss=stack.loss, stack.x)
> stack.fit <- lm(loss ~ ., stack)
> coef(stack.fit)
 (Intercept)  Air.Flow Water.Temp Acid.Conc.
  -39.91967 0.7156402   1.295286 -0.1521225
```

The terms in this case are separate vectors rather than one matrix.

Formulas may be kept as objects to save the effort of retyping them:

```
> fuel.f <- formula(Fuel ~ Weight+Disp.)
> fuel.f
Fuel ~  Weight + Disp.
```

This assigns the formula expression as an unevaluated S object. The `formula` function will also extract a formula from objects representing a fitted model; an equivalent way to get at the formula from the fitted model `fuel.fit` on page 106 would be

```
> fuel.f <- formula(fuel.fit)
```

Formula objects can be edited like other objects using S editor functions such as `vi()`. Usually, however, the `update()` function makes direct editing of formulas unnecessary, as in Section 4.2.5.

4.2.2   Basic Summaries

A wide range of plots and summaries can be applied to linear models. We will describe several of them, but you are definitely encouraged to design your own as well.

A plot method for `lm` objects makes two plots against the fitted values, one of the response and another of the absolute residuals.

```
plot(fuel.fit)
```

produces the plot in Figure 4.1. The left panel shows the general pattern of the

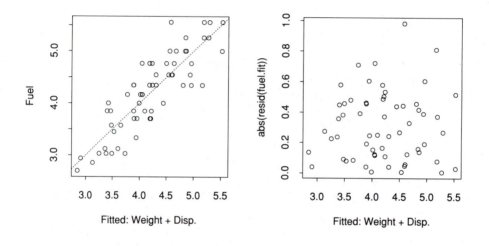

Figure 4.1: *Plot method for an* `lm` *object, applied to* `fuel.fit`. *The left panel shows response against fitted values, the right panel plots absolute values of residuals against fitted values.*

fit by plotting the response against the fitted values, with the $y = x$ line super-imposed. The right panel tries to point out patterns in the residuals by plotting their absolute values against the fitted values. Figure 4.1 shows, for example, one residual considerably larger than the rest. To find out which observation it is we can use `identify()` or look directly in the row names of the data frame. The high residual is the only residual with an absolute value greater than .9:

```
> row.names(fuel.frame)[ abs(resid(fuel.fit)) > .9 ]
[1] "Chevrolet Lumina APV V6"
```

The generic function `summary()` produces summary objects for fitted models intended to give more "statistical" information than comes from just printing the object. For a linear model, the summary contains estimated standard errors, correlations, and t-statistics for the individual coefficients, correlations for the coefficients, and summaries of the residuals:

```
> summary(fuel.fit)

Call: lm(formula = Fuel ~ Weight + Disp., data = fuel.frame)
Residuals:
   Min    1Q Median   3Q  Max
 -0.81 -0.26   0.02 0.27 0.98

Coefficients:
              Value Std. Error t value
(Intercept) 0.47897    0.34179    1.40
    Weight  0.00124    0.00017    7.22
     Disp.  0.00085    0.00157    0.54

Residual standard error: 0.39 on 57 degrees of freedom
Multiple R-Squared: 0.74

Correlation of Coefficients:
       (Intercept) Weight
Weight -0.90
 Disp.  0.47       -0.80
```

The `Coefficients` table gives the coefficients and their estimated standard error. The third column is the ratio of the estimated coefficients to the corresponding standard-error estimate, which could be compared to a Student's t distribution. The residual standard error is the sum of squared residuals, divided by the number of degrees of freedom for residuals (usually the number of observations less the number of coefficients). `Multiple R-squared` is the term for a quantity usually defined as the fraction of the total variation in the response accounted for by the variation in the fitted values. It can be a useful measure of the success in explaining the response by the current model, although it ignores the number of coefficients and so invites over-fitting. The table of correlations are those of the estimated coefficients; only the lower triangle is printed. Remember that these correlations are for the coefficients, *not* for the original variables.

The `summary()` function is normally used to produce printed output. However, the value of the function is an S object of class `"summary.lm"`, containing all the information printed. Computations needing the information printed above can extract it from the object. The printing is produced by a special method for `summary.lm` objects. Let's look at the components of the `summary.lm` object:

```
> fuel.summary <- summary(fuel.fit)
> names(fuel.summary)
[1] "call"          "terms"         "residuals"     "coefficients"
[5] "sigma"         "df"            "r.squared"     "cov.unscaled"
[9] "correlation"
```

The meaning of most of these will be obvious from the printed version of the summary. The cov.unscaled and df components are not reflected directly in the printing. The latter gives the degrees of freedom for each term. The discussion in Section 4.4.1 clarifies the use of the unscaled covariance, cov.unscaled.

The qqnorm() function ($\boxed{\text{S}}$, pp. 70–71) plots a vector of numbers, sorted, against corresponding expected values from a standard normal distribution. If the vector behaves like a sample from a normal distribution, the plot should look roughly linear. Since the standard statistical assumptions for linear models say that the residuals from the model are distributed as a normal sample, the plot is a useful way to look for patterns indicating that the candidate model needs to be modified. The residuals from the *fitted* model don't exactly follow this distribution, but the plot is still a reasonable way to look for problems. Figure 4.2 shows the result. The pattern looks reasonably linear in this example.

> qqnorm(residuals(fuel.fit))

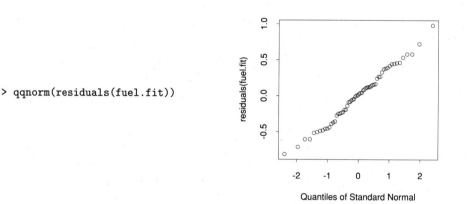

Figure 4.2: *Normal probability plot of residuals, for the fit of* Fuel $\sim$ Weight + Disp.

4.2.3   Prediction

After fitting a model to some data, we would often like to know the predicted response from the model for some *different* values of the predictor variables. This is provided by the generic function

```
predict(fit,newdata)
```

where `newdata` contains the data for which we want a predicted response. Following our general approach, `newdata` is a data frame including all the variables used in fitting the model. The response variable is ignored in `newdata` and need not be present.

The `fuel.fit` model was fitted to the automobile data for which the `Mileage` variable was present. One reason for fitting the model might well be to predict the fuel consumption for automobiles not road-tested by Consumers Union, for which `Mileage` would be `NA`. To obtain predicted values for a sample of those automobiles, we need to construct a data frame containing the `Weight` and `Disp.` variables. The data frame `car.all` contains all the variables for the automobile data. For prediction we want `Mileage` to be missing but `Weight` and `Disp.` to be present:

```
> attach(car.all)
> ok.for.predict <- is.na(Mileage) & !(is.na(Weight) | is.na(Disp.))
> sum(ok.for.predict)
[1] 55
> predict.rows <- row.names(car.all)[ok.for.predict]
```

Let's sample 10 of the possible rows for prediction:

```
> sample.rows <- sample(predict.rows,10)
> new.cars <- car.all[sample.rows,c("Weight", "Disp.")]
```

The data frame `new.cars` is now suitable as an argument to `predict()`:

```
> new.cars
                    Weight Disp.
   Volkswagen Golf    2215   109
   Volkswagen GTI     2270   109
         BMW 325i     2895   152
Pontiac Bonneville    3360   231
 Mitsubishi Precis    2185    90
    Hyundai Excel     2345    90
     Sterling 827     3295   163
Lincoln Continental   3695   232
        GEO Storm     2455    97
     Dodge Spirit     2940   181

> pred.fuel <- predict(fuel.fit,new.cars)
```

The predicted values have the same structure as the original response, either a numeric vector or a matrix. In our example:

```
> pred.fuel
Volkswagen Golf Volkswagen GTI BMW 325i Pontiac Bonneville
         3.3218          3.3901   4.2027             4.8475


Mitsubishi Precis Hyundai Excel Sterling 827 Lincoln Continental
           3.2684           3.467      4.7087               5.2642

GEO Storm Dodge Spirit
   3.6095       4.2834
```

Notice that the row names from `new.cars` have been retained as the `names` attribute of the predicted values. As for how *well* the model predicts, at least the names associated with low and high fuel consumption seem plausible.

The `predict()` function has a number of useful options. The option `se.fit=T` causes the prediction to include pointwise standard errors for the predicted values. The argument `type="terms"` causes the predictions to be broken down into the contributions from each term, returning a matrix whose columns correspond to the individual terms in the model. For examples of these options, see Section 7.3.3.

Prediction from a fitted model is usually straightforward, with one important exception. A problem arises whenever the expression for one of the predictors uses some overall summary of a variable. For example, consider the following two expressions:

```
x/3; x/sqrt(var(x))
```

The first expression is fine for prediction; the second is not. The problem is that while we could compute a subset of values for the first expression just from knowing the corresponding subset of `x`, the same is *not* true of the second expression. Ordinary prediction is precisely that: we try to compute values for the new data using only the formula and new values for the predictor variables. The functions to watch out for include:

```
poly(x) #orthogonal polynomials
bs(x)   #spline curves
```

and any function that uses overall summaries, such as `range()`, `mean()`, or `quantile()`. These are all fine expressions for linear models, but if you use them and want to do prediction, look ahead to Section 7.3 for a safe method.

Just in case you don't believe there is a problem, you can try a simple experiment. Fit the two models

```
fuel.fitq <- lm(Fuel ~ Weight + I(Weight^2) + Disp., fuel.frame)

fuel.fitq2 <- lm(Fuel ~ poly(Weight, 2) + Disp., fuel.frame)
```

These both fit a quadratic polynomial in `Weight` to the `Fuel` data. The coefficients will be different but the fitted values are equivalent; they represent different parametrizations of the same model. Now compute predicted values for the two models. The fitted values for `fuel.fitq` are not very different from those for `fuel.fit`; for example, the low and high values are:

```
Mitsubishi Precis Lincoln Continental
           3.209793              5.187772
```

which can be compared to `3.27` and `5.27` before. But for the fit using `poly()`, the corresponding values are

```
Mitsubishi Precis Lincoln Continental
           2.364744              6.642079
```

which have changed more than is plausible from the small contribution of the quadratic term.

4.2.4 Options in Fitting

Additional, optional arguments to the `lm()` function and special features of the formula language give a great deal of flexibility in fitting the model. Among other things, it is possible to select subsets of observations, provide weights for fitting, deal with missing values, fit parallel regressions, handle over-specified models, and update models to produce new models.

Fitting to Subsets of Observations

The `subset` argument allows the call to specify a rule for selecting a subset of the rows in the data to be included in the fit. For example, the data frame `car.test.frame` includes a factor, `Type`, that specifies one of six types of car:

```
> attach(car.test.frame)
> levels(Type)
[1] "Compact" "Large"   "Medium"  "Small"   "Sporty"  "Van"
```

Suppose we decide that cars of type "Van" should be excluded (perhaps because they tend to be big fuel guzzlers and we're really only interested in standard cars). Evaluating the expression

```
Type != "Van"
```

identifies the non-`Van` observations in factor `Type`. Including this expression as the `subset` argument causes `lm()` to fit the model only for this subset of the observations. Let's assume we added `Type` to `fuel.frame`, and fit the restricted model:

```
> fuel.not.van <- lm(Fuel ~ Weight + Disp., fuel.frame,
+      subset = (Type != "Van") )
> fuel.not.van
Call:
lm(formula = Fuel ~ Weight + Disp., data = fuel.frame,
       subset = (Type != "Van"))

Coefficients:
 (Intercept)     Weight       Disp.
      1.1528  0.00084507  0.0035165

Degrees of freedom: 53 total; 50 residual
Residual standard error: 0.36717
```

The original fit included seven cars of type van. We can see that the coefficients have changed somewhat and the residual standard error is reduced by about 10%. As an aside, suppose we want to ask also whether dropping out those seven vans from the data changed the pattern of the residuals. We can compare our two sets of residuals using qqplot, as in Figure 4.3. This function plots the quantiles of two sets of data and will give a roughly linear pattern if the two sets have the same distribution. In this case, the distribution of the residuals has changed very little,

```
> qqplot(residuals(fuel.fit),
+    residuals(fuel.not.van))
> abline(0,1)
```

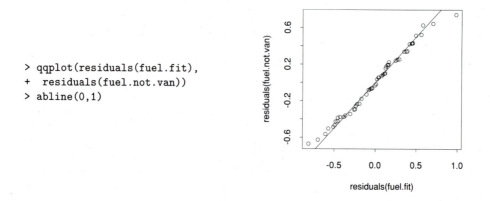

Figure 4.3: *Empirical quantile plot of residuals, the y-axis showing the fit with seven "Van" models excluded. The y = x line is included.*

despite the difference in the coefficients. The most discrepant point is the large residual in fuel.fit that appears also in Figure 4.1 on page 104.

Weighted Regression

A vector of non-negative weights can be supplied with the fit for applications where the contributions of the observations to the fitting criterion are not equal. The optional argument `weight=` to `lm()` allows weights to be supplied as an expression to be evaluated along with the formula. Note that the weights are defined to be the weights appearing in the sum of squared residuals; some discussions of regression use the square root of these instead.

Some data to illustrate this situation were reported by Baxter, Coutts, and Ross (1980). The data consist of the average cost of automobile insurance claims, along with the age of the policyholder, the age of the car, and the type of car, these predictors being recorded as factors with 8, 4, and 4 levels. The number of claims in each cell was also recorded. The data are in the data frame `claims`, as variables `cost`, `age`, `car.age`, `type`, and `number`. The model used in the published reference fits `cost` to the sum of terms in the age, the car's age and the type of car. Because the response is average cost, the rows should be weighted in the fitting by the number of claims. Also, there were five missing values in the reported cost data; we will explain how those were handled just a few paragraphs below.

```
> claims.fit <- lm(cost ~ age + type + car.age, claims,
+       weights = number, na.action = na.omit)
> coef(claims.fit)
 (Intercept)    age.L   age.Q   age.C  age ^ 4  age ^ 5  age ^ 6  age ^ 7
      250.64 -58.261 30.195  5.9625 -34.107    -33.5 -7.1807   18.667

   type1   type2   type3 car.age.L car.age.Q car.age.C
  2.6612 9.4708 24.269   -78.918   -54.769    -49.47
```

This model includes factors among the predictors, as discussed on page 20. The factor `age` has 8 levels, meaning there will be 7 linearly independent coefficients; similarly, for the factor `type` with 4 levels there will be 3 coefficients. Chapter 5 discusses appropriate summaries for linear models in this context in detail. In particular, the summary method used there groups together all the effects for a particular term. While we did not fit this example explicitly by the function `aov()`, as we would have in Chapter 5, the summary methods are compatible. By using the name of the method explicitly, we can produce the same summary information for the `lm()` fit:

```
> summary.aov(claims.fit)
             Df Sum of Sq   Mean Sq     F
      age     7   5618930   802704.3  2.18
     type     3  12110810  4036936.6 10.99
  car.age     3   6987861  2329287.1  6.34
Residuals 109  40056968   367495.1  1.00
```

The summary suggests that the `type` variable contributed the most to the fit, since it has the largest F-statistic value.

The use of the `summary.aov()` method on an `lm` object is worth remarking on. It illustrates a general principle throughout the book that the classes of model objects are linked as closely as makes sense, so that software for each kind of model takes advantage of work on other kinds of models. In particular, more advanced models are often designed to *inherit* from linear models. Many nice features of the computations result from this approach. Generalized linear models and additive models make particularly strong use of the style.

The `lm()` function allows weights to be exactly zero (but *not* negative). Zero weights are rather ambiguous, however, and discouraged: the problem is to decide whether they just happened as part of some numeric computation, or whether they really imply that the corresponding rows should be omitted from the computations. We assume the latter; for example, `summary()` does not count zero-weighted observations toward residual degrees of freedom. Expect to see a few warning messages if you use zero weights; on the whole, it's safer to use the `subset` argument to omit observations. In fact, the example just shown could have involved us in zero weights, since several observations had no recorded claims:

```
> sum(claims[, "number"] == 0)
[1] 5
```

One way of avoiding zero weights would be to supply as a `subset` argument the expression `number>0`.

Missing Values

The `lm()` function, like most other model-fitting functions in the book, cannot deal with `NA` values in either response or predictor; instead, it takes an argument `na.action` that allows the user to specify what technique should be used to remove the missing values. By default, `lm()` generates an error if there are any `NA`'s in the predictors, response, or weights. You can specify a method for removing missing values in a particular application, by giving an S function as an `na.action`, either in an argument to `lm()` or as an attribute of the data frame. The attribute is used if you want the `na.action` to apply to all models constructed from this data frame. This would be a reasonable approach in the example above: in the `claims` data, the average claim was, as it should have been, recorded as `NA` whenever there were no claims. A reasonable attitude would then be that *any* model that needs an `na.action` should omit such observations. This would be achieved by:

```
attr(claims,"na.action") <- na.omit
```

This function drops any row of the data frame for which any of the variables has a missing value. The function is applied to the data to be used in the actual fit;

missing values elsewhere in the data will be ignored. Giving an `na.action` attribute to a data frame automatically sets up the strategy for all models generated from that frame. If an action is supplied directly to `lm()` as the `na.action` argument, it will override the attribute, if any, of the data frame. Notice that since the `na.action` is a function, you can write your own function to take any action that is suitable for a specific application.

Section 3.3 discusses `na.action` functions in general. We have not tried to provide a sophisticated facility for imputing values for `NA`'s; the function `na.omit()` is the only really general one. The problem is difficult and to some extent depends on the kind of model as well, of course, as the assumptions that can be made about the missing values. Tree-based models deal rather better than linear models with missing values; Section 9.2 discusses a nice method for replacing missing values in tree models. Otherwise, we encourage users to write `na.action` functions appropriate to their own data: the essential requirement is that the function take a data frame as an argument and return one in which there are no missing values.

Another approach to removing the missing values is to work interactively to estimate them on a case-by-case basis, and then to work with the revised data frame. Keep in mind that only missing values in the observations and variables *included in the fit* will matter. Missing values anywhere else are irrelevant.

Fits Through the Origin; Parallel Regression

In (4.2) a constant or *intercept* term was included by default, as is usual. The coefficient for the intercept is labeled as `(Intercept)` in the fitted model. You can force a fit "through the origin"—that is, without an intercept—by including the term `-1` in the model formula:

```
Fuel ~ Weight + Disp. - 1
```

Used with a factor as the first predictor, this produces "parallel regressions," models in which a different intercept is included for each level of the factor but the coefficients of subsequent terms are estimated on all the observations. The following example produces a parallel regression of `Fuel` on `Weight` for each `Type` of automobile:

```
> lm(Fuel ~ Type - 1 + Weight, fuel.frame)
Call:
lm(formula = Fuel ~ Type - 1 + Weight, data = fuel.frame)

Coefficients:
 Type.Compact Type.Large Type.Medium Type.Small Type.Sporty Type.Van
       1.6721     1.7153      1.7743     1.2761      1.4816   2.2019

      Weight
 0.00088464
```

```
Degrees of freedom: 60 total; 53 residual
Residual standard error: 0.3634
```

Compare this with the use of `Type` with an intercept on page 102. In the parallel regression, there are 6 coefficients associated with each of the levels of `Type`. With an intercept there are 5 coefficients associated with the contrasts and not with individual levels.

Parallel regression can be used with any number of other terms in the model. If we want the slopes to be defined within each level of `Type` as well as for the intercept, then we are dealing with a *nested* model, as discussed in Section 5.2.1:

```
lm(Fuel ~ Type/Weight - 1, fuel.frame)
```

The separate slopes and intercepts will fit 12 coefficients; without the `-1` the model would fit contrasts for `Type` and then a coefficient for `Weight` within each level of `Type`.

Overdetermined Models

By default, `lm()` requires that the data be sufficient to estimate uniquely all the coefficients in the model. In numerical terms, the model matrix used in the fitting must be *nonsingular*. If the matrix is singular, the model is *overdetermined*; that is, there are (infinitely) many coefficient values that provide the same least-squares fit. In this case, computations using the coefficients may be meaningless. Our default approach is to treat overdetermined models as an error.

The application, the data, or the purpose of the analysis can suggest that overdetermined models should be allowed, or at least checked for by some nonstandard computations. Some designed experiments buy a smaller number of runs by arranging that not all the effects in the model will be estimable. A very different perspective comes from retrospective studies with predictor variables that can be highly correlated and at the same time not measured exactly. The statistical and numerical questions are subtle, and discussion of them is deferred to Section 4.4.3. Here we will just describe how to allow singularities and how to check for badly-determined models.

Fitted values and residuals will be well defined if the model is *exactly* singular. Such would be the case, for example, if the same predictor was effectively included twice in the model:

```
y ~ x1 + x2 + poly(x1,2)
```

The linear predictor in `x1` appears twice in this formula. There was no need to have the first `x1` term, but except for the coefficients, everything about the model should be well defined.

If the application suggests that such exact singularities are possible and not a problem, you can allow them by using the `singular.ok` argument to `lm()`:

```
> lm( Fuel ~ Weight + Disp. + poly(Weight, 2), fuel.frame,
+      singular.ok = TRUE)
Call:
lm(formula = Fuel ~ Weight + Disp. + poly(Weight, 2), data = fuel.frame,
     singular.ok = T)

Coefficients: (1 not defined because of singularities)
  (Intercept)    Weight     Disp. poly(Weight, 2)2
      0.50447 0.0012196 0.0011032        -0.45364

Degrees of freedom: 60 total; 56 residual
Residual standard error: 0.38894
```

As the printing warns, not all the coefficients will be defined. The value of `coef(fit)` will give only the estimated coefficients; those not printed are stored as `NA`. The full story is that many coefficients will give the same fit. In this case, the linear part of the third term is *completely aliased* with the first term, and a one-dimensional linear family of coefficients will give the same fit. Chapter 5 deals extensively with overdetermined linear models and presents techniques for studying the pattern of aliasing.

A somewhat different issue is that of models that are *nearly* singular, in that some small change to the data could make the model singular. Again, this is a complicated question, that we deal with in Section 4.4.3. The function `kappa()` returns an estimate of the *condition number* of the model matrix, large values of which indicate that the model is close to singularity. The argument to `kappa()` can be a fitted model or a matrix. A single number is returned; if it is large, there is a possibility that the numerical results of the fit are not well defined. For example, for the model on page 100,

```
> kappa(fuel.fit)
[1] 15494.84
> kappa(diag(5))
[1] 1
```

How large is large? The condition number of any orthogonal matrix is 1. Values that are approaching ε^{-1}, where ε is the relative precision of the computations are cause for numerical concern. On the machine we are using, ε is small enough that the condition estimate for `fuel.fit` is not troubling from a computational view:

```
> .Machine$double.eps
[1] 2.220446e-16
> .Machine$double.eps * kappa(fuel.fit)
[1] 3.440546e-12
```

However, if the data in the problem are subject to sizable error of measurement you ought perhaps to be concerned long before *numerical* inaccuracy is a problem. Section 4.4.3 gives some suggestions.

4.2.5   Updating Models

Developing statistical models nearly always involves modifying an existing model when something is seen to be wrong with it. The structural formula may not be right or some other aspect of the fit may need to be adjusted. The function `update()` allows a new model to be created from an old one by providing only those arguments that have to be changed. This function works on all the kinds of models to be fit in this book, taking advantage of some common structure in the fitted objects and, again, of the ability of methods to be inherited from one class of objects to another.

As an example, the model `fuel.not.van` that we computed above differed from the model `fuel.fit` only in the `subset` argument. Therefore, an equivalent but somewhat simpler way to generate it would have been:

```
fuel.not.van <- update(fuel.fit, subset = (Type != "Van"))
```

Updating also allows the formula to be updated so as to add, drop or change pieces of it while keeping the rest constant. For example, to add the variable `Type` to the `fuel.fit` model, as on page 102, we could update the formula:

```
fuel.fit2 <- update(fuel.fit, . ~ . + Type)
```

In giving the formula, we used "." both on the left and the right of the "~" to stand for "whatever was here before". On the left it stands for the old response, `Fuel`, and on the right for the old right side, `Weight + Disp.`. The same shorthand works for dropping terms. We could have gone back from `fuel.fit2` to the original model by:

```
fuel.fit.old <- update(fuel.fit2, . ~ . - Type)
```

A different response can be fit to the same predictors. If we wanted to fit `sqrt(Fuel)`,

```
sqrt.fit <- update(fuel.fit2, sqrt(.) ~ . )
```

will substitute for "." the original response from `fuel.fit2`:

```
> formula(sqrt.fit)
sqrt(Fuel) ~ Weight + Disp. + Type
```

The computations in `update()` attempt to simplify the new formula, so that adding and then dropping terms will work reasonably well:

```
> formula(fuel.fit.old)
Fuel ~ Weight + Disp.
```

The expressions to be simplified must match as expressions, not just numerically; for example, `Weight^2` and `Weight * Weight` are different.

Any arguments in the original call can be deleted from the new call by giving them as empty arguments to `update()`. To go from the fit with a `subset=` argument to one without:

```
> fuel.no.sub <- update(fuel.not.van,subset=)
> fuel.no.sub
Call:
lm(formula = Fuel ~ Weight + Disp., data = fuel.frame)
    ...
```

Finally, you can change from one kind of model to another by giving the `class=` argument to `update()`. The value of the argument is the name of the fitting function to be used, instead of that called to produce the current fit. To refit a model using `glm()` instead of `lm()`, call `update()` with

```
class = "glm"
```

along with whatever other new arguments and changes in the formula are needed.

The `update()` function's only assumptions about the fitting function are that it has `formula` as its first argument, and that the call that produced the old fit can be extracted from a `"call"` component in the fit. The new fit must be computable from the reconstructed call; in particular, the data must be available. If all the relevant data are in a data frame given as an argument, there should be no trouble. If the data frame is omitted or if we mix in variables from several sources, updating can fail—for example, because the search list or the contents of the working data have changed since the original fit was computed.

4.3 Specializing and Extending the Computations

The techniques illustrated up to this point don't require knowing what `lm()` does internally. Most statistical use of the linear model software should be at this level: the analyst wants to concentrate on asking the relevant and interesting questions rather than being diverted into computational details. One's perspective begins to change when the modeling software starts to be used to develop further techniques (for example, when a number of related linear models are to be produced at once). Since most of the computing time taken by `lm()` is in analyzing and setting up the proposed model and only a small part in the actual numerical fitting, repeated fitting can be made faster by iterating at a lower level than `lm()` itself.

Section 4.3.1 illustrates this in the context of general fitting of related models, and Section 4.3.2 discusses the special case of adding and dropping terms from the current model. Section 4.3.3 provides a framework for a variety of statistical techniques assessing the influence of observations.

4.3.1   Repeated Fitting of Similar Models

When fitting a single linear model, the natural way to proceed is as shown in subsection 4.2.1. The model and data are provided as a formula and a data frame; all the details from that point on are worked out by the `lm()` function. In some applications, however, you may want to fit related linear models, varying some aspect each time but otherwise reusing the previous model.

There are several ways to do this. Generally, they amount to a tradeoff between saving computing time and getting deeper into the computations. The use of `update()` as described in Section 4.2.5. is by far the easiest and most flexible approach. Anything can be changed (the new model doesn't even have to use `lm()`), and shorthand notation is provided for the common case of making changes in the formula. You pay for the generality in that `update()` just constructs a new call to `lm()` or to a similar function and then recomputes the fit from scratch. In order to save the computer some of this work, a human will have to work harder. If similar problems arise repeatedly, the investment will usually be worthwhile. The first requirement is to understand a bit more about how `lm()` and its cousins work. The main stages are as follows:

- From the formula, the data, and the optional arguments `weights`, `subset`, and `na.action`, `lm()` constructs an intermediate data frame, called the `model.frame`, containing just the data needed in the model. This includes the variables that appear in the formula, taking account of any subset selection, `NA` action, etc. The model frame has an attribute, `terms`, that summarizes in detail the terms of the model defined by the formula. See Section 3.3.3 for details.

- From the model frame, `lm()` constructs the response as a numeric vector or matrix, and the predictors as the `model.matrix`, a numerical matrix with some additional information about the model. See Section 2.4.3.

- The model matrix and the response are the arguments to `lm.fit(x, y)`, which does the numeric fitting.

- The basic fitted model returned by `lm.fit()` is augmented with components describing the terms, the call to `lm()`, and the response and/or model matrix, if the call asked for those.

A look at the definition of `lm()` will show how this works. The special techniques for repeated fitting proceed by doing the first step or the first two steps once only, and then repeating the rest of the calculations as needed.

Repeatedly modifying and using the model frame mainly saves manipulating the large data frame from which the (perhaps much smaller) model frame was constructed for this specific model. This technique is very general, using no special features of linear models. Therefore, it is a useful paradigm for repeated fitting of any models. Repeating the fitting using the model matrix, rather than the model frame, saves much more of the preparation and so is important for large-scale computations, such as simulation studies.

The other main question in repeated fitting is how to organize the results. Here too there is a tradeoff, this time between a simple list of ordinary `lm` objects and some more specialized organization. The list of fits has the advantages that the elements of the list are ordinary `lm` objects: all the usual summaries can be applied. Also, building the list needs no special knowledge of how `lm` objects are organized internally. Special structures can save on some space and can sometimes make later computations simpler. We will show one special organization, as a matrix of repeated fits by components of the fit. This simplifies access by component (for example, getting the residuals from all the models at once) and can also save some space by not including components of the fit that stay the same or that are not needed in the result.

Simple Repeated Fitting

As an example, suppose we want to include a power of one or more predictor variables, and would like to investigate the fit for a sequence of possible powers. In particular, suppose we want to fit a model of the form

```
Fuel ~ I(Weight^k) + Disp.
```

where `k` is to be chosen. This is not a linear model with `k` as a parameter, but rather than drag in a more powerful algorithm, we can refit with a sequence of values for `k` and compare the fits. We choose to study different powers of a predictor, rather than the more common case of powers of the response, because the latter can be done in a single call to `lm()`; however, for models in later chapters such as those fit by `glm()`, powers of a response would need to be handled as in the example to follow.

Let's begin with a purely lazy approach. Suppose `k.values` is a vector of values we want to try for `k`:

```
> fits <- list()
> for(k in k.values)
+    fits[[as.character(k)]] <-
+       lm( Fuel ~ I(Weight^k) + Disp., fuel.frame)
```

Notice that we assigned the elements of `fits` by name, not by position. This trick is recommended for many computations that create lists, because it creates named lists directly and avoids the confusion that can happen when a `NULL` value is assigned, deleting a component of the list. In this example, we called `lm()` each time, since the formula and data expressions were constant, with only the value of `k` changing.

If there are changes in the call *other* than values for data, the `update()` function can be used to construct the new call each time.

Once the list of models is created, we can pluck off specific results to study the fits. Several functions in S help, one of the most useful being `sapply()`. This applies a function to each element of a list and then attempts to simplify the result to a vector or matrix. The function supplied to `sapply()` can be any function suitable to be applied to the elements of the list. In this case, each element will be an `lm` object, and we can apply any of the summarizing functions in Section 4.2.1.

The "function" supplied to `sapply()` really is an S function object, even though the argument is usually just a name. As a result, if the precise function you want doesn't exist, you can simply define it in-line. For example, suppose we want to compare the variance of the residuals of the fits. All that is needed is to supply `sapply()` with the function definition

```
function(x) var(residuals(x))
```

This function will be called for each of the `lm` objects in `fits`:

```
sapply(fits, function(x)var(residuals(x)))
        0.5          1        1.5          2
   0.1445124  0.1470052  0.1514131  0.1574302
```

The four values of `k`, `c(.5, 1, 1.5, 2)`, appear as the names of the vector of variances. It seems that the square-root is the best choice, but there is not much difference over the chosen range. To look more closely, let's draw boxplots of the sets of residuals. The argument to the `boxplot()` function is a list with each set of residuals as one element. The expression

```
sapply(fits, residuals)
```

would simplify its result to a matrix since each set of residuals is of the same length. Using the function `lapply()` or giving the argument `simplify=FALSE` to `sapply()`, suppresses the simplification, so the resulting list can be passed to `boxplot()` directly:

```
> boxplot(sapply(fits, residuals, simplify = F))
```

Fitting from the Model Matrix

Even though `update()` makes no effort to save on space or time, the simplicity of this approach makes it the one to start with. But if the computations are to be done many times, a lower-level version is useful. As noted, `lm()` eventually calls a function `lm.fit(x,y)` or, for weighted fitting, a function `lm.wfit(x,y,w)`, where `x` and `y` are the model matrix and the response. In order to do more efficient fitting, we can construct these arguments and call the appropriate fitting routine directly.

The lm() function computes the model matrix and response, but by default does not return them as components of the lm object. If arguments x=T or y=T are supplied to lm(), it will return the corresponding component in its value.

Suppose we wanted to resample the data in a model with nrows observations, picking rows with replacement each time, to generate a new fitted model (a "bootstrap" resampling of the predictor and response). We can generate the model matrix and response for the resample by sampling with replacement from the observations of the original model. To begin, we create the original model, including the model matrix and response:

```
> fit1 <- lm(Fuel ~ Weight + Disp., fuel.frame, x=T, y=T)
> x <- fit1$x; y <- fit1$y
```

The resampling of rows can be applied to the model matrix x and the response vector y, and a new fit produced by lm.fit():

```
> rows <- sample(nrows, nrows, replace=T)
> fit2 <- lm.fit(x[rows, ], y[rows])
```

We can repeat the calls to sample() and lm.fit() to resample as many times as desired. For a substantial number of resamples, much less computation will be needed than when using update() or lm() each time. In this case, the computationally simpler form is simple for the human as well, once the necessary details of lm() are understood.

The real payoff, of course, comes from writing an S function to encapsulate the details. Let's write bootstrap.lm() to carry out the bootstrap sampling given a single initial fit, z, and the number of resamples wanted:

```
bootstrap.lm <- function(z, nsample) {
    x <- z$x; y <- z$y
    nrows <- dim(x)[1]
    value <- list()
    for(i in seq(nsample)){
        rows <- sample(nrows, nrows, T)
        value[[i]] <- lm.fit(x[rows, ], y[rows])
    }
    value
}
```

A user-friendly version of bootstrap.lm() should be a bit more careful than this; for example, if the original fit was missing the x or y component, bootstrap.lm() could use update() to redo the fit.

One caveat about proceeding at this lower level of numerical fitting is that not all of the information in the full lm object will have been generated by lm.fit(). A fairly simple modification of the function above would retain all the information available in the lm objects. Replace the line

```
value[[i]] <- lm.fit(x[rows,],  y[rows])
```

with the lines

```
z$x <- xx <- x[rows,]; z$y <- yy <- y[rows]
vi <- lm.fit(xx,  yy)
z[names(vi)] <- vi
value[[i]] <- z
```

The technique here, which is quite general, arranges for each element of the list to
have the following components:

- all the components returned by `lm.fit()` on the ith call;

- anything in the original fit not returned by `lm.fit()`;

- the data `x` and `y` used in the ith fit (these are not required and could be
  omitted if there was a large amount of data).

For large examples, the replication of identical information in the elements of the
list may waste a serious amount of space. An organization of the results in an
object of a special class (see, for example, page 124) avoids this wasted space at
the cost of more effort in organizing the results. Either approach only works on the
assumption that the information *other* than the components returned by `lm.fit()`
stays the same from one fit to the next. This is often correct, but needs to be checked
for each application. For example, `bootstrap.lm()` as written does not handle the
case that the sampled model is overdetermined (singular), as it could well be. A
possible approach is to use the argument `singular.ok=T` to `lm.fit()`, but comparing
a singular model with a nonsingular one is a bit ambiguous. Let's pass over these
details, however, and discuss an alternative organization of the results.

Matrix Organization for Repeated Fits

Organizing the results of repeated fitting as a list of fits is undoubtedly the best way
to start, and is probably the best choice in any case unless one expects to apply
some extensive calculations repeatedly across the different fits. However, notice
that it was necessary to use the function `sapply()` to get at all the residuals. This
does rather a lot of computing and efficiency might again be a consideration. In
some specialized applications, it may be important to make indexing symmetric and
efficient, either across the models or across the components of the models.

What kind of data organization in S makes subscripting in two different ways
easy? Obviously, a matrix. It is designed exactly for this purpose, and we should
consider organizing the fits as a matrix, indexed by the components of an `lm` object
along one dimension, and by the different circumstances giving rise to the model
along the other. This may sound a little strange at first since you are more likely

to encounter matrices of numerical data or of character strings than matrices of mode "list", as this one will be. However, nothing bizarre or specially constructed is necessary, and the technique will prove extremely useful. Therefore, let's pause to review a few things about matrices in S, to clarify what we are going to do. Sections 5.5.1 and 5.5.2 of $\boxed{S}$ (pages 126–135) are also useful reading. A matrix, or generally a multiway array, is any S object that has an attribute dim. This attribute gives the dimensions of the array; a matrix is just the special case in which the dim attribute is of length 2. S takes care to keep the dimension information meaningful and consistent with the length of the object, but otherwise it imposes no constraints on a matrix. In particular, the mode can be anything at all.

The meaning and the importance of matrices reside largely in the functions that understand how to deal with them. In this case, it is the [and [[operators that will be essential. These both understand how to subscript matrices ($\boxed{S}$, p. 127), and they also are written so that matrices can have any mode at all. You will not have used [[much with matrices, but when the mode of the matrix is "list", there is an important distinction. The result of [is always an object of the same mode as the original object; however, if you want to get a particular single element of a list, rather than the list of length 1 that contains that element, the [[operator is the one to use ($\boxed{S}$, p. 111). Arrays can also have a dimnames attribute that gives names to the rows and/or the columns, which can be used with the [or [[operators.

With this brief review in mind, let's turn the value of bootstrap.lm() into a matrix, whose rows correspond to the components returned by lm.fit() and whose columns correspond to the different samples. The calculations are identical, except that just before returning value:

```
rnames <- names(value[[1]])
value <- unlist(value, recursive = F)
dim(value) <- c(length(value)/nsample, nsample)
dimnames(value) <-  list(rnames, paste("Sample",1:nsample)))
class(value) <- "matrix"
```

The call to unlist() makes a single vector from the components of all the nsample fits. Both the dim and dimnames rely on the assumption that lm.fit() returns the same components each time; other than that, the paradigm used above applies to essentially any similar sequence of refitting. The dimnames attribute is computed from the component names of the object returned by lm.fit(). Setting the class to "matrix" will bring in some matrix methods, the most important being a special printing method for matrices of mode "list" (see page 126).

This same organization could have been used in the computation of the fits with different powers of the Weight variable on page 119.

```
fits <- array(fits, c(length(fit1),length(k.values)),
    list(names(fit1), k.values))
```

The boxplot function used on page 129 to produce Figure 4.5 expects as its argument a list of the data vectors to be plotted. With the matrix organization this does not require sapply():

```
boxplot(fits["residuals",])
```

Other functions might want, for example, a matrix whose columns are the residuals from the different fits. This is obtained by the expression

```
matrix(unlist(fits["resid",]), ncol = ncol(fits))
```

Remember also that the [[operator works for matrices of any mode. For example, a normal probability plot of the residuals for k=.5 is produced by:

```
qqnorm(fits[["resid", ".5"]])
```

If you get unexpected error messages about trying to use non-atomic or non-numeric data, chances are you meant to use [[, but used [instead.

The matrix form makes extractions simpler to write and computationally faster, but the extra programming involved in computing the object is not worthwhile for "one-off" analyses. For substantial new software efforts, where the objects created may themselves be used for further computing, the matrix form can be useful. The approach could be taken further by designing a *class* of objects for the matrix version of bootstrap.lm(), for example, so that methods could smooth over the details of the implementation. This would allow us to return complete lm objects when columns are selected, without replicating all the components in each fit, as we did previously. The extra information would be copied once as an attribute of the object, and a method for the "["() function would insert these in the value when a column subset was computed. The details would make an interesting, fairly advanced exercise. See Appendix A on how such classes and methods are designed, and the function "[.data.frame"() for an example.

If you have read Chapter 3 on data, all this discussion of lists versus matrices might suggest that data frames are likely to pop up. Indeed they do, in somewhat different uses of refitting. One example is shown in Section 4.3.2, in which new models will be formed by systematically adding or dropping one term from the current model. This example is sufficiently important that it rates a generic function and methods, but it is also interesting as another version of the general refitting techniques.

4.3.2   Adding and Dropping Terms

The analytic techniques to be considered next focus on individual terms that are candidates for inclusion in a linear model. The questions to consider include "Does this term appear to add a useful structural relationship to the model?", "What structural form seems to represent the relationship best?", and "Is the relationship strongly influenced by a few observations?". These questions hark back to the displays and summaries in Chapter 3, where we were studying the relationships between variables in the data frame. The focus now is sharper and narrower. We have chosen a response and are looking for a good linear model. A candidate, initial model has been fitted. We want to study some possible new models.

Computationally and statistically, two different situations arise. In one case, we begin with a model and ask what would happen if we *dropped* a term. In the other, we begin with a model and an additional set of candidate terms, and ask what would happen if we *added* one of those terms. The analytic techniques are essentially the same, but the context is different, and different computations are used to provide an efficient answer.

For the most part, we will talk about *terms* in this discussion, rather than coefficients. A distinction arises when the term being considered corresponds to a factor or a matrix, so that more than one coefficient is involved. Dropping single coefficients rather than terms is an option, but dropping some of the coefficients used to fit a categorical variable means quite a different thing from dropping the term, and is not always a meaningful operation. For example, if we coded an ordered factor by polynomial contrasts—linear, quadratic, cubic, etc.—then dropping the highest power would make sense, but dropping a lower power usually would not. So the discussions here will be of adding and dropping terms:

```
drop1(fit)
```

will return all the fits obtained from dropping one of the terms in `fit` and

```
add1(fit, scope)
```

will return all the fits from adding one term to `fit` from the possibilities in `scope`. These are generic functions. They find more frequent application in generalized linear models (Chapter 6), where adding and dropping terms involves somewhat more computations. The `glm` objects inherit from linear-model objects, however, and in this case the methods used for `glm`'s are applicable to *any* linear model.

As an example, suppose we consider the two terms in the original `fuel.fit`, and form an object representing this model and the two models that can be formed from dropping either `Weight` or `Disp.`:

```
fuel.drop1 <- drop1(fuel.fit)
```

The object `fuel.fit` is the fitted model computed on page 100. The `anova` object `fuel.drop1` is a summary of the changes resulting from dropping each of the terms:

```
> fuel.drop1
Single term deletions
```

```
Model: Fuel ~ Weight + Disp.

        Df Sum of Sq    RSS    Cp
<none>                 8.67   9.59
Weight   1     7.931  16.60  17.21
 Disp.   1     0.045   8.72   9.33
```

The printing method shows the initial model, and then a table with rows corresponding to the original and each of the deletions. The columns give the number of degrees of freedom in the deleted term, the sum-of-squares due to the deleted term, the residual sum-of-squares for the reduced model, and the C_p statistic for the subset of terms in the reduced model. For a discussion of the C_p statistic in this context, see page 233 in Chapter 7.

The object returned by drop1() or add1() summarizes the fits by the quantities Df, etc., as shown in the printing above. In contrast to the examples in the previous section, the object is not a list of the individual fits. Additional information for each of the models will be returned if you supply the argument keep=T to the drop1() function:

```
> fuel.keep <- drop1(fuel.fit, keep=T)
```

When printed out, this object will list some additional statistics kept for future computations:

```
> fuel.keep
$anova:
Single term deletions

Model: Fuel ~ Weight + Disp.

        Df Sum of Sq    RSS    Cp
<none>                 8.67   9.59
Weight   1     7.931  16.60  17.21
 Disp.   1     0.045   8.72   9.33


$keep:
       coefficients      fitted    residuals
Weight numeric, 2    numeric, 60  numeric, 60
 Disp. numeric, 2    numeric, 60  numeric, 60

       x.residuals      effects          R
Weight numeric, 60  numeric, 60  numeric, 4
 Disp. numeric, 60  numeric, 60  numeric, 4
```

In this version, the object returned is a list whose anova component is the table seen before. The other component, keep, is a matrix of mode "list", whose columns are indexed by different statistics and whose rows correspond to the dropped terms, here `"Weight"` and `"Disp."`. The printout `"numeric, 2"` means that the corresponding element of the matrix is a numeric vector of length 2.

The `keep` component is similar to the organization in Section 4.3.1, where repeated fits are represented as matrices. Elements of the matrix give single components for particular models:

```
fuel.keep$keep[[ "Weight", "residuals"]]
```

gives the residuals from the model in which `Weight` was dropped. The column `"x.residuals"` gives the residuals of the corresponding predictor when it is regressed against all the *other* predictors in the original model. This statistic arises in a number of plots and summaries used for studying added and dropped terms. The other columns should be self-explanatory. Since the individual elements are whole objects, not numbers or character strings, the printing method for class `"matrix"` just prints a brief summary of each element.

As an example of using the additional statistics from `keep=T`, suppose we want to look at the residuals of the model with `Weight` omitted. One useful plot shows the residuals of `Fuel` from this model plotted against residuals of `Weight` fitted against the same terms—in this case, regressed against `Disp.`. If the configuration of points in this plot shows a noticeable linear trend, this is evidence that `Weight` should be included; in particular, the simple linear regression of the points in this plot gives the same coefficient for `Weight` as in the full model `lmf`. For the statistical background, see Belsley, Kuh, and Welsch (1980, p. 30), Cook and Weisberg (1982, p. 44), and Chambers et al. (1983, pp. 268–277). All three references give a different name to the plot; we will call it the *added variable* plot, following the second reference, but with the caution that there are other reasonable plots for looking at added variables.

Figure 4.4 shows the plot for our example, along with a similar plot when `Displacement` is dropped. There is a clear linear pattern for `Weight`, indicating that it deserves to be in the model, but a much less obvious pattern for `Displacement`, reinforcing the message given from the two printed summaries. To produce these plots we extract the corresponding components from each of the two columns:

```
keepstuff <- fuel.keep$keep
for(x in  dimnames(keepstuff)[[1]])
    plot(keepstuff[[x,"x.resid"]],keepstuff[[x,"resid"]],
       xlab= paste("Residuals of",x,"from rest"),
       ylab=paste("Residuals without",x))
```

Note the use of double square-brackets: we want the vector in the corresponding element of the list, not a sublist (see **S**, p. 111, if this is an unfamiliar distinction). It should also be obvious that we could write an "added variables plot" function that did little more than the above, to produce this sort of plot for any drop-1 linear

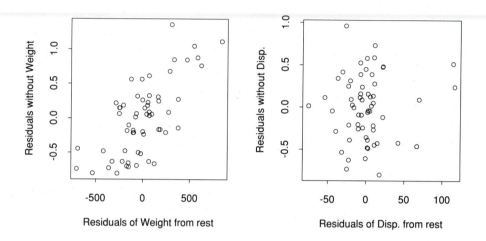

Figure 4.4: *Plots of the "drop-1" fits from the fit of* Fuel *on* Weight + Displacement

model summary. As in the previous subsection, the goal of functions like drop1()
is to produce an object from which a wide variety of specific diagnostics can be
generated easily.

The opposite approach starts from a model and adds on one term from each of
a possible choice of terms; for example,

```
> fit0 <- lm(Fuel ~ 1, fuel.frame)
> fuel.add1 <- add1(fit0, . ~ Weight + Disp. + Type)
```

The first statement creates an empty model, fitting only the intercept. The second
then investigates all the one-variable fits by adding each of them to the empty
model. The resulting object has the same structure as for drop1(), but this time
the individual models in the rows will all have one *more* term than the original
model.

```
> fuel.add1
Single term additions

Model: Fuel ~ 1
        Df Sum of Sq   RSS    Cp
<none>                33.86 35.00
Weight   1    25.14   8.72 11.01
 Disp.   1    17.25  16.60 18.90
```

    Type 5   24.24 9.62 16.50

The `keep=T` option is again available, with the same statistics. The `x.residuals`
statistic contains the residuals from fitting each of the new terms against the same,
original model. If we had computed `fuel.add1` with `keep=T`, boxplots of the three
sets of residuals could be produced easily. Figure 4.5 shows that the spread of the

> boxplot(fuel.add1$keep[, "residuals"])

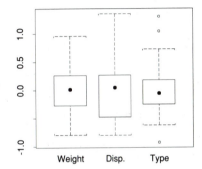

Figure 4.5: *Boxplots of residuals, fitting* `Fuel` *to three variables in the automobile data
using the* `add1()` *function*

residuals is much less when fitting `Weight` or `Type` than `Disp.` (the comparison is not
entirely fair, since `Type` uses 5 degrees of freedom and the numeric variables only 1).

4.3.3 Influence of Individual Observations

A weakness of least-squares models, in theory and sometimes in practice, is that
individual observations can exert a large influence on the fitted model. If those
observations were in some sense inconsistent with the bulk of the data, the con-
clusions drawn from the model could be misleading. A wide variety of statistical
techniques have been proposed to detect and analyze such situations: the books by
Belsley, Kuh, and Welsch (1980) and by Cook and Weisberg(1982) present a variety
of techniques, some of which we will illustrate. The questions that we will consider
center around what influence individual observations have on the fit. In particular,
what would happen to various aspects of the fitted model if individual observations
were omitted? Computing all the n models arising if one of the n observations were
omitted would be impractical for large datasets. Fortunately, numerical techniques
for linear least-squares models allow many relevant summaries of such models to be

computed all at once. (If this were not the case, the subject would have received much less attention than it has!)

The function lm.influence() takes an initial linear model and returns an object whose components contain various summaries of the n models obtained by omitting one observation. These summaries tend to be either vectors of length n or matrices with n rows, corresponding to the omitted observation. Specifically, lm.influence() returns a list with components coefficients, sigma, and hat. The first is a matrix whose rows are the coefficients for the model with the corresponding observation omitted, and sigma is a vector of the estimates for the residual standard error in the corresponding model. The definition of hat is a little more technical, and has to anticipate just a little the theoretical discussion of Section 4.4. The fitted values for the response, say $\hat{y}$, can always be written as the product of an n by n matrix times y:

$$\hat{y} = \mathbf{H} \cdot y$$

The component hat is the vector of the n diagonal elements of $\mathbf{H}$; large values of this vector indicate observations with a large influence on the fit. The matrix $\mathbf{H}$ depends only on the predictors, not on y. The utility of hat arises from its ability to summarize what is often called the *leverage* of the individual observations—that is, the effect on the fit arising from the corresponding row of the model matrix.

The components of the lm.influence object, or at least coefficients and sigma, are directly interpretable and of some interest. The real point in their design, however, is that they can be combined with some of the components of the lm object describing the original fit to compute a very wide variety of diagnostics. A number of these are summarized in Table 4.1. The S expressions in the second column of the table compute the specified diagnostics, using objects assumed to have been extracted from an lm object, say lmf, and from corresponding summaries and influence objects as follows:

```
lms <- summary(lmf)
lmi <- lm.influence(lmf)
e <- residuals(lmf)
s <- lms$sigma; xxi <- diag(lms$cov.unscaled)
si <- lmi$sigma; h <- lmi$hat
bi <-  t(coef(lmf) - t(coef(lmi)))
```

A typical use of this information would be to construct S functions that produce the particular statistics you want, using whichever of the quantities in the table are needed; for example,

```
dfbetas <- function(fit, lms = summary(fit), lmi = lm.influence(fit)) {
    xxi <- diag(lms$cov.unscaled)
    si <- lmi$sigma
    bi <-  t(coef(fit) - t(coef(lmi)))
```

Quantity	Expression	Meaning	Reference
Standardized Residuals	`e/(s*(1-h)`$^\wedge$`.5)`	Residuals with equal variance	BKW(20)
Studentized Residuals	`e/(si*(1-h)`$^\wedge$`.5)`	Use **si** as standard error	BKW(21), CW(18)
DFBETAS	`bi/(si %o% xxi`$^\wedge$`.5)`	The change in the coefficients, scaled by the standard error for the coefficients	BKW(13), CW(125)
DFFIT	`h*e/(1-h)`	The change in the fitted value when that observation is dropped	BKW(15)
DFFITS	`h`$^\wedge$`.5*e/(si*(1-h))`	Change in fitted values, standardized to variance 1	BKW(15), CW(124)

Table 4.1: *Computation of some diagnostic summaries for influential observations from the components of* `lm()` *and* `lm.influence()`. *References are to page numbers in <u>B</u>elsley, <u>K</u>uh, and <u>W</u>elsch (1980) and <u>C</u>ook and <u>W</u>eisberg(1982).*

```
    bi/(si %o% xxi^0.5)
}
```

The use of the outer product operator, `%o%`, produces a 60 by 3 matrix (observations by coefficients), matching the dimension of `bi`.

Similar functions could be created for the other statistics in Table 4.1, as well as for many other related quantities. The computations done by `summary.lm()` and `lm.influence()` provide the building blocks.

4.4 Numerical and Statistical Methods

This section discusses details of fitting linear models that underlie the functions in the previous sections. You should read the material if you want to understand *why* things work the way they do, or if you want to make them work in a seriously different way. You're welcome to read it anyway, if you're just curious. Section 4.4.1 gives a more formal discussion of the statistical regression model as we have used it in the chapter. Section 4.4.2 presents similar elaboration on the numerical methods used. Section 4.4.3 goes into some detail about the difficulties associated with over- or ill-determined models.

4.4.1   Mathematical and Statistical Results

Mathematical, computational, and statistical results on linear models are covered in many books and articles. We give some references for further reading at the end of the chapter. In this section, the main results will be given briefly, along with some special considerations for statistical modeling not discussed in the numerical-analysis literature. The mathematical and computational discussion of linear least-squares fitting rests on a few fundamental results. To express these compactly, we need to use matrix notation. Let X be the model matrix whose columns, x_j, are the predictors in the model, including a first column whose elements are all 1, if the intercept is included. Let β be the vector of coefficients, again including the intercept if one is fitted. Let p be the length of β. The least-squares estimate of the coefficients will be denoted $\hat{\beta}$.

The vector of n fitted values from the linear model can be written in matrix form, for any coefficients β, as

$$X \cdot \beta$$

where "$\cdot$" denotes matrix multiplication. As in equation (4.3) on page 97, the residuals are:

$$\varepsilon = y - X \cdot \beta$$

The least-squares fit chooses $\beta = \hat{\beta}$ to minimize the sum of squares of the residuals (4.2). Two characterizations of a least-squares fit arise most frequently. One can be stated in terms of an orthogonal transformation that takes X into an upper-triangular matrix. Suppose Q is an orthogonal n by n matrix; that is,

$$Q^T \cdot Q = I$$

where Q^T is the transpose of Q, and I is the identity matrix. For the orthogonal decomposition, Q is chosen so that

$$Q^T \cdot X = \left[\begin{array}{c} R \\ O \end{array} \right] \tag{4.5}$$

where R is a p by p upper-triangular matrix; that is,

$$R_{i,j} = 0, \; i > j$$

and O is a matrix of all zeros. If Q_1 is the first p columns of Q, then

$$X = Q_1 \cdot R$$

In geometric terms, Q_1 is a set of orthogonal vectors in n-space that span the columns of X; i.e., any linear combination of columns of X can be written as a linear combination of the columns of Q_1.

We now define the *effects* to be the vector c such that:

$$c = \begin{bmatrix} c_1 \\ c_2 \end{bmatrix} = Q^T \cdot y \qquad (4.6)$$

where c_1 is of length p and c_2 is of length $n - p$. Any $\hat{\beta}$ that satisfies the equations

$$R \cdot \hat{\beta} = c_1 = Q_1^T \cdot y \qquad (4.7)$$

can be shown to produce a least-squares fit. Equation (4.7) is the basis for practical numerical methods for least-squares solutions. The computations performed by lm() use the Householder decomposition for computing Q; this produces a complete and accurate definition of the transformation. Choosing R upper-triangular provides efficient, accurate solutions for (4.7). Mathematically, we have not used the upper-triangular property, and all we need is the ability to compute an orthogonal basis of X and then solve for $\hat{\beta}$. More about this when we come to discuss the singular-value decomposition. We are assuming in this section that the linear model is of full rank. The characterizations of the results in terms of an orthogonal decomposition carry over to the case that the model is over-determined, but the notation becomes more cumbersome, so we prefer to postpone this generalization to Section 4.4.3.

A second characterization of the least-squares fit begins from the cross-products and can be derived from (4.5) - (4.7). Any $\hat{\beta}$ satisfying the *normal equations*; namely,

$$(X^T \cdot X) \cdot \hat{\beta} = X^T \cdot y \qquad (4.8)$$

gives a least-squares fit. The normal equations follow directly by substitution; in particular,

$$\begin{aligned} R^T \cdot R &= X^T \cdot X \\ R^T \cdot c_1 &= X^T \cdot y \end{aligned} \qquad (4.9)$$

which provides a basis for the computations using the computed cross-products. The first equation of (4.9) is implemented by computing the Choleski decomposition of $X^T \cdot X$. The second equation is then used to solve for c_1; this is the solution of another triangular system of equations, and therefore can be computed quickly and accurately. Given c_1 and R, $\hat{\beta}$ is computed as before from (4.7).

The solution to (4.7) can be written as a linear transformation of c_1, say

$$\hat{\beta} = R^- \cdot c_1$$

If all the columns of X are linearly independent, as we are assuming in this section, then R^- is the inverse of R, is upper-triangular, and is uniquely defined. Substituting c_1 from (4.7), we can write $\hat{\beta}$ as

$$\hat{\beta} = X^- \cdot y \qquad (4.10)$$

where $X^- = R^- \cdot Q_1^T$. The relevance of (4.10) is not that one computes X^-, but that writing $\hat{\beta}$ in this way shows that it is a linear transformation of y. So also are the least-squares residuals and fitted values. The fitted values are

$$
\begin{aligned}
\hat{y} &= X \cdot \hat{\beta} \\
&= X \cdot X^- \cdot y \\
&= Q_1 \cdot Q_1^T \cdot y
\end{aligned}
$$

The last line follows from the definition of X^- and the fact that

$$R \cdot R^- = I$$

The n by n matrix

$$H = Q_1 \cdot Q_1^T$$

projects y into its fitted values $\hat{y}$. The diagonal elements of $\hat{y}$ are used in Section 4.3.3 as a measure of the leverage the individual observations have on the fitted values. The residuals are also a linear transformation of y,

$$e = y - \hat{y} = (I - H) \cdot y$$

Much of the formal neatness of linear least-squares results from the estimated coefficients, the fitted values, and the residuals all being linear functions of the response.

Statistical characterizations follow from making assumptions about the process that generated the data. In the usual treatment, the values of X are taken as fixed and the model assumes that

$$y = X \cdot \beta + \varepsilon$$

where the elements ε_i of ε satisfy the assumptions of independence, constant variance, and (usually) normal distribution on page 97. The combination of these assumptions with the algebraic characterizations given before leads to relatively precise theoretical results on the distribution of estimates from the fitted model.

In particular, because coefficients, fitted values, and residuals are linear transformations of y, the assumptions determine normal distributions for these quantities as well. Linear combinations of normally distributed quantities are also normally distributed. If z has a multivariate normal distribution with mean μ and variance matrix Σ, then any linear transformation of z, say $A \cdot z$, has a normal distribution with mean $A \cdot \mu$ and variance

$$V(A \cdot z) = A \cdot \Sigma \cdot A^T$$

From the characterization of (4.10), $\hat{\beta}$ is normally distributed with mean β and variance

$$
\begin{aligned}
V(\hat{\beta}) &= X^- \cdot X^{-T} \cdot V(y) \\
&= R^- \cdot R^{-T} \times \sigma^2
\end{aligned}
$$

The variances and covariances of the coefficients are estimated by replacing σ^2 by an estimate of the residual standard error, typically s^2, the sum of squared residuals divided by the number of degrees of freedom for residuals ($n - p$ if X is nonsingular). Notice that *whatever* estimate of σ^2 is used, it just multiplies the unscaled covariance, $R^- \cdot R^{-T}$. This is one reason why the unscaled covariance is included as a component of the summary object returned by `summary.lm()` (see page 106). In particular, the correlation matrix of $\hat{\beta}$ in the model is the corresponding cross-product when the rows of R^- are rescaled to have unit sum-of-squares.

Given some new predictor data, say x, the model predicts the corresponding response values to be $x \cdot \hat{\beta}$. Since this is a linear transformation of $\hat{\beta}$, the variance of the vector of the prediction is given by the formula for the variance of $\hat{\beta}$:

$$V(x \cdot \hat{\beta}) = x \cdot V(\hat{\beta}) \cdot x^T = x^- \cdot x^{-T} \times \sigma^2$$

where $x^- = x \cdot R^-$; that is, x^- satisfies

$$R \cdot x^- = x$$

Notice also that the prediction itself can be written as $x^- \cdot c_1$.

Let's summarize by noting what information from the fit we need in order to define various summaries. R and c_1 are essential: they are used to get the coefficients and, with an estimate of σ, are enough to estimate distributional properties, such as variances and correlations, for the coefficients and for predictions corresponding to new data. Some additional information is needed to answer similar questions about the n fitted values or residuals. This information comes from the orthogonal basis, Q_1, or through the projection matrix H defined from it.

4.4.2 Numerical Methods

One argument to `lm()` not discussed so far is `method`. This argument can specify a function to use for the numerical fitting. The method of *estimation* remains least squares; the motivation is to provide a more efficient or desirable numerical algorithm for special applications. Three methods are supplied, implementing the algorithms discussed in the previous section: `"qr"` implements the QR decomposition (in particular, the Householder method); `"chol"`, the Choleski decomposition method; and `"svd"`, the singular-value method. The default is `"qr"`.

```
> fuel.chol <- update(fuel.fit, method="chol")
> fuel.chol
Call:
lm(formula = Fuel ~ Weight + Disp., data = fuel.frame, method = "chol")

Coefficients:
```

```
(Intercept)    Weight        Disp.
      0.47897 0.0012414 0.00085436


Degrees of freedom: 60 total; 57 residual
Residual standard error: 0.39008
> effects(fuel.chol)
 (Intercept) Weight    Disp.
       32.611 5.0139 0.21169
```

Different methods will produce slightly different numeric results, as we will discuss, but only in situations where their particular numerical properties are relevant is this likely to matter to the user. For example, the Choleski method may be important in situations where computations using the cross-products are simpler or faster than those using the QR decomposition. The `"svd"` method may be important in some studies of ill-conditioned problems.

Other than these numerical issues, the different methods should produce essentially equivalent objects, in that the summary methods and other functions using the `lm` objects can function regardless of the method. This goal is largely, but not completely, achieved with the current methods. If the summary methods are themselves built around generic functions that do not depend on the particular numerical method, the results will be transparent. However, the information available is slightly different, even between `"qr"` and `"chol"`, which refer to the same underlying decomposition. As the example shows, one difference with the Choleski method is that the solution from the normal equations only determines the first p elements of the vector of effects. For most purposes, such as calling `summary()`, this makes no difference. Occasionally, you may encounter a diagnostic function that assumes all n elements of the effects have been computed, such as when doing plots of effects; for such diagnostics, the `"chol"` method will not be adequate.

The most commonly used numerical methods for finding an orthogonal basis, in the sense of equation (4.5) on page 132, are the orthogonal-triangular decomposition methods, usually just called orthogonal decompositions, and in particular the Householder decomposition. Solving linear equations in R is done by the efficient, accurate process of back-substitution. When the computations proceed from the normal equations rather than by decomposing X, this choice of R comes from the Choleski decomposition of $X^T \cdot X$. Either method works fine for most practical examples. The main argument in favor of the normal equations is usually speed, but it is necessary to understand, first of all, that in most applications the computer time taken in the *numerical* phase of solving linear least-squares problems is a small fraction of the total time spent preparing the model, displaying the results, and carrying out other tasks. A more likely reason for using the method might be that the cross-products can be accumulated or derived from some other computation, but that the corresponding full data are not easily produced. Historically, methods using cross-products came from a background of accessing data sequentially by rows, with the assumption that the full model matrix would take too much space.

The time taken by the numerical solution of least-squares problems can be estimated by counting "flops," floating-point operations. The counts can be expressed in terms of p and α, where

$$\alpha = n/p$$

is the ratio of observations to coefficients (hopefully $\alpha \geq 1$). The Householder solution takes $2p^3(\alpha - 1/3)$ operations, the solution from the normal equations, $p^3(\alpha + 1/3)$. This means that if α is large, the normal equations solution approaches half the operation count of the Householder method, whereas when $\alpha = 1$ the counts are equal. If you think such considerations might be important to your applications, you should study a careful account of the numerical methods, such as found in Golub and van Loan(1989), and also be prepared in most cases to do some detailed programming in FORTRAN or C. Just to make the situation more complicated (but also more interesting), if you *do* really need to worry about speed, you should probably consider the option of special hardware dedicated to the numerical processing. For example, parallel computation is a practical alternative that can alter the relevant time estimates in a fundamental way. The book by Golub and van Loan is again a good place to start studying such questions. In using lm() to fit linear models, more time goes into computing the terms, model frame, and model matrix objects than into solving the estimation numerically. Therefore, techniques to re-use the model matrix and response, as discussed on page 120, should be the first step in saving on computations. After that, considerations of the algorithm used may become relevant.

The other side of the comparison between normal equations and orthogonal decompositions concerns accuracy. To oversimplify a complicated topic, if the numerical rank of X is well-defined computationally, either method will produce accurate solutions to the least-squares problems, but as X becomes *ill-conditioned* computationally, two things happen. First, the solution itself becomes unstable, in the sense that small changes in X will substantially alter the least-squares coefficients and, perhaps, the fitted values or residuals. Second, the numerical accuracy of a solution using the normal equations will tend to degrade faster than that from an orthogonal decomposition. It is in this sense that the orthogonal decomposition can be said to be more accurate. But the question of non-unique or ill-conditioned solutions to the computation must be considered in the context of the statistical nature of the model, a context that usually dominates strictly numerical questions. We will say a little about this in the next section.

One last topic in comparing the two approaches is important, but subjective. It can be argued that orthogonal decomposition starting from the model matrix is a more informative and more elegant solution than computations based on normal equations. A basis for this statement is that, particularly with the Householder method, one obtains a simple geometric characterization of the solution, consid-

ered as the definition of a subspace spanned by the columns of X, within the
n-dimensional space of possible response values. The Householder QR method defines this subspace by p reflections. Stored in the form of these reflections, the
corresponding transformation can be applied to project vectors onto this subspace,
or onto its orthogonal complement. S has a variety of functions related to the QR
decomposition that implement the various computations in a simple manner, using
for example the `qr` component that can optionally be returned from a call to `lm()`.
For someone familiar with a little of the underlying mathematics, the computations
made possible by this form of the decomposition may simplify programming new
computations related to linear models.

The third method provided for solving linear least-squares problems corresponds
to choosing R in (4.5) of the form

$$R = D \cdot V$$

where D is a diagonal matrix and V is a p by p orthogonal matrix. Solving equations
in R is again easy: divide by the diagonal elements of D (assuming these are
positive), and multiply by V^T. These methods correspond to forming the singular
value decomposition of X. In using the singular values, it is important to carry out
the computations appropriately for least-squares problems: in particular, one can
avoid the portion of the decomposition requiring an n by p orthogonal matrix. See
again the comments in the next section, and the reference by Golub and van Loan.

The different numerical methods are integrated into the `lm` object by the component `R`. This component must exist, and there must be methods for each new class of
`R` objects to allow the functions treating `lm` objects to work with the new numerical
method. As far as the generic functions described in this book are concerned, the
only critical method is for solving (back-solving in the case of upper-triangular `R`),
as required by `summary.lm()`.

4.4.3   Overdetermined and Ill-determined Models

There is no guarantee that the formula and data in a model will define the least-
squares fit precisely, or even be numerically unique. Two situations need to be
considered, both arising quite frequently. First, if some column of X is exactly
equal to a linear combination of the preceding columns, then the coefficients β
are determined only up to a one-dimensional family. However, the fitted values
from the estimated model are unique: any least-squares coefficients are equivalent,
geometrically, to fitting y to the $p - 1$ linearly independent columns of X. This
sort of linear dependency arises all the time in models, such as those in Chapter
5, that include factors. In this case, it is quite natural that some variables will be
functionally equivalent to other variables; in particular, fitting such a factor always
generates columns in the model matrix that are linearly dependent on the vector of
ones representing the intercept term.

A somewhat different situation arises when some numeric terms are not exactly known, perhaps because of observation error. In this case, a column of X may be *approximately* equal to a linear combination of previous columns. The issue here is that a model may be statistically meaningless and substantively misleading because some linear combinations of variables are just noise. Hard decisions about which model to choose may not be possible in this situation; it may be necessary to look at several possible solutions. To make matters worse, in this case the fitted values may be affected by the choice as well as the coefficients. If we choose to include a questionable term in the model, the corresponding column of X will *not* be an exact linear combination of the preceding columns. Therefore, there is no guarantee what effect this inclusion will have on the fit. It could range from entirely negligible to being the only important effect in the model.

Clearly, the proper treatment of these questions involves both statistical and numerical questions. The two situations above correspond numerically to *singular* and *ill-determined* models, in numerical terminology. The distinction between these is clearly important statistically, but is not always made clearly in numerical discussions.

Attitudes to the Problem

To set the discussion in a realistic context, let's take the viewpoint of a user who needs to decide what precautions, if any, to take about these potential problems. Four plausible attitudes that one might have in presenting a linear model for fitting are:

1. "This model is not singular or ill-determined, and I don't want to waste time checking."

2. "This model shouldn't be singular or ill-determined, and if it is, treat that as an error. Errors in the predictor variables are not important."

3. "This model may well be singular, because the particular design makes it impossible to fit all the coefficients I'd like; if it is, just fit the nonsingular part and note the family of coefficients. The model should not be ill-determined once we take account of these design limitations."

4. "This model may be ill-determined; in particular, I realize that the predictor data are both correlated and not known precisely. Solve it in such a way that I can take account of these considerations."

These alternatives do not exhaust all the possible situations, but they do cover the majority of practical situations. Number 3 is the typical situation in designed experiments that are not "complete"; Chapter 5 will discuss this situation. Number 4, on the other hand, is a common situation in data where all the variables, response and predictors, are measured numeric quantities. Data in the social sciences particularly tend to be both subject to measurement error and strongly correlated.

Alternative number 2 is the default assumption made by the lm() function, so let's begin by considering that situation. The approach *only* protects against numerical singularity; the important, but very difficult, question of errors in the predictors is not considered. Numerical methods for solving least-squares problems allow us to compute a solution, given the number of linearly independent columns of X, and then examine that solution to see whether the problem seems to be nearly singular. If singularity or ill-conditioning is to be an error condition, we proceed to compute the numerical solution assuming that all the columns of X are linearly independent (this will generate an error if the algorithm being used concludes that X is singular). If not, the standard lm() computation returns the result, without further checking. If ill-determined models are a statistical concern, however, some additional checking should be done. Theoretical results about numerical linear least-squares solutions offer some useful help.

Estimating Numerical Sensitivity

The *condition number* $\kappa(X)$ of a matrix X is defined as the ratio of the largest to the smallest nonzero singular value. The essential qualitative property is that $\kappa(X)$ will turn out to measure how well linear models using X are determined. Small condition numbers mean well-determined models, while large condition numbers mean that either the coefficients or the residuals may be poorly determined. This statement can be made more meaningful in terms of the sensitivity of the fitted model to changes in the data (that is, to X or y or both). Suppose we let $\delta = \delta(X, y)$ stand for a small relative change in the data, and let δ_e and $\delta_{\hat\beta}$ be corresponding relative changes in the residuals and the coefficients of the fitted models. The fundamental sensitivity results are then

$$
\begin{aligned}
\delta_e &< C\,\kappa(X)\delta \\
\delta_{\hat\beta} &< (C_1\,\kappa(X) + C_2(e)\,\kappa(X)^2)\delta
\end{aligned}
\tag{4.11}
$$

Here C and C_1 are constants, and $C_2(e)$ depends on the residuals; specifically, it is small if the residuals are small relative to y but becomes arbitrarily large if the residuals are nearly equal to y. Equation (4.11) is the key theoretical result on computational sensitivity in linear models. Both the residuals and the coefficients can be ill-determined if the condition number of X is large. The coefficients, however, are more sensitive than the residuals in that they can be ill-determined even for moderate condition numbers if the residuals are large relative to y, indicating that the model is not doing a good job of fitting y. In addition, the sensitivity of the coefficients grows with the square of the condition number rather than the condition number itself, even if the model gives a moderately good fit.

Although we have left a few notions unspecified in setting out (4.11), the definitions can all be made precise. The important practical message is that one should estimate the condition number of the model matrix. This and the size of the residuals relative to the response will give guidance as to the degree of ill-determinacy possible in the problem. We could compute $\kappa(X)$ from the singular-value decomposition, but this takes substantially more calculation than the default method for the least-squares computations themselves. An adequate approximation is available, as the function `kappa()`, using the triangular factor, R, from an orthogonal or Choleski decomposition of X. This requires only on the order of p^2 floating-point operations. The resulting estimate of the condition number is not guaranteed as to accuracy, but experience with it has been that it gives reliable order-of-magnitude estimates. The underlying algorithm comes from the LINPACK library; the method is described in Golub and van Loan (1989, page 128). Let's examine the estimate on the model fitted to the `Fuel` data. Consider the `lm` object `fuel.fit`:

```
> kappa(fuel.fit)
[1] 15494.84
```

Roughly speaking, if the condition number times the relative precision of the computations is small compared to 1, then linear algebra calculations using the matrix should be *numerically* well-defined (which of course does not say they are *statistically* meaningful). In this case, computations are being done with a precision of around 10^{-17}. The product of this with κ is then around 2×10^{-12}, which seems quite small.

We can compare the estimated condition number with the actual value computed from the `svd` function applied to the matrix R. The condition number is the ratio of the largest to the smallest singular value:

```
> xx <- svd(fuel.fit$R)$d
> xx
[1] 22822.766404    280.651689      1.141294
> xx[1]/xx[3]
[1] 19997.27
```

So the cheap estimate from `kappa()` has underestimated $\kappa(R)$, but not very seriously.

An estimate of $\kappa(X)$ produces an estimate of the two kinds of sensitivity in the fitted model, via equation (4.11). The function `lm.sensitivity` takes an `lm` object and returns two numbers: the estimated sensitivities for the residuals and for the coefficients. In the example previously shown:

```
> lm.sensitivity(fuel.fit)
 residuals coefficients
   1766469      21546499
```

Despite the numerical accuracy implied by the value of κ, these numbers would be cause for worry, particularly about the values of the coefficients. To interpret the value 2×10^7 for coefficients, think of it as estimating how much the relative change in the coefficients might be as a multiple of some small relative change in the data. The term "relative change" is defined as follows: if we change some vector x to $x + \delta x$, the relative change is $||\delta x|| / ||x||$, with "$||$" standing for the Euclidean norm, that is, the square root of the sum-of-squares. Clearly, the estimated coefficients in this model are very sensitive, or can be, to changes in the data.

The definition of relative change in residuals is a little different: it is taken to be the ratio of the norm of the change in the residuals over the norm of y. Here again, the fit is clearly sensitive to changes in the data, although not quite so much so.

Using the Estimated Sensitivity

The sensitivity numbers are upper bounds, if the condition number is known exactly. If the inexpensive estimate is used, this tends to underestimate the condition somewhat, but usually by less than a factor of two. In any case, the bound from the condition number is likely to be much larger than the actual change in the fit due to a small change in the data. Nevertheless, the bounds are useful in that they point out models in which there is potential for the results to be influenced by such small changes. Very large values, such as those in the example above, should be cause to interpret the model cautiously, and to experiment statistically with changes in the data or the model.

The use of such diagnostics following a fit that is numerically nonsingular is recommended as a careful approach to situation 2 in our list. This is also a reasonable approach to situation 4; in this case, however, some explicit account should, if possible, be taken of the uncertainties in the predictors. A thorough treatment of this problem is far beyond the scope of this section, but the following procedure is one approach. Suppose we are willing to assume that the uncertainty or measurement error in the values of X can be adequately represented by a model in which the observed value x_{ij} has an error, say ε_{ij}^x, with the errors being independently distributed with mean 0 and standard deviation σ_j. Furthermore, assume we have an estimate, say, s_j of σ_j. The prescription is then:

- Divide the jth column of X by s_j. If your model includes an intercept term, first subtract means from y and from each column of X.

- Use the singular-value decomposition of X to compute the regression. Examine the singular values. If any of them are substantially smaller than $\sqrt{n}$, the suggestion is that these linear combinations of columns of X are essentially noise, under the assumptions made about errors in X. Therefore, you should not use these columns in computing a meaningful regression.

- From the examination of singular values above, select one or more possible ranks for the regression and compute corresponding fitted values and resid-

Chapter 5

Analysis of Variance; Designed Experiments

John M. Chambers
Anne E. Freeny
Richard M. Heiberger

This chapter provides for the modeling of numeric variables by *factors*, variables that take on one of a specified set of levels. Such data are often the result of a designed experiment, in which observations of one or more *responses* are made for changing values of several factors. The variation in the response is then studied, in the hope of understanding how it depends on the underlying phenomena that the levels of the factors represent. One of the most important contributions in the history of statistics was R. A. Fisher's notion that choosing the factors according to a suitable experimental design could provide more information on such dependencies than varying only a single factor at a time. Techniques for choosing such designs and for analyzing the results, particularly the analysis of variance, have become an important part of scientific studies in agriculture, biology, social sciences and, increasingly, in manufacturing and other applications of engineering and the physical sciences.

Our goal is a general computing capability for such data. This chapter draws on the ideas of Chapter 3 to represent the experiments, and on those of Chapters 2 and 4 to represent and fit the models. The classical models for designed experiments are very closely related to the linear models of Chapter 4; in fact, many of the modeling techniques are the same, but with important differences in how the models are viewed and used. In addition to modeling techniques, this chapter contains functions to generate common designs and informal ways to look at the results of the experiments without using a specific model.

While the motivation and statistical theory arising from experimental design are central to this chapter, it's worth noting that many of the techniques work for any data in which the behavior of a numeric response is studied in terms of some categorical predictors. Most of the computational techniques are organized to be valid for arbitrary experiments. However, the statistical techniques sometimes are invalid, or at least the results from them need to be interpreted carefully when the assumptions of balanced experiments are not met.

Since there are many software packages for the analysis of variance, we should emphasize what this chapter provides and what it does not. We are presenting a new approach based on the general ideas of model formulas, data frames, classes, and the techniques these ideas make possible. We stress generality, directness of expression, and an ongoing process of making powerful computations easily accessible. The emphasis is on computed objects describing the data and fitted models, to be used in an open-ended set of displays and diagnostic computations, rather than on fixed output reports of the analysis. The functions described here are far from a final package and we especially hope that users will extend them and adapt them to their own interests. Some topics, such as multiple error terms, are treated relatively thoroughly; for others, such as random-effects models, we provide only a fraction of the possible software.

5.1   Models for Experiments: The Analysis of Variance

An experiment, in the sense we will use the term, is described by the values of some chosen variables for each of a number of runs or observations. In a *designed* experiment, some of those variables will, in principle, have values from a finite set of *levels* specified by the experimenter in advance. In classical experimental design, these variables are categorical, and are usually called the *factors* in the design. Other variables are observed when the experimental runs take place. The data frames introduced in Chapter 3 provide a natural way to represent such experiments in S. The factors are represented as factor objects and the response as a numeric variable or matrix.

Complete factorial designs consist of one run for each possible combination of levels for each of the factors. For example, Table 5.1 shows a design with three factors, each having two possible levels, giving eight runs. In this experiment, as described in Box, Hunter, and Hunter (1978, p. 308), the experimenter plans to run a process using two catalysts, identified only as A and B, at two choices of concentration and with two choices for the temperature at which the process runs. Notice that two of the three variables are inherently quantitative, although they are treated as factors in the design by choosing two specific values. Such variables are `ordered` factors as described in Chapter 3, although with only two levels the distinction is not important.

```
    Temp  Conc  Cat  Yield
1   160   20    A    60
2   180   20    A    72
3   160   40    A    54
4   180   40    A    68
5   160   20    B    52
6   180   20    B    83
7   160   40    B    45
8   180   40    B    80
```

Table 5.1: *A factorial experiment with factors temperature, concentration of catalyst, and catalyst type, each having two levels. The response is the yield of the process.*

The scenario is that the experimenter now runs the process in some randomized order under each of the eight conditions given by the design, and records the `Yield` in the last column of Table 5.1. In many situations, the experimenter will choose to randomize the order of the runs (see page 175), to the extent practicable, in the hope of disentangling sequential effects from the factors of interest. The results of the experiment are now available for analysis. As always, some careful studies of the data, particularly through plots, should precede or at least accompany the formal modeling. One such graph is shown in Figure 5.1: the average of the values of `Yield` are shown for each level of each factor, with a horizontal line showing the overall average for comparison.

The analysis of variance, the principal classical model for factorial experiments, uses formulas and estimation that are formally special cases of the linear models discussed in Chapter 4. The emphases differ substantially, however. Discussion centers more on the contribution of terms in the model to the total variation of the response and less on individual coefficients. The categorical nature of the variables means that detailed study of the way in which the response depends on a single predictor variable is usually impossible—with only two levels of temperature, the form of dependence of yield on temperature will not be available. The compensation is that much broader exploration of the effects of several factors may be possible. New questions arise, as well, such as which factors should be assumed to interact.

An experiment on the quality of integrated circuit fabrication was reported by Phadke et al. (1983). The objective of the experiment was to investigate various factors that might affect the quality of the fabrication process, with the goal of

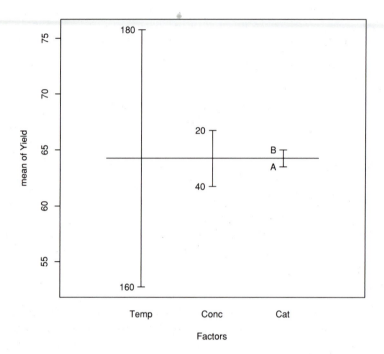

Figure 5.1: *Yield from the experiment in Table 5.1, averaged for each level of the factors. Vertical position shows the mean yield for the corresponding level; for example, the position labeled* 180 *is the mean of the observations with* Temp *at level* 180.

controlling both the average level and the variability of the quality. The experimental design had 18 runs, varying 8 factors based on an orthogonal-array design (p. 171). In principle, two wafers with 5 measurements each should have been manufactured for each run, but three wafers were broken. Two measures of quality are the pre-etch line width and post-etch line width. The line widths are numeric responses, measured five times on each wafer. The published analysis uses the mean and standard deviation for each experimental run for pre- and post-etch line width as responses, the goal of the experiment being to control both the average quality and the variability in quality. Table 5.2 shows a sample of 6 runs to suggest the form of the data. This experiment will be used several times in the chapter to illustrate some interesting techniques of analysis. See the reference for more discussion of the experiment and additional analysis.

	maskdim	visc.tem	spinsp	baketime	aperture	exptime	devtime	etchtime
2	2	204,90	normal	30	2	normal	45	13.2
5	2	206,90	normal	30	3	-20%	30	14.5
10	2.5	204,90	low	40	3	normal	45	14.5
14	2.5	206,90	normal	40	1	normal	30	15.8
15	2.5	206,90	high	20	2	-20%	45	14.5
18	2.5	204,105	high	30	1	normal	60	14.5

	pre.mean	pre.dev	post.mean	post.dev	N
2	2.684	0.1196	3.14	0.063	10
5	1.870	0.1168	1.72	0.400	5
10	2.660	0.1912	3.31	0.350	10
14	2.308	0.0964	3.14	0.160	10
15	2.464	0.0385	2.55	0.210	5
18	3.494	0.0473	4.34	0.078	5

Table 5.2: *Part of an experiment (6 of 18 runs) on wafer fabrication: 8 factors, 4 responses (the mean and standard deviation for* pre- *and* post-*etch line width), and the number* N *of measurements per run.*

Models for experiments may differ from other linear models also in the emphasis they give to structure among factors and in the error assumptions. Factors may be dependent on other factors, in that levels of the one factor are meaningful only within each level of the other factors, giving rise to *nested* or *hierarchical* models. The experimental situation may also imply that the errors of observation depend on the levels of some of the factors, giving rise to an *error model*. In our approach this situation can be expressed by including the error model as an additional term in the formula. The analysis of variance will then reflect the multiple sources of error, so that estimated effects are compared to the appropriate error estimates. These aspects of factorial experiments will be handled in this chapter by extensions to the basic computational techniques developed in earlier chapters, while keeping the organization built around formula, data, and fitted model.

The experiment in Table 5.1 is small and complete. Such experiments tend perhaps to appear mostly in textbooks. In practice, a large number of factors, more factor levels, unbalanced or incomplete designs, and other practicalities are likely to arise. The experiment in Table 5.2 is somewhat larger and, in its complexity and slight irregularity, more typical of practice. We will also deal with the wafer solder experiment introduced in Chapter 1. This, with its large size (900 observations) as well as its irregular features, is indeed a serious application. In choosing examples for this chapter, we will mix small examples for ease of illustration with these more realistic ones to emphasize the issues such experiments raise.

5.2   S Functions and Objects

This section describes the basic use of S functions to do analysis of variance, to display the results of experiments graphically, and to generate designs.

The modeling techniques of the analysis of variance apply the ideas of formulas and data from previous chapters, expanding on the range of formulas and introducing some summaries of particular relevance for designed experiments. Subsection 5.2.2 presents some summary and diagnostic techniques, chiefly graphical. These are particularly valuable, and we encourage you to use the plots as preliminary views of the data and as communication tools. Subsection 5.2.3 presents functions for generating some standard designs and running experiments. This subsection will not be relevant if you are only analyzing data after collection. The first two subsections, however, should be read by anyone interested in analyzing designed experiments.

5.2.1   Analysis of Variance Models

The basic expression for fitting an analysis of variance is

```
aov(formula, data)
```

where `formula` is a model formula, relating the response to appropriate factors, and `data` is an optional design object containing the data from the experiment. The object returned by the `aov()` function has class `"aov"`. It is very similar to a linear-model object, and inherits the use of functions `fitted`, `residuals`, `coef`, and `effects` to return the named components of the `aov` object.

Suppose the catalyst experiment in Table 5.1 is contained in design object `catalyst`. To fit a simple additive model in all the factors,

```
> aovcat <- aov(Yield ~ ., catalyst)
```

Notice that we used the shorthand "." to refer to all the variables in the data frame except for the response on the left of the formula. As with all fitted models, there are methods for printing and for a slightly more statistical summary:

```
> aovcat
Call:
   aov(formula = Yield ~ Temp + Conc + Cat, data = catalyst)

Terms:
                    Temp    Conc    Cat Residuals
Sum of Squares    1058.0    50.0    4.5     205.0
Deg. of Freedom        1       1      1         4

Residual standard error: 7.1589
```

```
Estimated effects are balanced

> summary(aovcat)
            Df Sum of Sq Mean Sq F Value  Pr(F)
    Temp    1      1058    1058   20.64 0.0105
    Conc    1        50      50    0.98 0.3782
     Cat    1         5       5    0.09 0.7791
Residuals   4       205      51
```

The printing and summary methods for analysis of variance objects reflect the different approach to what is essentially the same information as returned by lm(). While a printed linear-model object shows individual coefficients, printed aov objects show terms, which may correspond to several coefficients. In this, and in other ways we will demonstrate, the focus will shift somewhat in this chapter. Nothing has been lost, however, because methods for linear models can always be applied explicitly to aov objects, such as those for the coefficients or residuals:

```
> coef(aovcat)
 (Intercept) Temp Conc  Cat
       64.25 11.5 -2.5 0.75
> resid(aovcat)
   1   2   3    4  5 6   7 8
 5.5 -5.5 4.5 -4.5 -4 4 -6 6
```

We could also have previously attached the design object:

```
attach(catalyst)
```

to allow us to refer to the individual variables (Temp, etc.) in arbitrary S expressions. In this case, the second argument to aov() can be omitted, but at a price: without the data argument there is no context to define the meaning of ".", so the formula would have to be given in full.

Crossed and Nested Terms

In models involving factors as predictors, interactions among the factors are often important. When possible, good data analysis suggests that one should ask whether the effect of one factor depends on the levels of one or more other terms. This leads us to a richer use of formulas than is typical of Chapter 4.

The expression A * B in a model formula puts into the model the terms for A, B, and the interaction of A and B (represented as A:B). The notation can be used in any general way that makes sense: the operands of * can be any expression that evaluates to a factor, and crossed terms can be combined with other operators, and parenthesized to indicate grouping. This is needed to permit covariates to be crossed and nested. With a small factorial experiment, we may want to fit the full model including all possible effects. This model is specified by combining all the terms with * (an alternative specification avoids writing an explicit formula at all, as we show on page 153). In our previous example:

```
> aovall <- aov(Yield ~ Temp * Conc * Cat, catalyst)
> aovall
Call:
    aov(formula = Yield ~ Temp * Conc * Cat, data = catalyst)

Terms:
                    Temp    Conc    Cat  Temp:Conc  Temp:Cat  Conc:Cat
Sum of Squares    1058.0    50.0    4.5       4.5     200.0       0.0
Deg. of Freedom      1       1      1         1         1         1

                Temp:Conc:Cat
Sum of Squares           0.5
Deg. of Freedom            1

Estimated effects are balanced
```

The specified model fits three main effects, three two-way interactions and one three-way interaction. Notice that the term "Residuals" did not appear, nor the residual standard error. When all interactions in this experiment are included in the model, no degrees of freedom are left for residuals. Intermediate models can be specified, including whatever set of interactions make sense. Some short forms simplify writing common instances of such models, as we will illustrate shortly.

When the levels of factor B are meaningful only given the level of some other factor A, B is said to be nested in A. The notation in a formula is A / B, saying to fit first A and then the effects of B in each level of A. Main effects for B are not meaningful in this case. As with *, the operands to / can be arbitrary expressions.

An experiment on methods for firing naval guns was reported by Hicks (1973, page 194). Two methods were tested by gunners corresponding to three different physiques (slight, average, and heavy). Nine gunners of each physique were divided into three teams, and each team tested the two loading methods twice, for a total of 36 runs. The response was the number of rounds fired per minute. The data are in design object gun; a sample of six runs is as follows:

```
    Method Physique Team Rounds
 3    M1       A     T1    22.0
 4    M2       A     T1    14.1
14    M2       S     T3    12.5
25    M1       S     T2    26.9
29    M1       H     T2    23.7
34    M2       A     T3    16.0
```

For this experiment, the Team factor is only meaningful within each physique, so one possible model is:

Rounds ~ Method + Physique/Team

which will fit effects for Method, Physique and Team within Physique, as follows:

```
> gunaov <- aov(Rounds ~ Method + Physique/Team, gun)
> summary(gunaov)
                     Df Sum of Sq Mean Sq F Value  Pr(F)
           Method  1     652.0    652.0   316.8 0.0000
         Physique  2      16.1      8.0     3.9 0.0330
Team %in% Physique  6      39.3      6.5     3.2 0.0178
        Residuals 26      53.5      2.1     1.0 0.5000
```

Short Forms for Formulas

When an experiment involves a substantial number of factors, writing a formula to contain them all is tedious. An alternative is to generate a model specification including all the possible terms, up to a specified order of interaction. The "." convention introduced in Section 2.3 is a shorthand way of referring to all the variables, as an additive model. We used it to specify the main-effects model, but in fact it can appear anywhere. It is replaced by all the variables in the data frame, with the exception of those used in the expression for the response. In analysis of variance models, the notation is often conveniently combined with the "^" operator. This operator specifies all the main effects and interactions in the operand on its left, up to the limit defined by the "power" on its right. So:

aov(Yield ~ .^ 2 , catalyst)

says to fit the main effects and the two-way interactions of all the factors in the data.

If you plan to work for some time on a subset of the variables in a design, it may be worthwhile to create a new data frame containing only this subset, so that "." will have the desired meaning. The data from Table 5.2 are in the data frame wafer. It contains 8 factors, 4 responses, and the auxiliary variable N. To study pre.mean, the first response, we can create a new data frame, wpm:

```
> wpm <- wafer[, c(1:9)]
> waov1 <- aov( pre.mean ~ . , wpm )
> summary(waov1)
            Df Sum of Sq Mean Sq F Value  Pr(F)
   maskdim  1    0.652   0.6521   107.4 0.0092
   visc.tem 2    1.343   0.6717   110.6 0.0090
    spinsp  2    0.765   0.3827    63.0 0.0156
  baketime  2    0.002   0.0012     0.2 0.8380
```

```
 aperture  2      0.032  0.0158     2.6 0.2771
  exptime  2      0.545  0.2724    44.9 0.0218
  devtime  2      0.280  0.1401    23.1 0.0415
 etchtime  2      0.103  0.0517     8.5 0.1052
Residuals  2      0.012  0.0061
```

Model updating, discussed in Section 4.2.4, can also be used to refine the specification, by adding or dropping individual terms. For example, we can pool the contribution of the factor `baketime` into the estimate of the residual mean square:

```
> summary(update(waov1, . ~ . - baketime ))
           Df Sum of Sq Mean Sq F Value   Pr(F)
  maskdim   1     0.652  0.6521   180.0 0.00018
 visc.tem   2     1.343  0.6717   185.4 0.00011
   spinsp   2     0.765  0.3827   105.6 0.00035
 aperture   2     0.032  0.0158     4.4 0.09848
  exptime   2     0.545  0.2724    75.2 0.00067
  devtime   2     0.280  0.1401    38.7 0.00242
 etchtime   2     0.103  0.0517    14.3 0.01513
Residuals   4     0.014  0.0036
```

In the formula given to `update()`, the "." notation can be used on both sides of the "~"", referring to the response on the left and to all the terms in the original model on the right. See Section 4.2.4.

Multiple-Response Models

The examples described so far have dealt with single response variables. It is also possible to supply a numeric matrix as the response variable. Columns of the matrix are interpreted as the individual responses. The fit carries through as before, with the distinctions that the object now inherits from class `"maov"` and that the effects, coefficients, residuals, and fitted values will be matrices having as many columns as the response. These models should not be confused with the "manova" analysis, which studies the dependence of the covariance structure among the responses on one or more factors. Each response is modeled separately in our analysis.

Multi-response models can be described in the same way as single-response models. To illustrate this and also a useful way to organize multiple-response data, let's construct a new data frame from the `wafer` data with all four responses stored as a single matrix-valued response. While we're at it, we take logarithms of the standard deviations, which should be analyzed on the log scale:

```
> waferm <- wafer[, 1:8]
> attach("wafer")
> waferm[, "Line"] <- cbind(pre.mean, log(pre.dev), post.mean,
+        log(post.dev))
```

We first select the 8 factors, then insert in this data frame a single new variable, Line, set to all the columns of wafer containing responses. With a single column on the left of the assignment and a matrix on the right, the replacement method for data frames inserts the entire matrix as one variable. Now we can fit the multivariate anova simply:

```
> wmaov <- aov(Line ~ . , waferm)
```

Summaries for multi-response models repeat the univariate summary for each response:

```
> summary(wmaov)
Response: pre.mean
```

	Df	Sum of Sq	Mean Sq	F Value	Pr(F)
maskdim	1	0.652	0.6521	107.4	0.0092
visc.tem	2	1.343	0.6717	110.6	0.0090
spinsp	2	0.765	0.3827	63.0	0.0156
baketime	2	0.002	0.0012	0.2	0.8403
aperture	2	0.032	0.0158	2.6	0.2770
exptime	2	0.545	0.2724	44.9	0.0218
devtime	2	0.280	0.1401	23.1	0.0415
etchtime	2	0.103	0.0517	8.5	0.1052
Residuals	2	0.012	0.0061		

```
Response: log(pre.dev)
```

	Df	Sum of Sq	Mean Sq	F Value	Pr(F)
maskdim	1	0.661	0.6606	4.03	0.1825
visc.tem	2	0.165	0.0826	0.50	0.6667
spinsp	2	0.191	0.0956	0.58	0.6329
baketime	2	0.492	0.2460	1.50	0.4000
aperture	2	1.114	0.5570	3.40	0.2273
exptime	2	0.651	0.3255	1.99	0.3344
devtime	2	0.015	0.0074	0.05	0.9524
etchtime	2	0.313	0.1566	0.96	0.5102
Residuals	2	0.328	0.1638		

Etc.

Summaries for multi-response models (and also for models with Error strata) are lists of the summaries for the individual responses or error strata. If we wanted to print only the summary for one response, we could just select the corresponding element of the summary object for the whole model.

Aliasing; Over-specified Models

The specification of a model for the analysis of variance may include more terms than can actually be estimated from the design. In this case the fit goes through,

but some of the terms or coefficients specified in the formula may not appear in the
fit. A table of *aliasing* information can summarize the relation between effects that
can be estimated and those that cannot. Let's consider a small example. Suppose
that instead of the complete eight observations in the `catalyst` data shown in Table
5.1 on page 147, we could only afford four observations. By taking the half-replicate
consisting of rows 2, 3, 5, 8, and assuming the yields were the same as in Table 5.1,
we would obtain the following data, in design object `catalyst2`, say:

```
> catalyst2
     Temp Conc Cat Yield
2     180   20   A    72
3     160   40   A    54
5     160   20   B    52
8     180   40   B    80
```

What happens if we fit the complete model in the three factors?

```
> half.aov <- aov(Yield ~ Temp*Conc*Cat, catalyst2)
> half.aov
Call:
aov(formula = Yield ~ Temp * Conc * Cat, data = catalyst2)

Terms:
                   Temp Conc Cat
  Sum of Squares   529   25   9
  Deg. of Freedom    1    1   1

4 out of 8 effects not estimable
Estimated effects are balanced
```

The printing method warns us that some of the effects in the model formula can't
be estimated in the data. The fit includes three terms plus the intercept, using up
all 4 degrees of freedom available.

The function `alias()` defines the relation between the effects that could not be
estimated (the rows) and the effects that were estimated (the columns). Looking at
the alias pattern gives information about what happened to the interaction terms:

```
> alias(half.aov)
Complete
               (Intercept) Temp Conc Cat
     Temp:Conc                         1
      Temp:Cat                    1
      Conc:Cat               1
 Temp:Conc:Cat  1
```

For example, the first row shows that the estimate for the `Temp:Conc` interaction
was completely aliased to the `Cat` main effect, already included. Statistically, the

implication is that the design cannot distinguish these two terms. Only if one is willing a priori to treat the `Temp:Conc` interaction as known to be zero can the sum-of-squares assigned to `Cat` be attributed unambiguously to that term. The last row shows that the `Temp:Conc:Cat` interaction is aliased with the intercept. In the terminology of `fac.design()`, this says that the interaction is the defining contrast for the fractional design represented by the four rows of `catalyst2`. For more discussion of aliasing, see page 178.

Incomplete balanced, or fractional factorial, designs are widely used and useful when experiments are expensive to run or when many factors need to be studied simultaneously. It's perhaps preferable in these cases to understand beforehand what can be estimated, and to phrase the model to make the assumptions plain. In this case, for example, if we had specified only the main effects, as in the initial example on page 150, we would have got the same fit, but without aliasing. On the other hand, if you are unsure just what can be estimated from a particular balanced design, specifying a complete model and then examining the alias pattern is one way to find out. The aliasing depends only on the design and the structural form of the model, not on the response itself. Therefore, the aliasing can be studied before running the experiment; for example, if `halfdesign` were the half-replicate design in `catalyst2`:

```
alias(halfdesign, ~ .^3)
```

would show the same alias pattern as above. If the formula is omitted, `~.^2` is assumed. See Section 5.3.2 for more discussion of aliasing.

Error Terms

The analysis of experiments with multiple factors departs from standard linear models in an important way when the model includes multiple error terms. An example, described in Federer (1955, page 274), will suggest the idea. In an experiment on eight varieties of guayule (a Mexican plant yielding rubber), four different treatments were applied to the seeds. The question of interest was the effect of the treatments on the rate of seed germination. The experimenter reasoned that plants grown on different greenhouse flats were likely to have differences due to the flats. These differences are not of particular interest and can be modeled as a random quantity depending on the individual flat.

In order to gain the most information on the seed treatments, the experimenter divided each flat (*plot* in the classic terminology) into subplots and assigned all treatments to each subplot so as to allow estimation of the treatment effects orthogonal to flat effects and subject only to within-flat variability. Specifically, each flat was planted with seeds of one variety and each subplot contained 100 seeds treated with one of the four treatments. Each seed variety was planted in three flats, for a total of 24 flats. The response in this example consisted of 96 observations of the number of plants germinating per 100 seeds planted in each subplot. The first two treatments in eight of the flats are shown below:

```
> guayule[1:16, ]
   variety treatment reps plants  flats
1     V1        T1     1     66 1.V1
2     V2        T1     1     77 1.V2
3     V3        T1     1     51 1.V3
4     V4        T1     1     52 1.V4
5     V5        T1     1     45 1.V5
6     V6        T1     1     59 1.V6
7     V7        T1     1     56 1.V7
8     V8        T1     1     49 1.V8
9     V1        T2     1     12 1.V1
10    V2        T2     1     26 1.V2
11    V3        T2     1      8 1.V3
12    V4        T2     1      4 1.V4
13    V5        T2     1     20 1.V5
14    V6        T2     1      8 1.V6
15    V7        T2     1     12 1.V7
16    V8        T2     1     14 1.V8
```

In the analysis of variance, we want to specify an error term corresponding to the levels of flats, which we do by including the expression Error(flats) in the model formula:

```
> attach(guayule)
> gaov <- aov(plants ~ variety * treatment + Error(flats))
```

This model will produce a separate fit for the portion of the data corresponding to the effects of the variable flats and for the residuals from these effects—in classical anova terminology, the whole plot and subplot error strata.

As you might expect, since there is a separate fit for each error stratum, the overall fit is represented as a list of aov objects. Its class is aovlist, and methods exist for the usual functions such as print() and summary():

```
> summary(gaov)
Error: flats
            Df Sum of Sq Mean Sq F Value  Pr(F)
  variety  7       763   109.0   1.232 0.3421
Residuals 16      1416    88.5

Error: Within
                Df Sum of Sq Mean Sq F Value     Pr(F)
    treatment  3     30774   10258   423.4 0.000e+00
```

```
variety:treatment 21      2620      125      5.2 1.327e-06
        Residuals 48      1163       24
```

Notice that the experimenter chose a design such that `treatment` effects, which the experiment was particularly anxious to estimate accurately, were entirely orthogonal to `flats`. They do not appear in the first (less accurate) stratum at all. The experimenter's assumption of substantial flat-to-flat variation is supported by the residual mean square in the `flats` stratum being over three times that of the `Within` stratum.

The error model specified can include more than one term; the argument to `Error()` can be anything that would go on the right side of a formula. For example, depending on how the experiment was run, the experimenter might have wanted to reflect a situation in which flats within one replication would tend to be more alike than those in different replications. This reasoning suggests having two error terms: `reps` and `flats`. The formula

```
plants ~ variety * treatment + Error(reps + flats)
```

produces an analysis divided into three strata: `reps`, `flats` after removing `reps`, and residuals from both. In this case, `flats` are actually defined within `reps`, but nothing in the computations requires this. The model is as follows:

```
> gaov2 <- aov(plants ~ variety * treatment + Error(reps + flats))
> summary(gaov2)
Error: reps
          Df Sum of Sq Mean Sq F Value Pr(F)
Residuals  2     38.58   19.29

Error: flats
          Df Sum of Sq Mean Sq F Value Pr(F)
  variety  7       763   109.0   1.108  0.41
Residuals 14      1377    98.4

Error: Within
                  Df Sum of Sq Mean Sq F Value     Pr(F)
        treatment  3     30774   10258   423.4 0.000e+00
variety:treatment 21      2620     125     5.2 1.327e-06
        Residuals 48      1163      24
```

This experiment was designed so that treatments, varieties, and their interaction were balanced within each replication. As a result, there are no effects from `varieties*treatment` at all in the `reps` stratum.

Random-Effects Models

Some factors in experiments have the property that the levels chosen are not so much interesting in themselves but instead are examples of "typical" values for the

underlying variable. Samples of counties within a state, households on a block, or
animal or human subjects for testing may have been chosen in the hope that conclu-
sions made from the experiment can be generalized to the overall population from
which the samples are chosen. The *random-effects* model applied to such factors
looks at the variability in the effects for a particular term, not the individual effects
themselves. The standard distributional assumption is that the coefficients for individual levels are distributed with zero mean and some unknown standard deviation.
Large or small standard deviation then indicates important or unimportant terms.

We have included only a special case of this model, implementing the completely
random model for the balanced case, as described by Searle (1971, page 393). The
software tests for balance, by computing the number of replications for each of the
terms involved. To obtain a fitted analysis of variance including random-effects
information on all the factors, use

```
raov(formula, data)
```

instead of aov(). Let's do an example. Box, Hunter and Hunter (1978, page 572)
report an experiment in which 15 batches of a pigment were each sampled analytically twice and two repeated analyses were performed from each sample, to measure
the moisture content. Given this experiment, in a design object, pigment, we can
compute the random-effects analysis of variance as follows:

```
> pigment
   Batch Sample Test Moisture
1    B1    S1    T1      40
2    B2    S1    T1      26
3    B3    S1    T1      29
4    B4    S1    T1      30
5    B5    S1    T1      19
6    B6    S1    T1      33
7    B7    S1    T1      23
8    B8    S1    T1      34
9    B9    S1    T1      27

        ...

> praov <- raov(Moisture ~ Batch/Sample, pigment)
> summary(praov)
                      Df  Sum of Sq  Mean Sq Est. Var.
              Batch  14   1210.933  86.49524   7.12798
Sample %in% Batch   15    869.750  57.98333  28.53333
          Residuals  30     27.500   0.91667   0.91667
```

See the references cited for further discussion of the analysis.

Analysis of Unbalanced Experiments

The term *balance* describes a relation between a pair of factors in which every level of one factor appears with every level of the other factor the same number of times. If all the pairs of factors are balanced in this sense, the variables in the model matrix will be orthogonal, and interpreting the analysis will be more straightforward. On the other hand, data from an unbalanced experiment must often be analyzed. The nature of the unbalance can be studied, and alternatives to the usual summaries help with the interpretation. It should be emphasized that there is no problem in *fitting* the model, regardless of balance; the issues arise in summarizing and interpreting the fitted model.

The function `replications()` presents the relevant information about balance as part of a more general result. The arguments to `replications()` are a model formula and an optional data frame. The value returned has as many elements as there are terms in the model. Each element describes the pattern of replications for all the levels associated with the corresponding term. In general, these elements will each be a table of the number of replications of the levels (a one-way table for main effects, a two-way table for two-factor interactions, and so on). If the term is balanced in the design, all the numbers in this table will be equal. In this case `replications()` replaces the table by this single number. If *all* terms are balanced, `replications()` replaces the list by a numeric vector containing for each term the number of replications. The response in the formula given to `replications()` is not used in computing the replications, but may be helpful if we want to use "." in the formula (the response will then not be included incorrectly in the definition of "."").

An example of `replications()` on the solder data is

```
> replications(skips ~ . , solder.balance)
 Opening Solder Mask PadType Panel
     240     360  180      72   240
```

We stated that this subset of the data was balanced, and indeed it is. It came from a larger set of data from an experiment originally designed to be balanced, but not balanced as actually run. The result of `replications()` for the full set of data is

```
> rep.all <- replications( skips ~ . , solder)
```

The test for balance is

```
> is.numeric(rep.all)
[1] F
```

Given that there is some unbalance, a simple computation to find the unbalanced terms is

```
> sapply(rep.all, function(x)length(x)>1)
 Opening Solder Mask PadType Panel
       F      F    T       F     F
```

This calculation applies a function to each element of the list, in this case the in-line function

```
function(x)length(x)>1
```

which returns TRUE wherever the replications did not reduce to a single number. The result shows us that one term, that for Mask, caused the unbalance. We can look at the replication pattern for that term:

```
> rep.all$Mask
  A1.5  A3 A6  B3  B6
   180 270 90 180 180
```

Level A6 of the Mask factor is found in the full experiment, but not in solder.balance. You might be able to guess what happened from this printout. Level A6 was originally intended to be used on 180 runs, just like the other levels. In the actual experiment, however, half of those runs were done with level A3 instead. Level A6 turned out not to be a good one to pursue, but as a result, the full experiment left the data analyst with the dilemma of choosing between a larger, unbalanced experiment and a smaller, balanced one. With a properly randomized experiment, the balanced subset is still a legitimate one to study, but it does represent a loss of information.

Chapter 1 shows an analysis of sqrt(skips) over the balanced subset. For comparison, we now fit the main effects over the complete set of data. Since aov() is based on a linear model estimation, there is no problem with the fit, but the interpretation of the results is different. First, we fit and use the standard summary.

```
> aov.solder.all <- aov(sqrt(skips) ~ . , solder)
> summary(aov.solder.all)
              Df Sum of Sq Mean Sq F Value      Pr(F)
   Opening     2     740.8   370.4   527.7 0.000e+00
    Solder     1     295.1   295.1   420.4 0.000e+00
      Mask     4     548.7   137.2   195.4 0.000e+00
   PadType     9     161.3    17.9    25.5 0.000e+00
     Panel     2      22.7    11.3    16.2 1.272e-07
Residuals 881     618.4     0.7
```

With 881 degrees of freedom for residuals, and the F probabilities nearly all off-scale, the formal statistical analysis is not perhaps too relevant. However, if we were worried about its validity, we should take account of the unbalance. That is to say, the standard table above is to be interpreted sequentially. The contribution of each row must be interpreted as adding that term to the model containing the previous terms. A different analysis is obtained by the function drop1(), introduced in Section 4.3.2 for linear models. This function produces a table showing the effect of dropping each term from the complete model, and therefore is interpretable without the sequential considerations required for the standard summary.

```
> drop1(aov.solder.all)
Single term deletions
```

```
Model: sqrt(skips) ~ Opening + Solder + Mask + PadType + Panel
          Df Sum of Sq  RSS F Value       Pr(F)
   <none>                618
  Opening  2      684.3 1303     487.5 0.000e+00
   Solder  1      226.0  844     322.0 0.000e+00
     Mask  4      548.7 1167     195.4 0.000e+00
  PadType  9      161.3  780      25.5 0.000e+00
    Panel  2       22.7  641      16.2 1.272e-07
```

The summary here has sums of squares and F-statistics, but replaces the mean-square column with RSS, the residual sum-of-squares for the reduced model. This is useful in that it gives a direct comparison of the terms, showing that dropping Opening or Mask has the largest effect on RSS. The function drop1() is further discussed in the context of generalized linear models in Chapter 6.

Another way to investigate the lack of balance is with the function alias(). An example of this is given in Section 5.3.2. The aliasing information is derived from the numerical fit, not from the factors. In cases like the example here, where the unbalance is systematic, replications() generally gives more interpretable results. But if a few runs at random were omitted, no terms would be balanced. In this case, alias() gives direct numerical information about the pattern of unbalance.

5.2.2 Graphical Methods and Diagnostics

There are a number of useful graphical methods for displaying data that includes multiple factors. The plots shown in Figure 1.1 in Chapter 1 and Figure 5.1 in this chapter are examples. Figure 5.1 is produced by the default plotting method for designs:

```
plot(catalyst)
```

which shows a plot of the means at each level of each factor for the data given in Table 5.1. Similarly, Figure 1.1 was produced from the balanced subset of data from the experiment on wave soldering. The experiment, in design object solder.balance, has factors Opening, Solder, Mask, PadType, replication factor Panel, and response skips. So

```
plot(solder.balance)
```

plots the response, skips, showing its mean value for each level of each of the factors.

When the plotting method gets only one argument, as in the above examples, it looks for a numeric variable in the design to use as the response. If there are several such variables, separate plots will be produced for each of them. A second argument can be provided to define the variables to be plotted. Typically, this argument is a formula, to be interpreted in terms of the variables in the design; for example, to plot the square root of the skips in `solder.balance` against all the factors:

```
> plot(solder.balance,sqrt(skips) ~ .)
```

The plotting method for designs can use summary statistics other than means. In this case, a third argument is used to specify the function. Figure 5.2 shows the median skips for each level of each factor in the design, obtained by

```
plot(solder.balance, fun = median, ylim = c(0, 10))
```

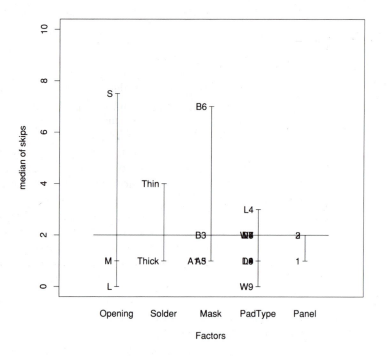

Figure 5.2: *Median number of solder skips at each level of each factor in the* `solder.balance` *data.*

Graphical parameters can be supplied in the call to the plot methods; in this case, the limits on the *y*-axis have been specified to include 0. The supplied function may be anything returning a single numeric value. Slightly more complicated cases—for example, when the function needs another argument—can be handled by providing an in-line function definition. See the example on page 167.

Two other specialized plotting functions are also available for use with designs:

- `plot.factor()` shows the distribution of a response for each level of one or more factors;

- `interaction.plot()` shows the interaction of two factors on some summary of the response.

The distribution of the response for different levels of one or more factors can be summarized using boxplots, shaded bar plots, and other plots, from `plot.factor()`. Figure 5.3 shows the first four of five boxplots produced by the expression:

```
plot.factor(solder.balance)
```

As its name implies, `plot.factor()` is a method for plotting factors or ordered factors. For example, the two calls

```
plot(Mask); plot(Mask, sqrt(skips))
```

would make a barplot of the factor `Mask` and a boxplot of `sqrt(skips)` for each level of `Mask`. Since the function is often called with a design or data frame rather than a factor as its first argument, we are emphasizing its direct use by calling `plot.factor()` as well as its automatic use as a plotting method for factors.

The arguments to `plot.factor()`, like those to the plot method for designs, can be a design plus an optional formula or response. If the second argument is omitted, as in this case, all the numeric variables in the design object will be plotted, split up by levels of each of the factors. One plot is produced for each combination of factor and response.

As another example of `plot.factor()`, suppose we want to produce a plot of the data in `solder.balance` by boards. The interaction of the factors `Opening`, `Solder`, and `Mask` uniquely identifies a board. We could plot boxes for each board, but let's try instead a *character* plot and identify the `Panel` within each board by plotting characters:

```
> attach(solder.balance)
> Boards <- interaction(Solder, Mask, Opening)
> plot(Boards, skips, character = Panel)
```

The function `interaction()` produces a new factor indexed by all combinations of levels of the arguments to `interaction()`. By giving the argument `character=` we tell the plotting method for factors to produce a character plot, using the levels of the factor `Panel` to label individual points. Figure 5.4 shows the result.

The function `interaction.plot()` summarizes graphically how pairs of factors interact in influencing the response. The function takes three arguments, the first two being factors and the third the response:

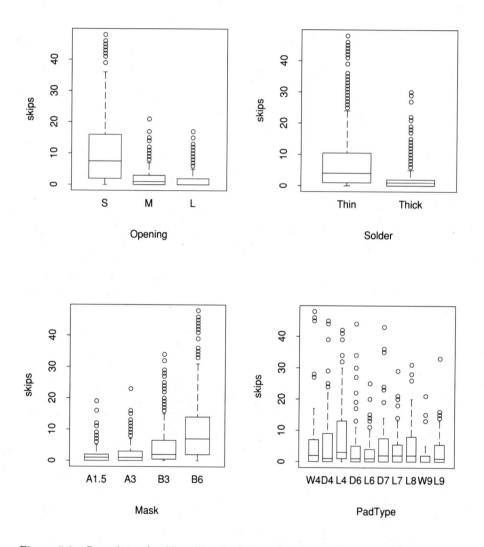

Figure 5.3: *Box plots of solder skips by factors for the* solder.balance *data. Each box pattern shows the distribution of the response* skips *for one level of a factor.*

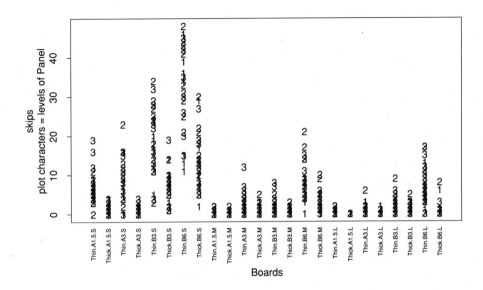

Figure 5.4: *Character plot of solder skips by boards, showing the levels of factor* `Panel` *as the characters* 1, 2, *and* 3. *The prevalence of* 2 *and* 3 *in higher values suggests that the second and third panels had more skips. Board* `Thin.B6.S` *tended to have the most skips overall.*

```
> interaction.plot(Opening, Solder, skips)
```

The left panel of Figure 5.5 is the result. The horizontal axis of the plot shows levels of the first factor. For each level of the second factor, connected lines are drawn through the mean of the response for the corresponding level of `Opening:Solder`. The style of the plots are related to the `matplot()` function ($\boxed{\text{S}}$, page 67). Like `plot.design()`, `interaction.plot()` uses a summary function, again the mean by default. Interesting plots for datasets with many factors may be suggested by looking at an analysis of variance table that includes two-factor interactions, or by previous knowledge of the factors. The right panel of Figure 5.5 shows the 95th-percent-point plot for `Mask` by `Opening`, produced by the expression:

```
> interaction.plot(Mask, Opening, skips,
+         fun = function(x) quantile(x, probs = .95) )
```

This example again shows the use of an in-line function definition to increase the flexibility of the plots. We wanted to give `quantile()` a second argument. Rather than assign a special object to be the new function, we can supply the expression

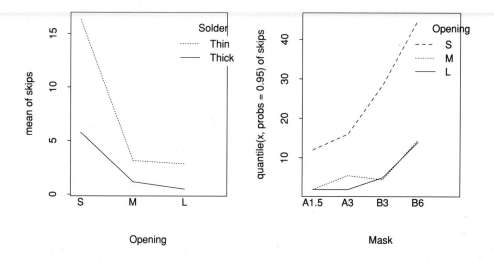

Figure 5.5: *Interaction plots of solder skips. The left panel plots mean skips at each combined level of* Opening *and* Solder, *and connects the points with the same level of* Solder. *The right panel does the same for* Mask *and* Opening, *using the 95th percentile rather than the mean.*

```
function(x) quantile(x, probs = .95)
```

The value of this expression is an S function object. The function takes one argument and returns the corresponding value from `quantile()`.

All of these plot functions take plotting parameters and can, as noted, sometimes take either factors or designs. In addition, `plot.design` and `plot.factor` can take formulas to specify the desired plot. See the on-line documentation for details.

Normal Quantile-Quantile Plots of Effects

Under the standard assumptions for linear models (independently normal errors with constant variance), the effects computed in fitting the model are also independently normal. If the true value of the coefficient for a particular contrast were zero, the corresponding effect would have mean zero. Thus, large values on a normal quantile-quantile plot of the effects suggest important effects. Since the signs of the effects are of secondary interest, the absolute values are traditionally plotted (a "half-normal" plot). A method for the S function `qqnorm()` for aov objects produces this plot. Effects corresponding to residuals or to factors having no contribution to the response should form a roughly linear pattern in the lower left of the half-normal plot; the slope of this pattern estimates the residual standard error in the fit. More importantly, points that lie well above this pattern suggest important contributions to the fit. Separate plots will be produced for each response in a multiple-response model and for each error term in a multiple-strata analysis of variance. Optional arguments can control the labeling of points on the normal or half-normal quantile-quantile plot of effects. If `label` is provided as a positive integer, the largest in absolute value `label` effects will be identified; if the option `identify=T` is included, the function prompts for interactive identification (e.g., with a mouse) of interesting points on the plot. In either case, the corresponding effect labels will be plotted.

As an example, we show in Figure 5.6 the half-normal plot of effects from the analysis of pre-etch mean in the `wafer` data discussed in the previous subsection, with the six largest effects identified:

```
qqnorm(waov1, label=6)
```

The effect names are abbreviated on the plot; typing `effects(waov1)` gives the full labels. For example, `vs1` is `visc.tem1`, the first effect for `visc.tem`, and `exp.L` is the linear contrast for the factor `exptime`.

5.2.3 Generating Designs

Design objects inherit from data frames, and thus any manipulations that can be done on data frames should also be appropriate for design objects. Designs can be created either by using some special functions or by taking any data frame and coercing it to be a design.

Two functions, `fac.design()` and `oa.design()`, are provided to generate commonly used experimental designs—namely, factorial designs (complete or fractional), and orthogonal array designs. In the simplest call, these functions take an argument `levels`, giving the number of levels for each of the factors to be included in the design:

```
fac.design(levels)
oa.design(levels)
```

A call to `fac.design()` produces a factorial design, with a row of the design object for each possible combination of the factors in the design. To generate a 2^3 design, as in Table 5.1 at the beginning of the chapter, we ask for a design with three factors, each at two levels:

```
> fac.design(c(2,2,2))
      A    B    C
1 A1   B1   C1
```

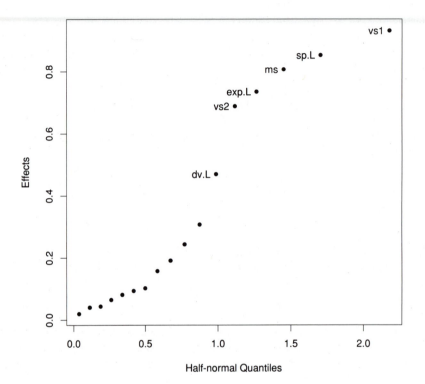

Figure 5.6: *Half-normal plot of pre-etch mean effects in the wafer experiment. The six largest points are labeled: "vs1" is the first effect for factor* visc.tem; *"sp.L" is the linear effect for ordered factor* spinsp, *etc.*

```
2 A2    B1    C1
3 A1    B2    C1
4 A2    B2    C1
5 A1    B1    C2
6 A2    B1    C2
7 A1    B2    C2
8 A2    B2    C2
```

A call to oa.design() produces an orthogonal array design. These designs aim to study a large number of factors in as few runs as possible. The designs returned by oa.design() select as small a design as possible from a given set of such designs, on the assumption that only main effects need to be included in the model, and allowing for an optional request to provide a specified number of degrees of freedom for residuals. The wafer experiment in Table 5.2 is an example of such a design. The underlying design had nine factors, three with two levels and six with three levels. Two of the two-level factors were combined in a special way to define a three-level factor, creating an experiment with one two-level factor and seven three-level factors.

```
> wafer.design <- oa.design(c(2,rep(3,7)))
> wafer.design
      A  B  C  D  E  G  H  I
1 A1 B1 C1 D1 E1 G1 H1 I1
2 A1 B1 C2 D2 E2 G2 H2 I2
3 A1 B1 C3 D3 E3 G3 H3 I3
4 A1 B2 C1 D1 E2 G2 H3 I3
5 A1 B2 C2 D2 E3 G3 H1 I1
6 A1 B2 C3 D3 E1 G1 H2 I2
          Etc.


Orthogonal array design with 2 residual df.
Using columns 1, 2, 3, 4, 5, 6, 7, 8 from design oa.18.2p1x3p7
> replications(wafer.design)
 A B C D E G H I
 9 6 6 6 6 6 6 6
```

The printing method for the design shows the number of degrees of freedom for residuals (the default minimum requested is 3). If this is less than 10% of the size of the design, a warning is given. The call to replications shows that we have a balanced design with 18 runs.

The oa.design() function uses a table of known designs of which the largest contains 36 runs to handle a limited range of requests for factors with two or three levels. Outside of that range, it gives up:

```
> oa.design(rep(2,100))
Error in oa.design(rep(2, 100)): Don't have an all 2 levels oa for > 32
    runs, use fac.design
```

Other factors can sometimes be accommodated by adjusting the design; for example, a four-level factor can be formed out of two two-level factors, provided the interaction of those factors is estimable.

By default, fac.design() and oa.design() generate standard names for factors and for the levels of the factors. These names may be provided by the factor.names argument. If supplied, factor.names can either be a character vector or a list; in either case, it should have length equal to the number of factors. A character vector provides names for the factors. The names for the levels are then constructed from an abbreviated form of the factor names:

```
> fac.design(c(2,2,2), factor.names = c("Temp","Conc","Cat"))
  Temp Conc  Cat
1   T1  Co1  Ca1
2   T2  Co1  Ca1
3   T1  Co2  Ca1
4   T2  Co2  Ca1
5   T1  Co1  Ca2
6   T2  Co1  Ca2
7   T1  Co2  Ca2
8   T2  Co2  Ca2
```

If `factor.names` is a list, its elements provide names for the levels of the corresponding factors, and the names attribute of the list gives the factor names:

```
> nlist <- list(Temp = c(160, 180), Conc = c(20, 40), Cat = LETTERS[1:2])
> fac.design(c(2,2,2), factor.names = nlist)
  Temp Conc Cat
1  160   20   A
2  180   20   A
3  160   40   A
4  180   40   A
5  160   20   B
6  180   20   B
7  160   40   B
8  180   40   B
```

The expression `factor.names(design)` returns a similar list giving the factor and level names of `design`. Factor names can be assigned to an existing design by using `factor.names()` on the left of an assignment. The same interpretation is given to the object on the right of the assignment as was outlined for the `factor.names` argument:

```
> cdesign <- fac.design(c(2,2,2))
> factor.names(cdesign) <- nlist
```

This produces the same design as in the previous example. A `factor.names` argument can be used in calling `oa.design()` as well, with the same interpretation.

The function `fac.design()` also takes an optional argument `row.names` to provide names for the rows of the resulting design object:

```
> cdesign <- fac.design(c(2,2,2), names = nlist,
+     row.names = paste("Run",1:8))
```

The row names are the attribute `"row.names"` of the resulting data frame, and can be extracted or assigned as this attribute. For `oa.design`, and by default for `fac.design`, the values of `row.names` are the sequence numbers of the rows. As with the factor names, the row names can be assigned after creating the design:

```
> cdesign <- fac.design(c(2,2,2))
> row.names(cdesign) <- paste("Run",1:nrow(cdesign))
```

For designs created by `oa.design()` this will generally be necessary, since the number of runs is not known in advance.

Factorial designs can be generated with multiple replications of each combination of factor levels by setting the optional `replications` argument to the number of replicates wanted. Fractional factorial designs can also be generated from `fac.design()`, but only for 2^k designs. The argument `fraction` defines what fraction of a design is wanted. The argument can be a numeric fraction:

```
> fac.design(rep(2,5),fraction=1/4)
    A    B    C    D    E
1   A1   B2   C1   D1   E1
2   A2   B1   C2   D1   E1
3   A2   B2   C1   D2   E1
4   A1   B1   C2   D2   E1
5   A2   B1   C1   D1   E2
6   A1   B2   C2   D1   E2
7   A1   B1   C1   D2   E2
8   A2   B2   C2   D2   E2

Fraction: y ~ A:B:D + B:C:E
```

In this case, the function provides a one-quarter replicate of a 2^5 design, using some standard choices to pick the fraction. For more control over how the fractionation is done, `fraction` can be one or more *defining contrasts*. In terms of our model formulas, a defining contrast is an interaction of factors, such as `A:B:C`. In a two-level design, each such interaction defines one contrast with two possible values. Therefore, each such contrast implicitly divides the design into two halves, one of which will be chosen in forming the fractional design. If one contrast is supplied, a half-replicate is produced; if two contrasts are supplied, a quarter-replicate, and so on. The half chosen is specified by the sign of the corresponding term in the formula. The following creates a specific quarter-replicate of a 2^5 design:

```
> fac.design(rep(2,5), fraction = ~ A:B:C - B:D:E)
    A  B  C  D  E
1  A2 B1 C1 D1 E1
2  A1 B1 C2 D1 E1
3  A1 B2 C1 D2 E1
4  A2 B2 C2 D2 E1
5  A1 B2 C1 D1 E2
6  A2 B2 C2 D1 E2
7  A2 B1 C1 D2 E2
8  A1 B1 C2 D2 E2
```

```
Fraction:   ~ A:B:C - B:D:E
```

It's important that we used ":", not "*" in defining the contrasts. We wanted *only* the three-factor interactions, not the expanded model with all the main effects, etc. The latter will produce a design, but not one we want. The printing method for designs will show the defining contrasts, whether chosen automatically or specified in the call.

Design objects are essentially just data frames in which the design itself is represented by the factor variables. The function `design()` can be used to make a design object from one or more arguments (data frames, matrices, etc.), each variable or column being coerced to be a factor. In particular, designs are sometimes printed as tables, with numbers 1, 2, ..., in each column to stand for the levels of the factors. For instance, the first five rows of the orthogonal array design on page 171 in this form and without row labels would look like:

```
1 1 1 1 1 1 1 1
1 1 2 2 2 2 2 2
1 1 3 3 3 3 3 3
1 2 1 1 2 2 3 3
1 2 2 2 3 3 1 1
```

The function `read.table()` will read in such a file and convert it to a data frame. Suppose the complete design was read in this form, as `waferd`. Then `design(waferd)` will convert the data frame `waferd` to a design object.

Once a design object has been created, the factor names can be assigned from a list. For example, we could make up the names for the factors in the `wafer.design` design as follows:

```
> fn <- list()
> fn$maskdim <- c(2, 2.5)
> fn$visc.tem <- c("204,90", "206,90", "204,105")
> fn$spinsp <- c("low", "normal", "high")
> fn$baketime <- c(20, 30, 40)
     .

     .

     .

> factor.names(wafer.design) <- fn
```

Alternatively, `fn` could have been given to `design()` as the `factor.names=` argument. In some designs, we want to indicate that some factors are ordered. Either single factors or selected factors in a design can be designated as ordered by using the function `ordered()` on the left of an assignment. The right side can give for each factor either a TRUE/FALSE value or the vector of levels in the order desired. With a logical value, the current levels are assumed to be in increasing order. For example, in `wafer.design`, all the factors except the second and fifth are ordered:

```
> ord <- rep(T, 8)
> ord[c(2,5)] <- F
> ordered(wafer) <- ord
```

Further control over the parametrization can be obtained by specifying contrasts for the factors. See Section 5.3.1.

Randomization

In carrying out an experiment, one may want to randomize the order in which the runs are to take place. A randomized order could be generated by

```
mydesign[ sample(1:nrow(mydesign)), ]
```

which permutes the rows of the design in a random order. Unfortunately, things may not be quite so simple. In practice, some factors may be difficult to vary. We want to restrict randomization to leave these factors alone, so they can be varied as infrequently as possible. The function `randomize()` takes a design and the names of some factors to be restricted, and returns an ordering, first by levels of the restricted factors, and then randomized within those levels. Running the experiment in the order given by this permutation would provide the restricted randomization requested. For example, suppose we wanted to randomize the design in Table 5.1, but not over the `Cat` factor:

```
> perm <- randomize(catalyst, restrict = "Cat")
> perm
[1] 1 3 4 2 8 6 5 7
> catalyst[perm,]
    Temp Conc Cat Yield
1    160   20   A     60
3    160   40   A     54
4    180   40   A     68
2    180   20   A     72
8    180   40   B     80
6    180   20   B     83
5    160   20   B     52
7    160   40   B     45
```

The result is to do all the `A` catalyst runs first, then all the `B` catalyst runs, while randomly permuting within each level of the catalyst factor. The `restrict` argument is essentially a subscript argument on the factors and can have any form suitable for such a subscript: character, numeric, or logical.

5.3   The S Functions: Advanced Use

In this section, we present some more advanced techniques for using S functions to parametrize contrasts in more customized ways, to investigate aliasing, and to compute projections.

5.3.1   Parametrization; Contrasts

If we think of the anova model as implying coefficients for each level of each factor or interaction, then the model is inherently over-parametrized. The sum of all the coefficients for a main effect is trying to estimate the same thing as the intercept term, and the sum of the coefficients over one factor in a two-factor interaction is estimating the same thing as a single level of the other factor. No matter how many observations we take, these parameters are functionally aliased. In addition, if the experiment does not include all the possible combinations of levels for the factors in the model, even parameters that are not functionally aliased may be aliased in the design.

The resolution of these ambiguities is done in two steps. First, a parametrization of the model is constructed according to a choice of *contrasts* for each factor appearing in the model, to eliminate functional aliasing. Second, the numerical fitting of the model checks for and identifies any further design-dependent aliasing.

The need to parametrize factors to fit linear models is common to all models in Chapters 4 to 7, although the details may vary a little depending on the emphasis in the modeling. For most analysis of variance, the default choice of contrasts is adequate: unordered factors use the Helmert contrasts, and ordered factors use orthogonal polynomials. In particular applications, you may want to set the contrasts or study the fit in terms of particular contrasts. In doing any of these computations, two tools are particularly useful:

- The function C() takes a factor and a chosen set of contrasts, and returns a factor with those contrasts inserted. You typically use C() directly in the formula of a fit, to set the contrasts for that fit.

- The function contrasts() returns or sets the contrasts of a factor. If you want to modify the contrasts, you can start by getting the current contrast matrix and then make any changes you want to that matrix. The matrix can then be used with C() in a formula or assigned as the contrasts of the factor.

Chapter 2 presents the techniques for defining and modifying contrasts in Section 2.3.2 with the rules examined in detail in Section 2.4.1. In addition to C() and contrasts(), a number of other techniques in this chapter are handy when studying contrasts. For example, the projections of the fit for individual terms are useful as responses in fits designed to study the effects of different contrasts for the corresponding factors. The remainder of this section illustrates some of these techniques.

It is possible to choose fewer than $k - 1$ contrasts for a factor with k levels, if the remaining contrasts are of no particular interest. There are two distinct ways to do this:

```
contrasts(x) <- value
contrasts(x, how.many) <- value
```

The first version value and fills it out to give a complete parametrization. The second version assigns only how.many contrasts, allowing a partial term in the analysis, in the sense that at most how.many degrees of freedom will be given to this term.

Suppose we wanted to make sure the first effect for a four-level factor contrasted the average of the first and third against the average of the second and fourth levels:

```
> attach(state, 1)
> contrasts(region) <- c(1,-1,1,-1)
> contrasts(region)
                [,1] [,2] [,3]
    Northeast     1 -0.7 -0.1
        South    -1  0.1 -0.7
North Central     1  0.7  0.1
         West    -1 -0.1  0.7
```

The assignment function for contrasts has appended two additional, orthogonal columns to the supplied contrast. The rows of the contrast matrix, as always, are labeled by the levels of the factor.

For an example of omitting degrees of freedom from a term, we look again at the wafer data in Table 5.2. The factor called visc.tem was a three-level factor concocted from three of the four possible levels constructed by combining two levels each for viscosity (204 and 206) and baking temperature (90 and 105). In analyzing this factor, one needs to choose contrasts specially. If one chose to assume that the temperature factor had no effect, then the quadratic effect from three-level orthogonal polynomial contrasts turns out to be equivalent to the single contrast for viscosity. One way to install this assumption in the anova fitting would then be:

```
> attach(wafer, 1)
> contrasts(visc.tem,1) <- contr.poly(levels(visc.tem))[, 2]
> contrasts(visc.tem)
              [,1]
 204,90   0.40825
 206,90  -0.81650
 204,105  0.40825
> detach(1, save = "wpm1")
```

This is a fairly typical computation; let's examine it a step at a time. We want
to create a new data frame with the modified contrasts for `visc.tem`; to start this
off, we attach `wafer` as the working data. Now we generate the quadratic contrast
for `visc.tem` (since it is not an ordered factor, its default contrasts would not use
`contr.poly()`). Because we assigned this with `how.many=1`, only one degree of freedom will be used for this term. Detaching and saving gives us our new data, in
`wpm1`. Now we fit the `pre.mean` response to the new data:

```
> wvaov <- aov(pre.mean ~ ., wpm1)
> summary(wvaov)
          Df Sum of Sq Mean Sq F Value  Pr(F)
  maskdim  1     0.652   0.652    67.1 0.0038
 visc.tem  1     1.326   1.326   136.4 0.0013
   spinsp  2     0.765   0.383    39.4 0.0070
 baketime  2     0.002   0.001     0.1 0.8904
 aperture  2     0.032   0.016     1.6 0.3318
  exptime  2     0.545   0.272    28.0 0.0115
  devtime  2     0.280   0.140    14.4 0.0289
 etchtime  2     0.103   0.052     5.3 0.1033
Residuals  3     0.029   0.010
```

Only one degree of freedom goes to `visc.tem`, the other being included in the residuals. The same technique could have been used in a formula, replacing `visc.tem`
by

    C(visc.tem, contr.poly(levels(visc.tem))[, 2], 1)

The third argument to `C()` is again `how.many`, with the same interpretation as when
setting contrasts. In this example, permanently setting the contrasts seems more
straightforward than having to include such a complicated expression in formulas.

5.3.2   More on Aliasing

The term *aliasing* is used in the analysis of variance to refer to the inability in some
circumstances to talk about the estimate of an effect in the model without reference
to other effects, either:

- complete or full aliasing, in which the estimate is identical to (completely
   aliased with) some previously estimated effect or linear combination of effects;
   or,

- partial aliasing, in which all the effects can be estimated but with correlation
   between estimates of the coefficients.

There are many ways to phrase the definition of aliasing, and other terminology
is sometimes used (the term *confounding* is used in some cases for partial aliasing,

particularly between block and treatment factors). The wording we use reflects our linking of analysis of variance with linear models. In fact, complete aliasing corresponds to singularity of the model matrix and partial aliasing to non-orthogonality of the columns of that matrix.

Complete aliasing will happen in fractional factorial designs, and may be deliberate if the design has been chosen to alias some higher-order interactions in the hope that not all of these will be important. Numerically, an effect or coefficient that would have been estimated is found to be equal to some linear combination of previously estimated parts of the model. That is, a column in the model matrix representing a contrast for one of the terms is found to be linearly dependent on previous columns of that matrix.

Complete aliasing is represented by a matrix of dependencies, with rows for inestimable effects and columns for estimated effects. Each row equates an inestimable effect to a linear combination of previously included effects. The expression `alias(fit)` returns the aliasing pattern appropriate to the object `fit`: for a linear model or an anova with a single error stratum the result is a matrix or `NULL`; for a model with multiple error strata the result is a list of the non-null alias matrices for the strata. The alias pattern in the fractional factorial analyzed on page 156 is:

```
> alias(half.aov)
Complete
               (Intercept) Temp Conc Cat
    Temp:Conc                         1
     Temp:Cat                    1
     Conc:Cat              1
Temp:Conc:Cat          1
```

In this example the interpretation of the pattern is very simple, since each inestimable contrast is exactly aliased with one earlier contrast. For example, one can say that the two-factor interaction `Temp:Conc` is aliased with the main effect `Cat`. When several nonzero coefficients appear in single rows of the alias matrix, interpretation may be more difficult.

Partial aliasing is used to refer to the situation when two contrasts are correlated but not exactly linearly dependent. Where complete aliasing relates inestimable effects to estimable ones, partial aliasing is a relationship *among* the estimable effects. The measure we use is the correlation matrix of the coefficient estimates, with the diagonal terms set to zero.

As an example of partial aliasing, we return to the wafer-processing example in Table 5.2 on page 149. The analysis in the referenced paper, and our own analysis so far, have ignored the fact that 3 of the 18 runs were done on one wafer, meaning that they had 5 instead of 10 repeated observations from which to compute the mean and standard deviation of the line widths. Under these circumstances, linear model computations should weight the observations proportionally to the number of repeated observations, since the variance of the mean line width is inversely proportional to this number. The weighting is easy to do with the `aov()` function. Arguments to `aov()` can include optional arguments to `lm()`, such as `weight=` to provide weights to be used in the fitting.

```
wwaov <- aov( post.mean ~ maskdim + visc.tem + spinsp + baketime +
       aperture + exptime + devtime + etchtime, data = wafer, weight = N)
```

The weighted linear model is not balanced, and we can examine the alias pattern:

```
> alias(wwaov)
Partial
              (I)  m v1 v2 sL sQ bL bQ a1 a2 eL eQ dL dQ eL2 eQ2
(Intercept)     3  6 -1  6  1 -4 -5  1 -1 -6  1 -1 -1     -9
   maskdim         4  6  6 -3  1 -1 -6    -4  6 -3     -4
  visc.tem1       -5  4 -3 -4 -2  4  3 -7  5 -4 -3     -8
  visc.tem2        2  4  2 -3 -6 -4  5 -7  6  4      1
 spinsp.L          4 -4 -3  1 -4 -4 -2  4 -2     -8
 spinsp.Q            -2  3 -1 -7  3 -4  6 -3     -1
baketime.L              -2 -4  2  4 -2  1  4      5
baketime.Q               5 -3  2  3 -1 -6      6
 aperture1                  1 -4  6 -4 -6     -1
 aperture2                    -3  4 -6  3      1
 exptime.L                      -5  4  3      8
 exptime.Q                         -6 -4     -1
 devtime.L                             1      1
 devtime.Q                                    1
etchtime.L
etchtime.Q

Notes:
$"Max. Abs. Corr.":
[1] 0.184
```

Most effects are partially aliased with most other effects. The object returned to represent partial aliasing is a table simplified by coding the substantially nonzero correlations from -9 to +9 relative to the maximum absolute correlation, which is added as a note to the table. In addition, since the table is symmetric, only the upper triangle is shown, with the column labels abbreviated since they are the same as the row labels. There are many other ways to try to simplify this sort of pattern. The function `pattern()` is used to produce the simplified form:

```
alias(wwaov, pattern = F)
```

suppresses the call to `pattern()` and returns just the numerical alias pattern. Some nested designs tend to produce less global partial aliasing patterns.

In the examples shown, we either had complete aliasing (`half.aov`) or partial aliasing (`wwaov`). It's possible, of course, to have both, in which case `alias()` returns both as elements of a list. Arguments `Complete=F` or `Partial=F` will suppress one part of the report if it is not of interest.

So far, we have shown `alias()` as applied to a fitted model. You might well want to study the aliasing properties of a proposed design *before* fitting the model. In particular, studying the alias pattern of a proposed design and model may be a useful step in selecting a design. For this reason, `alias()` has a method for design objects:

```
alias(mydesign)
```

This method returns the aliasing pattern of the design, with respect to a model that fits all possible main effects and second-order interactions. Generally, the numerical aliasing pattern is determined by the design *and* the structural form of the model (but not by the response data). For this reason, an optional model formula can be added to the above call (a response is not needed). The default model is equivalent to

```
alias(mydesign, ~ .^2)
```

5.3.3 Anova Models as Projections

The description of a model formula as the sum of terms suggests an analogous way of looking at the fitted model. The model formula for the gun example on page 152 is

```
Rounds ~ Method + Physique/Team
```

which expands to

```
Rounds ~ Method + Physique + Team %in% Physique
```

This additive model can usefully be related to the sum of five vectors, one for each of the terms in the model (including the intercept) and one for the residuals. These vectors identically sum to the response:

$$\text{Rounds} = y_1 + y_2 + y_3 + y_4 + y_5$$

where y_1 is the *projection* of the response on the intercept, y_2 is the projection on the `Method` term, and so on. These projections are useful diagnostics and summaries of the fit. They can be computed, either during the fit or later. The projections are represented as a matrix. If computed during the fit (the most efficient approach) they are returned as the `projections` component of the fit. In the `gun` example, the computations would be as follows:

```
> gunaovp <- aov(Rounds ~ Method + Physique/Team, gun,
+    projections = TRUE)
> gunproj <- proj(gunaovp)
> dim(gunproj)
[1] 36  5
> dimnames(gunproj)[[2]]
[1] "(Intercept)"        "Method"            "Physique"
[4] "Team %in% Physique" "Residuals"
```

The argument `projections=T` to `aov()` causes projections to be computed. If we had previously computed a fit and just wanted to produce the projections, this can be done by

```
> gunproj <- proj(gunaov)
```

This usually does as much work as the previous version, since `proj()` must refit the model unless a `qr` component was requested in the original fit.

The projection has one column for each term and one for the residuals. The linear equation above corresponds to asserting that the sum of the columns of the projection is equal to the response (up to rounding error):

```
> sum.of.projections <- gunproj %*% rep(1,5)
> all.equal.numeric(sum.of.projections, gun$Rounds)
[1] T
```

Since the columns of the projections are defined by the corresponding factor (main effect or interaction), all the elements of the column that correspond to the same level of the factor are equal. Thus, for all rows with the same level for `Method`, the values of `gunproj[,"Method"]` are the same. For example, with `Method=="M1"`, the common value is `4.256`. The same is true for the `Physique` effect. For the `"Team %in% Physique"` nesting, all cells indexed by each unique combination of `Team` and `Physique` have a common value.

The sums of squares of each of the columns of `gunproj` are the sums of squares listed in the anova table (with, as usual, the sum of squares for the `"(Intercept)"` suppressed from the anova table).

```
> apply( gunproj^2, 2, sum)
 (Intercept) Method Physique Team %in% Physique Residuals
       13456 651.95   16.052             39.258    53.499
```

Compare this with the summary on page 152.

Indeed, the anova table is a systematic way of recording the quadratic equation that expresses the total sum-of-squares as the sum of the five values above. This quadratic equation is often called *Cochran's theorem* in the statistics literature. It is a multidimensional analogue of Pythagoras's theorem. The degrees of freedom associated with the sums of squares are kept in the computed projection, as `attr(gunproj,"df")`. The sums of squares for single-degree-of-freedom projections are `effects(gunaov)^2`. We could store the sums of squares as an attribute of `gunproj`, but have chosen not to do so. Recording the degrees of freedom in `gunproj` is necessary. The row sums of the linear equation recover the response variable `Rounds`. The sum of the sums of squares equals the sum-of-squares of `Rounds`, and the sum of the sums of squares excluding `"(Intercept)"` is the *corrected* total sum of squares often seen in anova tables.

Each column of `gunproj` is the projection of the response variable into the linear space spanned by the columns of the model matrix corresponding to that term of the model. The standard notation for the model expansion

```
Rounds ~ Method + Physique + Team %in% Physique
```

gives the linear equation:

$$
\begin{aligned}
R &= \overline{R} + \widehat{R}_{\mathrm{M}} + \widehat{R}_{\mathrm{P}} + \widehat{R}_{\mathrm{T\%in\%P}} + e \\
&= \mathbf{1}\beta_0 + X_{\mathrm{R}}\beta_{\mathrm{R}} + X_{\mathrm{M}}\beta_{\mathrm{M}} + X_{\mathrm{T\%in\%P}}\beta_{\mathrm{T\%in\%P}} + e
\end{aligned}
$$

where the R_{term} columns are projections of R onto the linear space of the model terms, the X_{term} matrices are subsets of the columns of the model matrix, and the β_{term} coefficients are subsets of `gunaov.qr$coef`. In this notation Cochran's equation is:

$$
R^t R = \overline{R}^t \overline{R} + \widehat{R}_{\mathrm{M}}^t \widehat{R}_{\mathrm{M}} + \widehat{R}_{\mathrm{P}}^t \widehat{R}_{\mathrm{P}} + \widehat{R}_{\mathrm{T\%in\%P}}^t \widehat{R}_{\mathrm{T\%in\%P}} + e^t e
$$

Projections of the vector `gun[,"Rounds"]` onto each of the single-degree-of-freedom columns of the model matrix are also possible:

```
> gunproj1 <- proj(gunaov, onedf = T)
> dim(gunproj1)
[1] 36 11
> dimnames(gunproj1)[[2]]
 [1] "(Intercept)"    "Method"
 [3] "Physique.L"     "Physique.Q"
 [5] "PhysiqueSTeam1" "PhysiqueATeam1"
 [7] "PhysiqueHTeam1" "PhysiqueSTeam2"
 [9] "PhysiqueATeam2" "PhysiqueHTeam2"
[11] "Residuals"
```

The sums of squares of the columns of `gunproj1` are the single-degree-of-freedom sums of squares. They are identical to the squared effects, `effects(gunaov)^2`.

Projections with Multiple Sources of Variation

Projections in the multiple-stratum models form a list whose elements are the projections for each of the `Error()` terms. They are created by the same generic `proj()` function. We illustrate using the guayule example:

```
> gaov.proj <- proj(gaov)
Refitting model to allow projection
> names(gaov.proj)
[1] "(Intercept)" "flats"       "Within"
```

Each element of the list of projections is the projection matrix for the corresponding error term. The number of columns is the number of non-empty terms in the analysis for that error term:

```
> sapply(gaov.proj, dim)
     (Intercept) flats Within
[1,]          96    96     96
[2,]           1     2      3
```

In the analysis on page 158 that produced gaov, there are two terms for the flats strata, with two corresponding columns of gaov.proj$flats:

```
> dimnames(gaov.proj$flats)[[2]]
[1] variety Residuals
```

Each of the projection matrices has its own df attribute, recording the degrees of freedom for the projections:

```
> attr(gaov.proj$flats,"df")
 variety Residuals
       7        16
```

The projections in a multi-stratum model represent two steps of projection. First, the response is decomposed into projections onto each term of the error model. Then, each of these projections is decomposed into projections onto each term of the treatment model.

As with the single-stratum projections, we could use design.table() to examine each of the projection matrices in the multiple-stratum case. The value of each projection is constant within each level of the interaction factor corresponding to that term.

Single-degree-of-freedom projections are again available, with the same interpretation.

5.4   Computational Techniques

This section presents some background on the computations shown in previous sections, including numeric and statistical results to justify the computations. The purpose is to make clearer the computations available and how they might be further extended. The topic is a rich one and our discussion here is necessarily brief and incomplete. However, it should help to clarify what is going on and why. Section

5.4.1 relates the effects and coefficients to the underlying linear model. Section 5.4.2 discusses what happens when aliasing prevents estimating all the effects in the model. Section 5.4.3 adds the results needed when the model has an `Error()` model. Computations for projections are discussed in Section 5.4.4.

5.4.1 Basic Computational Theory

This section states some results that link the computations for `aov` models to the results for linear models. Specifically, we state the algorithm for generating the model matrix and give the essential properties of the effects. A thorough treatment would require a fairly extensive excursion into linear algebra or geometry. Fortunately, the numerical algorithms themselves provide a natural way to motivate these results in an informal way, extending the corresponding results from Section 4.4.

We start by rephrasing the linear model appropriately for this chapter. Suppose there are m terms in the formula, with expressions involving * or / expanded, the result simplified to eliminate duplicate terms, and with the intercept, if any, included as the first term. The n observations on the response y can be written:

$$y = T_1 + T_2 + \cdots + T_m + \varepsilon \tag{5.1}$$

where each T_j represents the expected contribution from the jth term and ε is the error, conventionally assumed independently normal with zero mean and variance σ^2, as outlined in Chapter 4, page 97. In this chapter, the emphasis is on terms and on single-degree-of-freedom effects. The algorithms used to fit the models, however, work by finding a model matrix, X, that is equivalent to (5.1). The term T_j corresponds to X_j, a submatrix of X. The columns of X_j are generated from contrast definitions for all the factors appearing in the term T_j. These contrasts are chosen so as to represent all the linear combinations of the dummy variables for the factors that are not redundant (in the sense of being functionally dependent on previous terms).

The rules for constructing the columns of X_j were given in Section 2.4.1, where a recursive rule was given. For a factor, F, appearing in T_j, the computation generates a matrix with n rows and either k or $k-1$ columns, where k is the number of levels of F. If T_j is an interaction of two or more factors, then X_j is formed by taking all possible products of columns from the matrices generated for each of these factors. Section 2.4.1 has an informal proof that this is a valid coding of the model.

We can write the model matrix as:

$$X = [X_1 X_2 \cdots X_k] \tag{5.2}$$

and let X_j have d_j columns. Then d_j is the maximum number of degrees of freedom for T_j, as determined from the form of the model (5.1). If the design is complete or the model is chosen so that all terms are estimable, then X will be of full rank.

Otherwise, the number of degrees of freedom for some T_j will be less than d_j. Since the notation is a bit simpler, let us assume to begin with that X has full rank, $d = \sum d_j$, and then consider the general case in Section 5.4.2.

Given a valid coding of the model in the form (5.2), we can use the orthogonal decomposition discussed in Section 4.4 to state the properties of the effects. The decomposition expresses X in terms of an n by n orthogonal matrix Q and an n by d upper-triangular matrix R such that

$$
\begin{aligned}
Q^t \cdot X &= R \\
Q^t \cdot y &= c \\
&= [c_1\, c_2\, \cdots\, c_m\, c_{m+1}]
\end{aligned}
\tag{5.3}
$$

where c_j is the set of effects associated with term T_j (d_j of them by the assumption of full rank), and c_{m+1} is the set of residual effects, if any. These effects are the key computational tool for standard anova summaries. Their essential properties follow directly from the construction of the orthogonal decomposition:

1. The elements of c are all uncorrelated, both within and between terms, regardless of whether the design is balanced or not.

2. The distribution of c_j is unaffected by any of the preceding terms $T_{j'}$, $j' < j$; in particular, $T_{j'}$ contributes nothing to the expected value of c_j. Again, this is true regardless of balance.

3. In addition, *if* a following term $T_{j'}$, $j' > j$, is balanced with respect to T_j, the same is true of it; that is, the distribution of c_j is unaffected by $T_{j'}$.

By "unaffected" in properties 2 and 3 we mean that the mean of c_j does not involve the coefficients from term $T_{j'}$. These three properties justify the use of standard summaries, such as the anova tables and the association of effects in probability plots with the corresponding terms.

To complete this section, we outline how the three properties can be derived from the linear model. Equation (5.3) and the model assumptions imply the first property, since the elements of y are independent and c is an orthogonal transformation of y. To see the other properties, we write Q in columns corresponding to the terms,

$$
Q = [Q_1\, Q_2\, \cdots\, Q_k\, Q_{k+1}]
$$

In the Householder algorithm, Q is not stored explicitly as an n by n matrix, but the matrix form is nevertheless fully defined. We also write out the linear model for y in terms of coefficients corresponding to the chosen contrasts,

$$
y = X_1\beta_1 + X_2\beta_2 + \cdots + X_k\beta_k + \varepsilon
\tag{5.4}
$$

and partition the upper-triangular matrix R into rows and columns corresponding to the terms; that is, R_{11} is the first d_1 rows and d_1 columns, R_{21} the next d_2 rows of the same columns, and so on. Since R is upper-triangular,

$$R_{jj'} = 0, \; j' < j$$

In addition, if the j'th term is balanced with respect to the jth term,

$$R_{jj'} = 0, \; j' > j$$

as well. The second and third properties then follow from writing out c_j:

$$
\begin{aligned}
c_j &= Q_j^t \cdot y \\
&= Q_j^t \cdot (X_1 \beta_1 + X_2 \beta_2 + \cdots + X_k \beta_k + \varepsilon) \\
&= \sum_{j'=1}^{k} R_{jj'} \cdot \beta_{j'} + Q_j^t \cdot \varepsilon \\
&= R_{jj} \cdot \beta_j + \cdots + R_{jk} \cdot \beta_k + Q_j^t \cdot \varepsilon \tag{5.5}
\end{aligned}
$$

and $c_j = R_{jj} \cdot \beta_j$, if all the terms are balanced.

As in the linear model, all the basic results extend directly to the case of a multivariate response—that is, to the case that y is an n by q matrix. Then c, c_j, β and β_j all become matrices with q columns.

5.4.2 Aliasing; Rank-deficiency

This section derives the matrix returned by the `alias()` function to represent aliasing in an over-determined model. When the computed rank r of X is less than $d = \sum d_j$, the decomposition pivots columns so that the first r columns are linearly independent. In this case, the numerical decomposition is written:

$$
\begin{aligned}
Q^t \cdot X \cdot P &= R \\
&= \begin{bmatrix} R_+ & R_a \\ 0 & 0 \end{bmatrix} \tag{5.6}
\end{aligned}
$$

Here P is a permutation matrix, representing the permutation of columns described by the component `pivot` returned from the decomposition, and R_+ is an r by r upper-triangular, nonsingular matrix. The computation has decided that the diagonal elements in the triangular matrix corresponding to the pivoted columns are all effectively zero.

The matrix R_a contains the aliasing information. To see this, let X_0 be the columns of X found to be linearly dependent, the last $d - r$ columns of $X \cdot P$ in

(5.6). Similarly, let X_+ be the first r columns of $X \cdot P$, and Q_0, Q_+ the corresponding columns of Q. Then

$$Q_+^t \cdot X_0 = R_a$$

By analogy with (5.3), R_a contains the effects from fitting X_0 to X_+. In the sense developed in Section 4.4, the elements in R_a define the contribution of the contrasts in X_+ to predicting the contrasts in X_0. But since the fit is exact, the effects actually define X_0 in terms of X_+. The `alias()` method returns R_a^t as the definition of complete aliasing as illustrated in Section 5.3.2.

This discussion of aliasing shows clearly the extension to partial aliasing. When complete aliasing occurs, the contrasts in X_0 are regressed on X_+ with effectively zero residuals. However, when the jth contrast is not fully aliased, we can still consider fitting it to the preceding $j-1$ columns of X, and it remains true that the jth column of R represents the effects of this fit. If the first $j-1$ elements of this column are effectively zero, then the jth column is orthogonal to (unaliased with) the preceding columns. Otherwise, partial aliasing exists, and can be summarized in various ways. The result returned by `alias()` uses the above-diagonal elements of the correlation matrix of the coefficients. This is a relatively easy summary to interpret. However, there is no single definitive numerical summary. In terms of theory, the matrix R itself is attractive. It clearly has all the information and connects fairly smoothly between complete and partial aliasing. It is not scale-independent, on the other hand, and probably is not as easy to explain intuitively.

5.4.3   Error Terms

This section explains the computations used with an explicit error model, supplied as an `Error()` expression in the formula. The method is nicely simple and general, and uses the basic linear model computations in a neat, recursive way. The general problem is to fit an analysis of variance model when the error is assumed to be the sum of errors due to the levels of certain factors occurring in the error model, as well as a common residual error.

To begin, we need to write such a model explicitly. Let's submerge the specifics of the terms and just let μ stand for the sum of all the terms in (5.1) on page 185, except for the `Error()` expression. This is often called the treatment model. Suppose the error model has s terms. Then (5.1) can be rewritten for observation i:

$$y_i = \mu_i + \sum_{j=1}^{s} \varepsilon_{l(i,j)} + \varepsilon_i \tag{5.7}$$

where the notation $l(i,j)$ just means that the jth error term is at level $l(i,j)$ for the ith observation.

The key to the computations is to write the element-wise model (5.7) in terms of the equivalent model matrix, X_E, for all the terms in the `Error()` expression. In matrix form:

$$y = \mu + X_E \cdot \varepsilon_E + \varepsilon$$

But now we can compute an orthogonal matrix Q_E from the decomposition of X_E. As in (5.3), Q_E applied to y produces $s + 1$ sets of orthogonal effects, which we write y^*:

$$
\begin{aligned}
Q_E^t \cdot y &= y^* \\
&= (y_1^* y_2^* \cdots y_s^* y_{s+1}^*) \\
&= \mu^* + R_E \cdot \varepsilon_E + Q_E^t \cdot \varepsilon
\end{aligned}
\tag{5.8}
$$

where y_j^* are the effects for error term j, and

$$\mu^* = Q_E^t \cdot \mu$$

Since Q_E is orthogonal, $Q_E^t \cdot \varepsilon$ has the same distribution as ε under the normal error assumptions. The analogy with (5.3) extends to equation (5.5). In particular, the j'th error term contributes nothing to the variance of y_j^*, for $j' < j$. Statistically, this says that the jth term of the error model produces a set of effects, say b_j of them, constituting the components of y affected by the jth error term (but not by previous error terms), and a set of $n - \sum b_j$ components affected only by the `Within` (intra-block) errors. The appropriate analysis then applies the treatment model separately to these sets of effects, producing analyses for each that can be related to the corresponding error estimates.

Computationally, this works out quite simply. The quantity we have written as μ is really the sum of the components from each of the terms in the treatment part of the model (5.1) and can be written in terms of the model matrix (5.2):

$$\mu = X \cdot \beta$$

Therefore

$$\mu^* = X^* \cdot \beta$$

where $X^* = Q_E^t \cdot X$. Equation (5.8) fits y_1^* to the first b_1 rows of X^*, y_2^* to the next b_2 rows, and so on. Each fit may produce estimates for any of the terms in the treatment model, which are to be compared to the relevant error estimates for that fit, assuming there are degrees of freedom left for residuals in that fit. It is these separate fits that correspond to the error *strata*.

To summarize the computations:

1. Compute the orthogonal transformation, Q_E, for the error model.

2. Apply this to the response and to the model matrix for the treatment model, producing y^* and X^*.

3. For each term in the error model, fit the elements of y^* assigned to that term to the corresponding rows of X^*.

4. The object representing the complete analysis of variance is the list of these models.

5.4.4   Computations for Projection

Section 5.4.1 extends the QR calculations for linear models to handle the single-stratum analysis of variance. In essence, we partitioned the response variable y in equation (5.4):

$$y = X\beta + \varepsilon = X_1\beta_1 + X_2\beta_2 + \cdots + X_k\beta_k + \varepsilon$$

The columns of the X_j components are not necessarily orthogonal, neither within nor between X_j. The projections are a reparametrization of the same equations with the substitution $X = Q \cdot R$. In terms of computed quantities:

$$y \approx X\hat{\beta} + e = QR\hat{\beta} + e = Q_1\hat{\gamma}_1 + Q_2\hat{\gamma}_2 + \cdots + Q_k\hat{\gamma}_k + e \qquad (5.9)$$

where $\approx$ means equal up to computational error, the $\hat{\gamma}_j$ are the elements of the product $\hat{\gamma} = R\cdot\hat{\beta}$ and e is the computed residuals. Each term $Q_j\hat{\gamma}_j$ in equation (5.9) is one column of the n by m projection matrix returned by the generic function `proj()`. For ordinary linear models, and so for anova models without an `Error()` model, the result follows directly from the QR method of fitting. If single-degree-of-freedom projections are not wanted, all the columns associated with each term of the model formula are summed into one column. By supplying the argument `onedf=T`, this summation is suppressed. The column of residuals is appended to the projection matrix, so the columns sum to the response.

The multiple-stratum projections are more interesting. They begin with the objects constructed as described in Section 5.4.3. The fit in stratum j b_j, generally much smaller than n. Specifically, the computations solved the linear model:

$$y_j^* = X_j^*\beta_j^* + \varepsilon_j^*$$

where ε_j^* is the part of $Q_E\cdot\varepsilon$ associated with the jth term in the `Error()` model. The projection computations in this case begin by producing the projections of column-length b_j for the ith error term. They then reconstruct the n-row projection matrix for each error term by embedding the length b_j projection into an n-row matrix of zeros and premultiplying by Q_E.

To make this concrete, consider again the `guayule` example. Recall that the `Error()` model forms a stratum from the `flats` factor, specifically,

```
plants ~ variety * treatment + Error(flats)
```

The `assign` component of `gaov.qr` assigns 1 degree of freedom to the `Intercept` stratum, 23 degrees of freedom to the `flats` stratum, and the remaining 72 degrees of freedom to the `Within` stratum. Suppose Q_E is the orthogonal transformation defined by the QR decomposition of the `Error()` model. For any $(n = 96)$-vector x, $Q_E^t \cdot x$ projects the first element onto the `Intercept` stratum, the next 23 to the `flats` stratum, and the remaining 72 to the `Within` stratum. In particular, if M is the treatments model matrix, $Q_E^t \cdot M$ projects each column of M in this way. This is the key step in computing the analysis of variance (and the projections) with multiple error strata.

The anova for the `flats` stratum is computed by fitting elements 2 through 24 of $Q_E^t \cdot y$ to the corresponding rows of $Q^t \cdot M$, and the anova for the `Within` stratum fits elements 25 through 96 to rows 25 through 96. The computational efficiency is realized because each of these steps has computational cost proportional to ν^3, where ν is the number of degrees of freedom in the stratum. For example, $\nu = 23$ in the `flats` stratum. The projection process reverses the subsetting process. The column-length 23 projection in the `flats` stratum is embedded into rows 2 through 24 of an otherwise 0-valued 96×23 matrix, which we will call P. The column-length 96 projection returned in the component `gaov.proj$flats` is calculated as $Q_E \cdot P$.

As we noted, there is no need to specify separate treatment models for the different strata; with a balanced design the balance will cause the columns of the transformed model matrix to be zero (within computational accuracy) in strata where the corresponding terms drop out of the analysis. For example, the `flats` in this experiment are orthogonal to the `treatment` factor, so that the corresponding `treatment` columns of $Q_E^t \cdot M$ will be zero in rows 2 through 24.

The computations are designed to be general. An error structure with more than two strata would correspond to additional factors in the `Error()` model. Also, there is no computational requirement for balance, either in the error strata or between the error factors and the treatment factors. Of course, statistical use and interpretation of analysis of variance models fitted to unbalanced data are adversely affected and need to be much more carefully stated.

Generalized Yates Algorithm for Direct Projection

We derive here an algorithm for fitting the models of this chapter from projections, without explicitly using linear model calculations. The function `aov.genyates()` implements this algorithm. The relevance is that for some large, balanced designs the calculations may be much more efficient than those based on linear models. In addition, the method illustrates a more general notion; namely, that our approach to these models can be adapted to any algorithm capable of producing the essential information used for computing summaries and derived quantities. While we do not discuss them, methods exist that solve a wider range of models without using linear model calculations. These too could, in principle, be used to generate `aov` objects.

The generalized Yates algorithm is based on knowing that the columns of the X matrix in the regression setting are orthogonal. Hence the fundamental mathematical specification for the projections,

$$\hat{\beta} = (X^t X)^{-1} X^t Y$$

is numerically simplified because $X^t X$ is diagonal. By starting with an orthogonal X matrix we are able to avoid the cross-product and inversion in the cross-product algorithm, or the orthogonalization process in the QR algorithm:

```
xx <- apply(X*X,2,sum);
xy <- crossprod(X,Y)
coef <- xy/xx
```

From this we can calculate the projections directly:

```
# element by element multiplication
proj <- X * matrix(coef,nrow(X),ncol(X),byrow=T)
Yhat <- apply(proj,1,sum)
# alternatively:
Yhat <- X %*% coef
residual <- Y - Yhat
```

We collect in function aov.genyates() the direct calculation of the single-degree-of-freedom projection information from the model formula and the data frame. The result is an aov object with an added proj component.

This process gives us projections onto single-degree-of-freedom contrasts that correspond to the columns of the specific contrast matrix used in the construction of the model matrix. Often we care about the projection into the subspace, but not onto the arbitrary individual degrees of freedom. We therefore collect them by summation of all the columns of the projection that have been assigned (with the assign attribute) to the same model term. Since the columns are orthogonal, their sums by the assign groups retain the orthogonality.

Generalized Yates Algorithm with Multiple Error Strata

The error strata correspond to a model matrix similar to that used for treatments. The calculations for the multiple-strata designs are also similar to those for the simpler one-stratum designs. We describe them in terms of the direct projection algorithm.

First, the response variable Y is partitioned via aov.genyates() into a set of projections onto the terms of the Error() model formula. For the gun example, there are the "(Intercept)", flats, and Within terms. Each carries the entire sum of squares and degrees of freedom of what will become its stratum.

Second, each column of the projection onto the error space and its residual is partitioned, by recursive use of the aov.genyates() function, into columns corresponding to the terms of the treatment model. Treatment terms whose model matrix columns are orthogonal to the columns of the error model matrix that define a stratum have projections with 0 sum-of-squares and 0 degrees of freedom. Should a stratum be entirely orthogonal to the treatment model, the entire sum-of-squares for the stratum appears in the stratum residual. To avoid redundancy in terminology, the stratum to which the residuals from the fit to the Error() model are assigned is called the Within stratum.

The direct projection algorithm is efficient but not immediately generalizable to situations where the error and treatment models do not define orthogonal model matrices. Common designs such as partially confounded designs are therefore excluded. In addition, designs for which there are missing values on some observations are excluded.

We can partially compensate for the lack of generality by including an optional argument in the multiple-stratum function. When it is known that the error and treatment model matrices are not orthogonal, setting the argument causes the treatment model matrix to be projected onto the space of the error model matrix before the recursive anova of the current partition of the response variable.

Bibliographic Notes

There are many good introductions to the analysis of variance. Box, Hunter, and Hunter (1978) is widely used in industrial applications. Searle (1971) gives a somewhat less applied approach, connecting the methods to linear models as we have done here. Computational methods for the analysis of variance are presented in detail in Heiberger (1989).

Chapter 6

Generalized Linear Models

Trevor J. Hastie
Daryl Pregibon

Linear models and analysis of variance are popular. Many phenomena behave linearly (at least over small ranges) and have errors that are Gaussian. Often even nonlinear phenomena can be modeled linearly by transforming or "bending" the response. This simplifies the computations, but it can lead to interpreting the model in unnatural scales.

This chapter is devoted to a class of models that is as tractable as classical linear models but does not force the data into unnatural scales. Instead, separate functions are introduced to allow for nonlinearity and heterogeneous variances. Generalized linear models are closer to a reparametrization of the model than to a reexpression of the response.

Take, for example, binary response data, where the outcome variable takes one of two values, say "success" or "failure." If the response is coded 1 for success and 0 for failure, the mean is the probability of success and is a natural candidate for modeling. We might want to investigate the effect of the predictor variables on this probability. Is the linear regression model still appropriate for binary response data? Probably not, unless we are willing to constrain the fitted values to be in $[0, 1]$. Furthermore, the variance of a binary response depends on the mean, which we are modeling as a function of the predictors, so we should account for that as well. The *logistic regression model*, described in this chapter, is a generalized linear model (GLM) that is specially designed for modeling binary and more generally binomial data.

The family of generalizations includes log-linear models for contingency tables and count data, models for multinomial responses, gamma models for positive data with long-tailed error distributions, and many more. These models share several features:

- They can be described concisely in terms of a *link function*, which describes the relationship between the mean and the *linear predictor*, and a *variance function*, which relates the variance to the mean.

- They can be fitted by iteratively-reweighted least squares. Apart from the computational convenience, the accompanying quadratic approximation to the log-likelihood makes for simple approximate inferences.

- The fitted nonlinear models can be summarized by statistics, tables, and plots that are all natural generalizations of their linear model counterparts.

S is a natural environment for exploring generalized linear models and for creating suitable data structures for their representation and fitting. For example, the link and variance functions are packaged up in the *family* argument to the `glm()` function, and are themselves S functions; typically they are very simple, and as such they can be easily modified and new link and variance functions can be created.

6.1   Statistical Methods

The classical linear model

$$y = \beta^T x + \varepsilon \tag{6.1}$$

postulates that ε is (normally) distributed with zero mean and constant variance. This model serves a large variety of data situations very nicely, and it has seen a lot of use. In some situations, (6.1) is not appropriate for one or more reasons:

- If y assumes values over a limited range, the model $E(y) = \mu = \beta^t x$ for the mean does not incorporate this restriction.

- For many types of data a change in the mean of y is accompanied by a change in its variance.

For example, binary response data have their mean $\mu \in [0, 1]$ and a variance $\mu(1-\mu)$ that changes with the mean. The problem seems to be that a mean linear in the parameters/predictors is not restricted to $[0, 1]$ without additional assumptions, and that a constant variance is not always realistic.

Generalized linear models deal with these issues in a natural way by using reparametrization to induce linearity and by allowing a nonconstant variance to be directly incorporated into the analysis. Specifically, a generalized linear model requires two functions:

- a *link* function that describes how the mean depends on linear predictors, $g(\mu) = \boldsymbol{\beta}^{\mathrm{T}}\boldsymbol{x}$, and

- a *variance* function that captures how the variance of y depends upon the mean, $\mathrm{var}(y) = \phi V(\mu)$, with ϕ constant.

Link functions are monotone increasing, and hence invertible; the inverse link $f = g^{-1}$ is an equivalent and often a more convenient function for relating μ to the predictors:

$$\mu = f(\boldsymbol{\beta}^{\mathrm{T}}\boldsymbol{x}).$$

For convenience, it is customary to denote the linear predictor by $\eta = \boldsymbol{\beta}^{\mathrm{T}}\boldsymbol{x}$. For example if the response is binary, the so-called *logit* link

$$\eta = \log\left(\frac{\mu}{1-\mu}\right)$$

or

$$\mu = \frac{e^{\eta}}{1+e^{\eta}}$$

guarantees that μ is in the interval $[0, 1]$, which is appropriate since μ is a proportion in this case. The logit link, together with the binomial variance function $V(\mu) = \mu(1 - \mu)$ defines the popular logistic regression model.

Although binary data problems are an important application, the following table summarizes other commonly used generalized linear models, along with their default link and variance functions:

Distribution	Link Function	Variance Function
Gaussian	μ	1
Bernoulli	$\log\{\mu/(1-\mu)\}$	$\mu(1-\mu)$
Binomial	$\log\{\mu/(1-\mu)\}$	$\mu(1-\mu)/n$
Poisson	$\log(\mu)$	μ
Gamma	$1/\mu$	μ^2
Inverse Gaussian	$1/\mu^2$	μ^3
Quasi	$g(\mu)$	$V(\mu)$

Apart from the last entry, all the distributions in the table belong to the one parameter *exponential family* of distributions. The last entry in the table refers to the *quasi-likelihood* model. While all the other entries are generated by a specific distribution or likelihood, this need not be the case for quasi-likelihood models; they are specified entirely by the mean and variance functions. A thorough account of quasi-likelihood inference would be out of place here; McCullagh and Nelder (1989) is a good reference for all the material in the chapter.

Generalized linear models are an alternative to response transformation models of the form

$$\psi(y) = \boldsymbol{\beta}^{\mathrm{T}} \boldsymbol{x} + \varepsilon \qquad (6.2)$$

which are also used for enhancing linearity and homogeneity of variance. In fact, certain choices of g and V above lead to analyses very similar to the class of response-variable reexpression models, but in fact they are more general due to their flexibility in allowing separate functions to specify linearity and variance relationships.

Reexpressions, although very useful at times, suffer from several defects:

- Familiarity of the measured response variable is sacrificed in the analysis of $\psi(y)$.

- A single reexpression $\psi(y)$ must simultaneously enhance both linearity and homogeneity of variance.

- Often the preferred transformations are not defined on the boundaries of the sample space; e.g., the *logit* transformation is not defined for observed proportions exactly equal to zero or one.

Generalized linear models finesse both these problems.

The next question is how to estimate the model. We use *maximum-likelihood estimation.* For the class of response models we consider here, maximum-likelihood estimation has a particularly convenient structure. An iteratively reweighted least-squares (IRLS) algorithm is used to compute the model parameter estimates, and weighted least-squares plays a central role in the asymptotic inference. We give an outline of the methodology in Section 6.4.1. Readers not familiar with this area might even read that section first, because we draw on the concepts throughout the chapter.

Another popular arena for GLMs is the analysis of cross-classified count data, or contingency tables. The margins of a table are indexed by factors, and the cell counts are very similar to the response in a balanced, multiway, designed experiment. For example, an entry in a three-way table n_{ijk} is the number of individuals at level i of factor I, level j of factor J and level k of factor K. The most popular models are linear in the logarithm of the expected cell count $\mu_{ijk} = En_{ijk}$. For example, the main-effects model has the form

$$\log \mu_{ijk} = \alpha + \beta_i^I + \beta_j^J + \beta_k^K$$

where the superscript refers to the factor and the subscript to the level of the factor. Alternatively, we can write the log-linear model as a multiplicative model for the expected cell count

$$\mu_{ijk} = \mu^0 \mu_i^I \mu_j^J \mu_k^K$$

which is the complete independence model for the table. All the models for various forms of independence, such as conditional independence and marginal independence, have a log-linear representation. Although the multinomial distribution is usually appropriate as the sampling model for the cell proportions, appropriate Poisson models for the cell counts produce identical estimates and inference. Despite the similarity with a fully-balanced designed experiment, the estimated effects are not independent because of the nonlinearity of the mean and the changing error variance. Nevertheless, the analysis is similar, and an *analysis of deviance* table is used to compare nested sequences of models. We return to this member of the GLM family in Section 6.2.5, where we fit a log-linear model to the wave-soldering data of Chapters 3 and 5.

In the next section we describe S functions for fitting GLMs. Since weighted least squares is used iteratively to fit the models, the final fit is also a weighted least-squares fit. Indeed, we have made sure that objects produced by the `glm()` functions inherit as many of the properties of the `lm` and `aov` objects as possible. This means that summaries and diagnostics described in Chapters 4 and 5 can also be used here, typically with minor modifications.

Section 6.3 describes the functions in more detail, and examines more advanced functions for model selection, diagnostics, and creating private families. Section 6.4 gives details on the statistical concepts associated with maximum-likelihood inference, as well as algorithmic details.

6.2   S Functions and Objects

Here we describe some S functions for fitting generalized linear models, and for printing, summarizing, and working with the fitted `glm` objects. Many of the functions here resemble those encountered in Chapter 4; indeed, the `glm` object inherits all the properties of an `lm` object. Readers familiar with the earlier chapters will not have to learn too many new names here; functions such as `summary()`, `plot()`, and so on work as before, with suitably modified effect. We show in the examples how this inheritance can also be exploited to provide additional views of a fitted generalized linear model.

We encourage you to read Chapter 4 on linear models before reading this chapter in order to become acquainted with the basic functions associated with linear models. Although this chapter is self-contained, the pace is quicker than in Chapter 4.

6.2.1   Fitting the Model

A call to the S function `glm()` in its simplest form looks like

```
glm(formula, family)
```

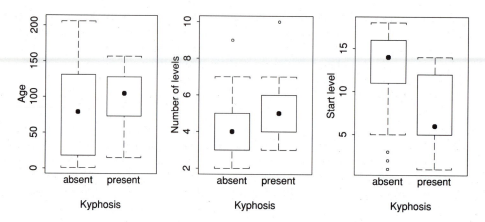

Figure 6.1: *Boxplots of the kyphosis data*

The argument `family` captures all the relevant information about the link and variance functions. Without the family argument, the `glm()` function is equivalent to `lm()`, modulo some additional returned components.

Let us explore `glm()` using some binary data. The data frame `kyphosis` consists of measurements on 81 children following corrective spinal surgery. The first few observations are

```
> kyphosis[1:13,]
   Age Number Kyphosis Start
1   71 3        absent     5
2  158 3        absent    14
3  128 4       present     5
4    2 5        absent     1
5    1 4        absent    15
6    1 2        absent    16
7   61 2        absent    17
8   37 3        absent    16
9  113 2        absent    16
10  59 6       present    12
11  82 5       present    14
12 148 3        absent    16
13  18 5        absent     2
```

where the binary outcome variable `Kyphosis` indicates the presence or absence of a postoperative deformity (called *Kyphosis*). The other three variables are `Age` of the child in months, `Number` of vertebrae involved in the operation, and `Start`, the beginning of the range of vertebrae involved.

```
> class(kyph.glm1)
[1] "glm" "lm"
```

This means that we could have printed `kyph.glm1` using `print.lm()` instead, although the merit of doing this is not obvious at this stage.

There are several functions for extracting single components from the fitted objects, which we list here and explain as we go along:

- `residuals()` or its abbreviation `resid()`: produces residuals, with an argument specifying the `type` of residual;

- `fitted()` or `fitted.values()`: extracts the vector of fitted values;

- `predict()`: has several arguments, and by default extracts the linear predictor vector;

- `coef()` or `coefficients()`: extracts the coefficient vector;

- `deviance()`: extracts the deviance of the fit;

- `effects()`: returns the vector of labeled 1 degree-of-freedom effects;

- `formula()`: extracts the model formula that defines the object;

- `family()`: returns the family object used or implicitly used in producing the object.

All of these functions are generic, which means they should produce sensible results for a number of different classes of models, provided specific methods exist.

The details of a `glm` object are described in the documentation section of the appendix, under `glm.object`, and can also be obtained online using the expression `help(glm.object)`. Among the components are both

```
fitted.values
linear.predictors.
```

The former is on the scale of the mean, while the latter is the linear parametrization obtained from the fitted values via the link function. A simple way to extract these is to use the functions `fitted()` and `predict()`. For example, the first three values of the linear predictor for `kyph.glm1` are

```
> predict(kyph.glm1)[1:3]
      1       2        3
 -1.061  -1.969  -0.02822
```

while the first three fitted values are

```
> fitted(kyph.glm1)[1:3]
      1      2      3
 0.2571 0.1225 0.4929
```

which lie in $[0, 1]$ as expected. These are obtained in this case by applying the inverse logit transformation

```
> TT <- exp(predict(kyph.glm1)[1:3])
> TT/(1+TT)
      1      2      3
 0.2571 0.1225 0.4929
```

The summary() function gives a more detailed description of the fitted model:

```
> summary(kyph.glm1)

Call: glm(formula = Kyphosis ~ Age + Start + Number,
        family = binomial, data = kyphosis)
Deviance Residuals:
    Min     1Q  Median      3Q    Max
 -2.398 -0.5702 -0.3726 -0.1697 2.197

Coefficients:
               Value Std. Error t value
(Intercept) -2.03491   1.432240  -1.421
        Age  0.01092   0.006353   1.719
      Start -0.20642   0.067061  -3.078
     Number  0.41027   0.221579   1.852

(Dispersion Parameter for Binomial family taken to be 1 )

    Null Deviance: 83.23 on 80 degrees of freedom

Residual Deviance: 61.38 on 77 degrees of freedom

Number of Fisher Scoring Iterations: 4

Correlation of Coefficients:
        (Intercept)      Age    Start
    Age -0.4552
  Start -0.3949      -0.2755
 Number -0.8456       0.2205   0.1236
```

Residuals for GLMs can be defined in several different ways. The summary() method produces *deviance residuals*, and prints a five-number summary of them. These are different from the residuals component of a glm object, which are the so-called *working* residuals from the final IRLS fit. The `residuals()` method has a `type=` argument, with four choices:

- `"deviance"` (the default): Deviance residuals are defined

$$r_i^D = \text{sign}(y_i - \hat{\mu}_i)\sqrt{d_i} \tag{6.3}$$

  where d_i is the contribution of the ith observation to the deviance. The deviance itself is then $D = \sum_i (r_i^D)^2$. These are presumably reasonable residuals for use in detecting observations with unduly large influence in the fitting process, since they reflect the same criterion as used in the fitting.

- `"working"`: We name these working residuals because they are the difference between the *working* response and the linear predictor at the final iteration of the IRLS algorithm. They are defined

$$r_i^W = (y_i - \hat{\mu}_i)\frac{\partial \hat{\eta}_i}{\partial \hat{\mu}_i} \tag{6.4}$$

  They are an example of why it is safer to use the *extractor* functions such as `residuals()` rather than accessing the components of a `glm` object directly;

  `residuals(kyph.glm1)`

  would produce the deviance residuals, while

  `kyph.glm1$residuals`

  would give the working residuals. Working residuals are used to construct *partial residual* plots; we give an example in Section 6.2.6.

- `"pearson"`: Pearson residuals are defined by

$$r_i^P = \frac{y_i - \hat{\mu}_i}{\sqrt{V(\hat{\mu}_i)}} \tag{6.5}$$

  and their sum-of-squares

$$X^2 = \sum_{i=1}^n \frac{(y_i - \hat{\mu}_i)^2}{V(\hat{\mu}_i)}$$

  is the chi-squared statistic. Pearson residuals are a rescaled version of the working residuals, when proper account is taken of the associated weights: $r_i^P = \sqrt{w_i}\,r_i^W$.

- `"response"`: These are simply $y_i - \hat{\mu}_i$.

For Gaussian models these definitions coincide.

The `summary()` function is generic; the method `summary.glm()` is used for `glm` objects. One can assign the result of `summary()` rather than printing it, in which case it produces a `summary.glm` object. The summary object includes the entire deviance residual vector and the asymptotic covariance matrix, which are often useful for further analysis.

Along with the estimated coefficients, the `summary()` method produces standard-error estimates and t values. The standard errors are square roots of the diagonal elements of the asymptotic covariance matrix of the coefficient estimates. More simply stated, they are the standard errors appropriate for weighted least-squares estimates, if the weights are inversely proportional to the variance of the observations. We discuss the quadratic approximation that leads to these estimates in Section 6.4. The t values are the estimated coefficients divided by their asymptotic standard errors and can be used to test whether the coefficients are zero.

The printed summary reports the value for the *dispersion* parameter ϕ. For the binomial and Poisson families, the dispersion parameter is identically 1, and is not estimated in these cases. For other families, such as the Gamma or Gaussian, ϕ is estimated by X^2/ν, the Pearson chi-squared statistic scaled by the residual degrees of freedom. Of course, for the Gaussian-error model, this is the usual procedure. The dispersion parameter is used in the computation of the reported standard errors and t values for the individual coefficients. These defaults can be explicitly overridden by specifying the value for the dispersion parameter using the `dispersion=` argument to the `summary()` method; `dispersion=0` will result in the Pearson estimate, irrespective of the family of the object.

6.2.2   Specifying the Link and Variance Functions

The `family` argument is the main difference between calls to `glm()` and calls to `lm()`. The family argument expects a family object, which is a list of functions and expressions that are needed to define the IRLS algorithm and calculate the deviance. In our example above we used `family=binomial`.

Actually, `binomial()` is itself not a family object but a family *generator* function that evaluates to a family object. It can have arguments of its own, as we see later.

Let's explore a binomial family object:

```
> fam1 <- binomial()
> names(fam1)
[1] "family"     "names"      "link"       "inverse"     "deriv"
[6] "initialize" "variance"   "deviance"   "weight"
```

The `family` component is simply a vector of names used for printing, and is used by the `print()` method for families

```
> fam1
Binomial family

     link function:    Logit: log(mu/(1 - mu))
   variance function:  Binomial: mu(1-mu)
```

Three components of `fam1` are S functions for computing the link function, its inverse, and its derivative:

```
$link:
function(mu)
log(mu/(1 - mu))

$inverse:
function(eta)
1 / (1+exp(-eta))

$deriv:
function(mu)
1/(mu * (1 - mu))
```

Similarly, the remaining components are also S functions and expressions that enable the IRLS algorithm to fit the logistic regression model.

For the binomial model, link functions other than the *logit* are also possible. For bioassay problems, the *probit* link, defined by $g(\mu) = \Phi^{-1}(\mu)$, where Φ is the Gaussian distribution function, is popular. To invoke the probit link, use

```
binomial(link = probit)
```

or, in a call to `glm()`,

```
glm(formula, binomial(link = probit))
```

Some of the components of the binomial family object are changed:

```
$link:
function(mu)
qnorm(mu)

$inverse:
function(eta)
pnorm(eta)

$deriv:
function(mu)
1/dnorm(qnorm(mu))
```

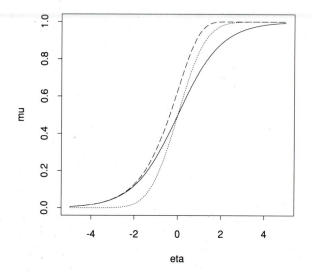

```
> eta <- seq(from = -5, to = 5, length = 200)
> plot(range(eta), c(0,1), xlab = "eta", ylab = "mu", type = "n")
> lines(eta, binomial(link = logit)$inverse(eta))
> lines(eta, binomial(link = probit)$inverse(eta), lty = 2)
> lines(eta, binomial(link = cloglog)$inverse(eta), lty = 4)
```

Figure 6.2: *A plot of three commonly used (inverse) link functions for the binomial family. The functions are the* logit *(solid line), the* probit *(dotted line) and the* complementary log-log *(broken line).*

Figure 6.2 displays the inverse *logit, probit,* and *complementary log-log* link functions, three commonly used links for binomial data. The expressions used to create the plot are also displayed, if only to convince the reader that these components of the family objects are legitimate functions.

Other family generator functions are gaussian() (the default), poisson(), Gamma(), inverse.gaussian(), and quasi() (we use uppercase Gamma to distinguish this family function from the S probability-distribution function gamma()). All but the last are special families implied by an error model with the same name. Since the error model determines the variance function, these generator functions do not have a variance= argument. The quasi() function has both a link= and variance= argument, and is used for constructing arbitrary link/variance combinations. In

Section 6.3.3, we give some more details on families. We show how to construct special-purpose families and modify existing ones. A great deal of flexibility can be achieved through the family argument to `glm()`. Two examples described later are

- the `power()` link function, which allows the link to be parametrized;

- the function `robust()`, which converts a family into a *robust* version of itself.

6.2.3 Updating Models

We seldom know in advance what model will be appropriate, so we typically fit a number of different models and explore various combinations of predictors. Usually, the new model will differ from the preceding one in a simple way:

- one additional predictor is included in the model formula;

- the response in the model formula is transformed;

- a subset of the data is used in the fit;

- a slightly different family is used in fitting the model, and often all the other arguments to the original call are held fixed.

It is convenient in situations such as these to have a function for updating a model.

Suppose we want to drop the term `Age` from our fitted model `kyph.glm1`. The call

```
kyph.glm2 <- update(kyph.glm1, ~ . - Age)
```

results in

```
> kyph.glm2
Call:
glm(formula = Kyphosis ~ Start + Number, family = binomial,
        data = kyphosis)

Coefficients:
 (Int.)    Start Number
 -1.029 -0.1849 0.3574

Degrees of freedom: 81 total; 78 residual
Residual Deviance: 64.54
```

Apparently `Age` is not a very important predictor (if modeled linearly), since the deviance increased by only 3.16 when `Age` was dropped.

The first argument to `update()` is a model object, and the second an updating formula. A "." (on either side of ~) is replaced by the corresponding left or right formula of the model object. In this case, the response is the same, and the linear predictor has the term `Age` removed. For convenience, the "." on the left of $\sim$ can be omitted. Any additional named arguments to `update()` are used to replace the corresponding arguments of the object being updated. For example,

```
update(kyph.glm1, subset = -79)
```

simply augments the existing call with a subset argument to repeat the fit, with observation 79 deleted.

Arguments can also be *removed* from the call. For example, suppose `kyph.subset` was created using a `subset=-79` argument; then

```
update(kyph.glm1, subset =)
```

will remove the subset argument, and fit the same model to all the data.

Updating is also described in some detail in Section 4.2.5, and is used several times in Chapter 1. Although we emphasize its simple uses, some more exotic applications include:

- The `class=` argument can be used to change the fitting mechanism from, say, `glm()` to `lm()`, or even to `tree()` or `loess()`.

- By supplying a `data=` argument to `update()`, we can refit the model to an entirely new dataset (as long as the variables named in the formula and other arguments in the call are to be found in the new data).

- Using `evaluate=F` causes update to return the call corresponding to the new model, without actually evaluating (fitting) the model. This can be useful if the original attempt at updating an object caused an error of some kind.

6.2.4   Analysis of Deviance Tables

More often than not, we fit more than one `glm` model. It is convenient to summarize a series of fitted models in an *analysis of deviance* table. An analysis of deviance table is simply the analogue of an analysis of variance table, such as that produced by `summary.aov()`.

There are many ways to arrange a series of models into such a table, and more importantly, there are many ways of *generating* an appropriate series of models. In this and subsequent sections we describe several functions that produce anova tables, or more precisely objects of class `"anova"`:

- `anova(...)` takes an arbitrary number of fitted models as arguments, and makes sequential pairwise comparisons in the order the fitted models are listed.

- `anova(object)`: given a single object, `anova()` fits a sequence of models by successively dropping each of the terms (from last to first), and produces a table to summarize the changes.

- `drop1()` and `add1()` produce tables by making a series of single term deletions or additions. These functions are also encountered in Chapters 4, 5 and 7, and we describe them in some detail in the next section. *pp. 235-236   also pp. 125-*

- `step()` is a stepwise model selection function that builds a model by sequentially adding or dropping terms (Sections 6.3.5 and 7.3.1).

Let's look at a simple example of the use of the `anova()` function:

```
> anova(kyph.glm1, kyph.glm2, kyph.glm3)
Analysis of Deviance Table

Response: Kyphosis
```

	Terms	Resid. Df	Resid. Dev	Test	Df	Deviance
1	Age + Start + Number	77	61.38			
2	Start + Number	78	64.54	-Age	-1	-3.157
3	Start	79	68.07	-Number	-1	-3.536

If adjacent models are nested with respect to the terms in the linear predictor, as is the case here, then the terms comprising the difference are named in the `Test` column. Either way, the column labeled `Deviance` reports the difference in deviances between each model and the one above it, and `Df` is the difference in degrees of freedom.

The output of `anova()` is an S object with class `"anova"` that inherits from the class `"data.frame"`. There are several advantages to arranging the output in this form. As a data frame, the numbers in the individual columns can be accessed, and columns can be subscripted out. For example, for large models the `Terms` column can be rather wide, so we might choose to omit it when printing the table. The table could then be printed in pieces, with the row numbers used for cross-referencing.

```
> kyph.anodev <- anova(kyph.glm1, kyph.glm2, kyph.glm3)
> kyph.anodev[,-1]
Analysis of Deviance Table

Response: Kyphosis
```

	Resid. Df	Resid. Dev	Test	Df	Deviance
1	77	61.38			
2	78	64.54	-Age	-1	-3.157
3	79	68.07	-Number	-1	-3.536

Similarly, if we want to use the numbers in the "Deviance" column, we can simply extract them:

```
> kyph.anodev$Deviance
[1]      NA -3.157 -3.536
```

When given a single model as an argument, anova() behaves a bit differently:

```
> anova(kyph.glm1)
Analysis of Deviance Table

Binomial model

Response: Kyphosis

Terms added sequentially (first to last)
        Df Deviance Resid. Df Resid. Dev
  NULL                      80        83.23
   Age  1      1.30         79        81.93
 Start  1     16.63         78        65.30
Number  1      3.92         77        61.38
```

Notice the table header is different, as are the row labels and columns in the table. The table reports the effect of sequentially including each of the terms in the original model, starting from the NULL model. The NULL model is a constant, and is the mean of the response if an intercept is present in the model, as is the case here; if there is no intercept, the NULL model has a linear predictor that is all zeros. The same table is obtained if we drop terms sequentially from the full model, from right to left in the formula. The formula in our example is rather simple; for more complicated formulas with interaction terms, the formula is first expanded and then terms are dropped while honoring the model's hierarchy. This version of anova() mimics the table produced by summary.aov() when applied to an aov or lm object. In fact, the anova() method for these two classes is summary.aov(). The distinction is worth noting, however. For a balanced aov model, the table is the same no matter what the order of the terms. For an unbalanced aov or lm model, the order is relevant, as it is for glm models. For large models, anova.glm(object) might take a while to compute, since it has to fit each of the submodels of object iteratively; for lm models no refitting is needed.

It is apparent that the contents of an anova object are rather general; any data frame can be transformed into one by attaching a character vector header attribute. There is a print method for anova objects, and all it does is print the header and hand it over to the printer for data frames. Typically each row corresponds to a term in a model, and there will be columns labeled "Deviance" or "Sum of Sq" and "Df".

The `anova()` methods all have a `test=` argument. The default is `"none"` for `anova.glm()`, and other choices are `"Chisq"`, `"F"`, and `"Cp"`. For a binomial model, the changes in deviances between nested models are typically treated as chi-squared variables, so `test="Chi"` is appropriate here (notice abbreviations are allowed):

```
> anova(kyph.glm1, test = "Chi")
Analysis of Deviance Table

Binomial model

Response: Kyphosis

Terms added sequentially (first to last
        Df Deviance Resid. Df Resid. Dev Pr(Chi)
  NULL                        80      83.23
   Age  1     1.30            79      81.93  0.2539
 Start  1    16.63            78      65.30  0.0000
Number  1     3.92            77      61.38  0.0477
```

The additional column labeled "Pr(Chi)" gives the tail probability (p-value) of the chi-squared distribution corresponding to the values in the "Df" and "Deviance" columns. The `test="F"` option is suitable for Gaussian GLMs, Gamma models with a dispersion parameter, and perhaps for overdispersed binomial and Poisson models. The choice `test="Cp"` is discussed in Section 6.3.5.

One can directly augment an `anova` object with one or more *test* columns using the function `stat.anova()`. For example, the sequence

```
> anova1 <- anova(kyph.glm1)
> stat.anova(anova1, test = "Chisq")
```

produces the same table as above. This function is useful in situations where the original table is expensive to compute.

6.2.5   Chi-squared Analyses

A `glm` object looks very similar to a fitted `lm` object, as it should. We say that it *inherits* the properties of an `lm` object. The fitting algorithm uses iteratively reweighted least squares, which means that the final iteration is a weighted least-squares fit. This linearization is not only a coincidence of the numerical algorithm, but is the same linear approximation that drives the first-order asymptotic inference for generalized linear models. So, for example, the usual covariance matrix from this linear fit is the same as the inverse of the expected Fisher information matrix for the maximum-likelihood estimates. This is precisely what is used as the asymptotic covariance matrix of the coefficients and is what is usually reported along with the fit. Section 6.4 has more details.

Apart from simplifying the algorithms for fitting these models, this linearization allows us to use many of the tools intended for linear models and models for designed experiments. To illustrate this, we return to the wave-soldering data introduced in Chapter 1 and revisited in subsequent chapters. Recall that in this experiment the response is the number of defects or `skips`, an integer count taking on values $0, 1, \ldots$. In some of the earlier analyses we reexpressed the response using the square-root transformation, and modeled the transformed data using `aov()` as though the errors were Gaussian. An alternate and perhaps more justifiable approach is to model the response directly as a Poisson process. The natural link for the Poisson family is the logarithm of the expected counts. The effect of the log link is similar to that of the square-root transformation, but in addition it guarantees that the fitted means are positive.

The call looks similar to the `aov()` call:

```
paov <- glm(skips ~ ., family = poisson, data = solder.balance)
```

Once again we can print and produce summaries of the fitted object using either the methods appropriate for `glm` objects or else, in this case, those appropriate for `aov` or `lm` objects:

```
> summary(paov)

Call: glm(formula = skips ~ Opening + Solder + Mask + PadType + Panel,
      family = poisson, data = solder.balance)
Deviance Residuals:
    Min    1Q   Median    3Q    Max
 -3.661 -1.089 -0.4411 0.6143 3.946

Coefficients:
                 Value Std. Error t value
(Intercept)  0.735680   0.029481  24.955
 Opening.L  -1.338898   0.037898 -35.329
 Opening.Q   0.561940   0.042005  13.378
    Solder  -0.777627   0.027310 -28.474
    Mask.1   0.214097   0.037719   5.676
    Mask.2   0.329383   0.016528  19.929
    Mask.3   0.330751   0.008946  36.970
 PadType.1   0.055000   0.033193   1.657
 PadType.2   0.105788   0.017333   6.103
 PadType.3  -0.104860   0.015163  -6.916
 PadType.4  -0.122877   0.013605  -9.032
 PadType.5   0.013085   0.008853   1.478
 PadType.6  -0.046620   0.008838  -5.275
 PadType.7  -0.007584   0.006976  -1.087
 PadType.8  -0.135502   0.010598 -12.786
```

```
PadType.9 -0.028288   0.006564  -4.310
  Panel.1  0.166761   0.021028   7.931
  Panel.2  0.029214   0.011744   2.488
```

(Dispersion Parameter for Poisson family taken to be 1)

    Null Deviance: 6856 on 719 degrees of freedom

  Residual Deviance: 1130 on 702 degrees of freedom

  Number of Fisher Scoring Iterations: 4

  Correlation of Coefficients:
```
              (Intercept) Opening.L Opening.Q    Solder  Mask.1
  Opening.L   0.4472
  Opening.Q  -0.1082       0.3844
     Solder   0.3277       0.0000    0.0000
     Mask.1  -0.1350       0.0000    0.0000   0.0000
     Mask.2  -0.3095       0.0000    0.0000   0.0000  0.1605
     Mask.3  -0.4050       0.0000    0.0000   0.0000  0.1483
```

plus many more correlations which we omit. Using `summary.aov()` instead we get

```
> summary.aov(paov)
            Df Sum of Sq Mean Sq F Value     Pr(F)
  Opening    2      2101    1050     706  0.00e+00
   Solder    1       811     811     545  0.00e+00
     Mask    3      1429     476     320  0.00e+00
  PadType    9       473      53      35  0.00e+00
    Panel    2        66      33      22  4.89e-10
Residuals  702      1045       1
```

The default summary method for `glm` objects concentrates on the individual parameters, while `summary.aov()` concentrates on the fitted terms. The F values and their tail probabilities are only valid under some special assumptions, in particular that the Poisson model is over- or underdispersed. In this case, the scaled chi-squared estimate of ϕ is 1.49, which is somewhat larger than 1 and does indicate overdispersion. Of course, the default `summary()` for a Poisson GLM makes the even stronger assumption that $\phi = 1$, which justifies the use of chi-squared rather than F-statistics. In practice, it seems that this often is not the case, the above example being a case in point.

It is interesting to note that the asymptotic correlations (and hence covariances) between the parameter estimates are zero in the same places as they are for the balanced analysis of variance model. This is a special feature of the main-effects Poisson model, a result of the multiplicative structure of the fitted values (they create weights that are similar to those arising from a proportionally balanced design).

The sums of squares in the analysis of variance table are appropriately weighted versions of the usual sums of squares.

If paov is summarized by summary.lm(), the residuals reported are the Pearson residuals

```
    Min      1Q  Median      3Q     Max
 -3.382  -0.852  -0.402   0.664   5.259
```

and what is reported as the residual standard error estimate is $\sqrt{X^2/\nu}$, the square root of the scaled Pearson chi-squared statistic.

The "balance" property referred to above also means that the weighted sums of squares decompose orthogonally. This is not the case in general for other GLMs, where orthogonality is rare. For example, we can use summary.aov() on the kyphosis model

```
> summary.aov(kyph.glm1)
          Df Sum of Sq Mean Sq F Value Pr(F)
     Age   1       0.1     0.1     0.1 0.738
   Start   1      11.1    11.1    12.5 0.001
  Number   1       3.4     3.4     3.9 0.053
Residuals 77      68.5     0.9
```

but the components here are not as meaningful. The entries would differ depending on the order of the variables, which is the case for any "unbalanced design." The last effect, Number in this case, approximates the change in deviance resulting from dropping this variable from the model formula. The approximation in effect takes one step toward the solution, and as such is similar to Rao's score test. The functions add1() and drop1() are also based on score tests, and are described in Section 6.3.5.

6.2.6   Plotting

We have already seen some plots of fitted GLMs in Figure 1.6, where we plot the observed versus fitted values (on the square-root scale) for the wave-soldering data. Many plots are possible for GLMs, so our selection for a plot method is bound to be subjective and not please all. Figure 6.3 shows the result of plot(paov), a plot of our Poisson model. The left panel graphs the response variable against their fitted values, while the right panel graphs the absolute deviance residuals against the linear predictor values. This is similar in spirit to the plot() method for lm objects introduced in Chapter 4. In fact, plot.lm() can be used on glm objects to produce a variation of the two plots in Figure 6.3. In that case, the axis labeled response would actually be the working response, and the fitted values would be the linear predictor.

By using the *extractor* functions fitted(), predict(), and residuals(), we can easily produce a large variety of plots. For the wave-soldering data, we plot *response* residuals against the fitted values in Figure 6.4:

plot (fitted(paov), skips)

abline (0,1)

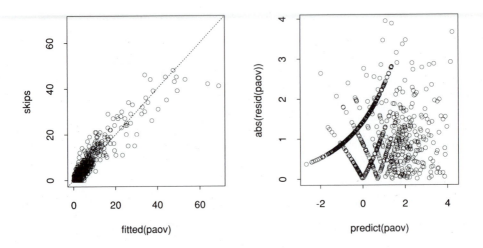

Figure 6.3: *A plot of the* `glm` *object* `paov`. *The left panel shows the response plotted against its fitted values; the broken line is at 45 degrees. The right panel shows the absolute deviance residuals plotted against the linear predictor. The discrete nature of the response introduces strange striations in the residuals.*

```
> attach(solder.balance)
> plot(sqrt(fitted(paov)), skips - fitted(paov))
> abline(h=0)
```

We could have used the `residuals()` function with `type="response"` to extract the residuals, although our usage here produces more informative plot labels. We actually plot against the square root of the fitted values to expand the horizontal scale.

The mean of the residuals is approximately zero, with the most striking feature being the increase of variability with the mean. This is, of course, expected for Poisson data, since the variance is supposed to increase linearly with the mean. The plot on the left uses the main-effects model, and we see some large negative residuals for high values of the fitted values. This indicates lack of fit for some regions of factor space, and we need to fit some interaction terms. The plot on the right is the residual plot for the model

```
paov2 <- glm(formula = skips ~ . + (Opening + Solder + Mask)^2,
        family= poisson, data = solder.balance)
```

which is the same model selected in the analysis in Chapter 1. The large neg-

label obsns on plot

plot(x,y, type="n")

text(x,y, labels=ch)

      character vector (not factor)

      same length as x+y

      contains character to plot at each point

plot(x, fitted(glm), text="n")

text(x, fitted(glm), labels=sex)

     — can do as.character (factor)

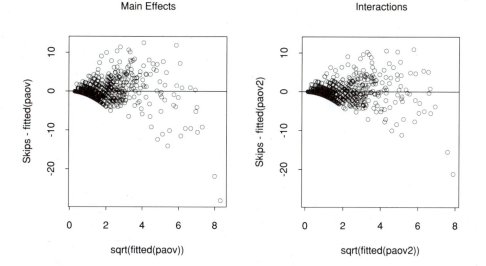

Figure 6.4: *The residuals for two Poisson models fitted to the wave-soldering data, plotted against the square root of the fitted values. The residuals on the left correspond to the main-effects model, while those on the right are from a model that includes some interactions.*

ative residuals are not as pronounced as for the main-effects model. Both plots exhibit rather strange curved bands due to the constrained and discrete nature of the response and fitted values. This behavior is far more extreme for binary data, rendering plots of this kind almost useless in that case.

Other plots such as the half-normal plot of Figure 5.6 can also be generated for a `glm` object, although one should exercise caution in making interpretations not necessarily valid in the nonlinear context.

Partial residual plots are useful for detecting nonlinearities and for identifying the possible cause of unduly large residuals. The partial residuals for variable x_j are defined to be

$$r_i^j = x_{ij}\hat{\beta}_j + (y_i - \mu_i)\partial\eta_i/\partial\mu_i. \tag{6.6}$$

The term to the right of the plus sign in (6.6) is simply the working residual, while the term on the left is the jth fitted term. As mentioned earlier, the working residual is available using the `type="working"` argument in a call to `residuals()`, and is in fact the `"residuals"` component of a fitted `glm` object. The individual fitted terms are available from the `predict()` method, with the argument `type="terms"`. A matrix of values is returned, with a column for every term in the model. We remind the reader that a term is often composite, involving several coefficients. So,

in fact, the expression

```
predict(kyph.glm1, type = "terms") + kyph.glm1$residuals
```

will produce a matrix of partial residuals, one column for each term. The expression takes advantage of the fact that matrix addition *recycles* the values in the vector-valued residuals term (fortunately columnwise).

Curvature in the pattern of these residuals (plotted against x_j) can suggest nonlinear transformations of the variables, which might improve the fit. A smooth curve fitted to these partial residuals enhances this display and allows one to detect the nonlinearities more easily. Figure 6.5 shows such plots for the kyphosis model.

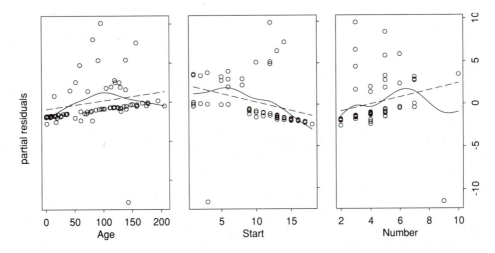

Figure 6.5: *Partial residual plots for the kyphosis data. In each plot the residuals are summarized by a* loess() *smooth curve, which suggests nonlinear transformations. One large negative residual seems to have a strong local effect on the curve for* Number.

After some exploration with these partial residual plots, and experimenting with various nonlinearities, we eventually arrived at the model

```
Kyphosis ~ poly(Age, 2) + I((Start > 12) * (Start - 12))
```

which has a quadratic effect for Age, and a piecewise-linear effect for Start. The functions poly() and I() are both special, and are described in Chapter 2. For convenience, we describe them again here.

- poly() is an S function that takes one or more vector arguments (or a matrix argument), as well as a degree= argument, and generates a basis for polynomial regression. Since this is itself a matrix, it is a valid object for inclusion in a formula. The more common application uses a single, vector argument (or an expression that evaluates to 1), as is the case here, and the resulting columns of the basis are orthonormal. For more than one argument or a matrix argument, the bases are no longer orthonormal. In either case, the polynomials defining the columns have maximum degree given by the `degree=` argument.

- `I()` is the identity function, and protects its argument from the formula parser. Such protection is clearly needed here since its argument uses the "`*`" and "`-`" operators, which have a different interpretation in a formula. In this case, the expression evaluates to the truncated linear function $(\text{Start} - 12)_+$, where the $()_+$ notation refers to the *positive part*.

If the positive part function were to be used frequently, it would make sense to write a special function, say `pos.part()`, for future inclusion in formulas. Any function with arbitrary arguments can be used in a formula, as long as it evaluates to a numeric vector, a matrix, or a factor. The use of compound expressions such as `poly()` and `I()` in formulas is discussed in some detail in Chapter 2, as well as in Chapter 7. We saved the fitted model in `kyph.glm4`:

```
> kyph.glm4
Call:
glm(formula = Kyphosis ~ poly(Age, 2) + I((Start > 12) * (Start - 12)),
        family = binomial, data = kyphosis)

Coefficients:
 (Intercept) poly(Age, 2)1 poly(Age, 2)2 I((Start > 12) * (Start - 12))
   -0.684961       5.77193      -10.3248                        -1.35101

Degrees of Freedom: 81 Total; 77 Residual
Residual Deviance: 51.9533
```

We can plot the fitted terms by separating out the relevant columns and multiplying by their fitted coefficients. If we want to see pointwise standard-error curves, we need to do a similar partitioning of the covariance matrix of the parameter estimates. Both these operations can be performed by the `predict()` method; the expression

```
predict(paov, type = "terms", se = T)
```

will return a list with two components, `"fit"` and `"se.fit"`. Both of these will be matrices, the former having as columns the fitted terms, the latter, the pointwise standard errors for each term. Section 6.3.6 has more details.

There is an even easier way out. Chapter 7 focuses on the fitted terms in *additive models*, and the `plot.gam()` method produces exactly the type of plot outlined above. The pair of plots in Figure 6.6 were created by the call

```
plot.gam(kyph.glm4, se = T, residuals = T)
```

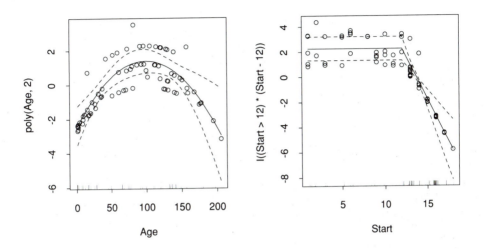

Figure 6.6: *The fitted polynomial and step function for* `kyph.glm4`, *constructed using* `plot.gam()`. *The broken lines are pointwise 2 × standard-error curves, and the points are the partial residuals.*

It can also be useful to plot factor terms and even linear terms in a similar fashion, especially if the factors are ordered. Figure 6.7 displays such a plot, created by the expressions

```
> preplot.paov <- preplot.gam(paov, terms = c("Mask", "PadType"))
> plot(preplot.paov, se = T, rug = F, scale = 2.1)
```

If all the plots are on the same scale (achieved by using the scale argument), the fitted effects or slopes can be easily compared visually. Detailed discussion of `plot.gam()` and `preplot.gam()` is given in Section 7.3.5.

6.3 Specializing and Extending the Computations

6.3.1 Other Arguments to `glm()`

In addition to the `formula` and `data` argument, `glm()` shares other arguments with `lm()` and `aov()`. These include the `subset=` and `weights=` arguments. The latter allows one to specify prior weights for the observations. A common situation requiring prior weights is when the responses are themselves averages over homogeneous groups of independent responses. The weights would then be the number

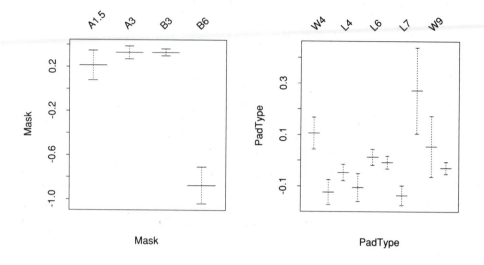

Figure 6.7: *The fitted effects for two of the factors in the main-effects model for the wave-soldering data. The broken bars indicate two standard errors.*

in each average. Binomial proportions have this form, in which case the n_i or numbers of trials corresponding to each proportion can be passed as weights. Binomial responses can also be presented as a two-column matrix; we discuss this in Section 6.3.3.

A commonly used device in GLM models is the *offset*, a component of the linear predictor that is known and requires no coefficient. An offset is redundant for standard Gaussian linear models, since one can simply work with the residuals. An offset allows a form of "residual" analysis for GLMs; we can evaluate the contribution of additional terms while holding fixed those already fit. In some stratified sampling situations, offsets are required to correct the sampling imbalance. An offset term is specified directly in the model formula by including it as the argument to the `offset()` function, as in

```
y ~ dose + age + offset(prior.fit)
```

There will be no coefficient fit for the offset term; it is added as is into the linear predictor.

The `start=` argument allows initial values for the linear predictor different from the default given in the `"initialize"` component of the `family` object. The `control=` argument sets algorithmic constants, and expects a named list. The default values are given by the expression

```
> glm.control()
$epsilon:
[1] 0.0001

$maxit:
[1] 10

$trace:
[1] F
```

and `glm.control()` can be used to adjust any of the values as well:

- `epsilon=` gives the convergence threshold, described Section 4.

- `maxit=` sets the maximum number of IRLS iterations.

- If `trace=T`, then iteration information is printed during the execution of `glm()`.

Any of the three arguments to `glm.control()` can optionally be given directly as named arguments in the call to `glm()`.

We list on page 202 some shortcuts in constructing the call to `glm()`. The main candidate for shortcuts is the `formula=` argument. The function `formula()`, which has methods for a variety of object classes, is used inside `glm()` (and other modeling functions) to extract a formula from the given argument, in the event that it is not an explicit formula. In particular, there is a `formula()` method for data frames, and a slightly different method for design matrices (as described in Chapter 5). The expression `formula(data.frame)` assumes the first column of `data.frame` is the response, the remaining columns are predictors, and constructs a simple linear model formula. A consequence of this is that `glm(data.frame)` will fit a Gaussian GLM (and implicitly use the argument `data=data.frame` as well). The expression `formula(design.matrix)`, on the other hand, uses the first numeric (nonfactor) column as the response, discards all the remaining numeric columns, and constructs a formula additive in all the factors.

6.3.2 Coding Factors for GLMs

In Sections 2.3.2 and 5.3.1, we describe how factors are coded in forming the model matrix. By default, unordered factors are coded using Helmert contrasts, and ordered factors by orthogonal polynomials. Both these choices lead to uncorrelated parameter estimates for balanced designs, both within and between factors. A lot of this appeal disappears for GLMs, where balanced designs rarely lead to uncorrelated parameter estimates, except in special cases. The easiest way to see this is through the information matrix

$$X^T W X$$

which will typically be full due to the presence of the diagonal weight matrix $\boldsymbol{W}$.

Two other popular coding schemes are described in Section 5.3.1, each accompanied by a contrast function to generate them. Suppose a factor f has k levels:

- contr.sum() produces $k - 1$ contrasts, which compare each level of the factor to the last level. This is equivalent to constraining the k original coefficients corresponding to each level of the factor to sum to zero.

- contr.treatment() simply produces the $k - 1$ dummy variables corresponding to all but the first level of the factor. GLMs are frequently used for analyzing medical data, where it is common to compare a number of new procedures, drugs, or, more generally, treatments. Usually there is a control treatment that can be considered the baseline for comparison, which we assume to be at the first level of the factor. Each coefficient then measures the difference between a treatment and the control.

One can explicitly attach a contrast attribute to a factor using the C() function, if a one-time special coding is desired. For example,

```
C(f, treatment)
```

creates a new version of f, with the appropriate contrast matrix attached. A more convenient and permanent approach is to reset the default using the options() function; for example,

```
options(contrasts = c("contr.treatment", "contr.poly"))
```

makes contr.treatment() instead of contr.helmert() the default for factors. The effect remains for the duration of the session.

Let's look at a simple example using contr.treatment() as the default:

```
> f <- factor(rep(1:3, 3))
> x <- -4:4
> model.matrix(~ f * x)
   (In) f.2 f.3  x  f:x.2  f:x.3
1    1   0   0  -4     0      0
2    1   1   0  -3    -3      0
3    1   0   1  -2     0     -2
4    1   0   0  -1     0      0
5    1   1   0   0     0      0
6    1   0   1   1     0      1
7    1   0   0   2     0      0
8    1   1   0   3     3      0
9    1   0   1   4     0      4
```

Using contr.helmert() as the default we get:

use model.matrix(data.frame) to get codings used

name of

```
      (In) f.1 f.2  x f:x.1 f:x.2
  1     1  -1  -1 -4    4     4
  2     1   1  -1 -3   -3     3
  3     1   0   2 -2    0    -4
  4     1  -1  -1 -1    1     1
  5     1   1  -1  0    0     0
  6     1   0   2  1    0     2
  7     1  -1  -1  2   -2    -2
  8     1   1  -1  3    3    -3
  9     1   0   2  4    0     8
```

Setting the default to `"contr.sum"`, on the other hand, produces:

```
      (In) f.1 f.2  x f:x.1 f:x.2
  1     1   1   0 -4   -4     0
  2     1   0   1 -3    0    -3
  3     1  -1  -1 -2    2     2
  4     1   1   0 -1   -1     0
  5     1   0   1  0    0     0
  6     1  -1  -1  1   -1    -1
  7     1   1   0  2    2     0
  8     1   0   1  3    0     3
  9     1  -1  -1  4   -4    -4
```

In examples like this, where we have a quantitative predictor as well as a factor, we may well prefer one of the alternative codings. For instance, it is harder to interpret a contrast of slopes than a contrast of simple mean effects.

6.3.3   More on Families

In Section 6.2.2, we introduce the `family=` argument of `glm()`, and looked at some of the components of the binomial family in detail. In this section we explore their flexibility and power in extending the capabilities of the GLM functions.

The `initialize` component of a family object is an expression that sets up internal variables before the iterations begin. At face value it simply assigns initial values for the linear predictor, on which everything else depends. The potential is far greater, however. Since it is simply an expression involving variables local to the main frame of `glm()`, it allows an opportunity to insert additional code in the `glm()` function at the point at which it is evaluated. The binomial family illustrates some of this potential:

```
> binomial()$initialize
expression({
    if(is.matrix(y)) {
        if(dim(y)[2] > 2)
```

```
            stop(
                "only binomial response matrices (2 columns)"
                )
        n <- as.vector(y %*% c(1, 1))
        y <- y[, 1]
    }
    else {
        if(is.category(y))
                y <- y != levels(y)[1]
        else y <- as.vector(y)
        n <- rep(1, length(y))
    }
    y <- y/n
    w <- w * n
    mu <- y + (0.5 - y)/n
}
)
```

In the above example, the local variables that are used in the body of `glm()` are `y`, `w`, and `mu`.

If the response is a matrix with two columns, say m_1 and m_2, they are assumed to be of the form m_{1i} successes and m_{2i} failures. The `initialize` expression converts to $y_i = m_{1i}/(m_{1i} + m_{2i})$ and incorporates the $m_{1i} + m_{2i}$ into the weight vector. If the response is a category, factor or ordered factor, the first level is assigned the value 0, and all other levels are 1.

In general, the `initialize` expression allows an arbitrary amount of user-defined preprocessing of the data. Although expressions such as this are more flexible than functions in that no arguments need to be specified, they are more dangerous since careless use of them can disturb local variables unintentionally.

The `variance` and `deviance` functions form a logical group in that a variance function implies a deviance function [see (6.7)]. Let's look at the Poisson family:

```
> poisson()[c("variance","deviance")]
$variance:
function(mu)
mu

$deviance:
function(mu, y, w, residuals = F)
{
    nz <- y > 0
    devi <-  - (y - mu)
    devi[nz] <- devi[nz] + y[nz] * log(y[nz]/mu[nz])
    if(residuals)
        sign(y - mu) * sqrt(2 * w * devi)
```

```
            else 2 * sum(w * devi)
    }
```

The variance function gives the variance of y_i as a function of the mean μ_i, while the deviance function computes the residual deviance. Notice that the deviance function has a `residuals=` argument that is convenient for computing the deviance residuals; it is mainly there for the `residuals()` function. Although variances and deviances usually arise from a likelihood corresponding to a particular error model, quasi-likelihood models are more general. They allow any variance function $V(\mu)$ to determine a corresponding deviance element:

$$D(y, \mu) = 2 \int_\mu^y \frac{y - u}{V(u)} du. \tag{6.7}$$

There are two auxiliary lists for families: `glm.links` and `glm.variances`. Their separate presence is indicative of the disjoint contributions of link and variance functions to a family. Each of them is stored as a matrix of mode `"list"`, and their `dimnames` are instructive:

```
> dimnames(glm.links)
[[1]]:
[1] "name"       "link"         "inverse"     "deriv"        "initialize"

[[2]]:
[1] "identity" "logit"     "cloglog"  "probit"    "log"         "inverse"
[7] "1/mu^2"   "sqrt"

> dimnames(glm.variances)
[[1]]:
[1] "name"       "variance" "deviance"

[[2]]:
[1] "constant" "mu(1-mu)" "mu"        "mu^2"       "mu^3"
```

We see that each column of `glm.links` is a link subfamily with five elements, and each column of `glm.variances` is a variance subfamily with three elements. The family generator functions, such as `binomial()` and `poisson()`, protect the user against bad choices; for example, only `logit`, `probit`, and `cloglog` are permissible links when constructing a binomial family.

There are several ways to modify the families and construct private ones:

- The `quasi()` function can be used to build a family from the supplied links and variances whose names appear in the two lists above.

- Users can build their own link or variance subfamilies (by mimicking any of the supplied ones). These can then be used to construct a family, either using `quasi()` or the function `make.family()`.

- An entire family object can simply be constructed from scratch.

- Functions such as `robust()` can be used to modify existing families.

The second approach is probably safer than the third. The function `make.family()` (type `?make.family` to see detailed documentation) forces the user to include the appropriate components together with suitable naming information, and builds an object with class `"family"`. This new object would then be passed via the `family=` argument to `glm()`. Of course, it is the responsibility of the creator to ensure that all the components behave as they should.

As a start to creating a family object, it might seem reasonable to simply type `poisson` to see an example; the Poisson family generator function would be printed out! Typing `poisson()` is not much better; this evaluates to the `poisson` family, but there is a `print()` method for families that simply prints the names of the family, link, and variance (we give an example on page 206). This feature was built in partly to protect the user from unwittingly having to see all the family function definitions. One can avoid the `print()` method for `family` objects (and more generally for objects of any class), by using either of the following expressions:

```
print.default(binomial())
unclass(binomial())
```

The link and variance subfamilies do not have `print()` methods that need to be side-stepped, so

```
glm.links[,"logit"]
```

would extract the `logit` link subfamily and print out the list of functions.

A slightly more advanced modification in family construction is to parametrize the link and/or the variance subfamilies. We provide a function `power(lambda)` that creates a link subfamily with components `link`, `inverse`, `deriv`, and `initialize`, each of which depends on the value of `lambda`. A call to `glm()` might have the form

```
> glm(formula, family = quasi(link = power(0.5)))
```

where the first few components of `quasi(link = power(0.5))` are

```
> quasi(link = power(0.5))[c("link","inverse","deriv")]
$link:
function(mu)
mu^0.5

$inverse:
function(eta)
eta^(1/0.5)
```

```
$deriv:
function(mu)
0.5 * mu^(0.5 - 1)
```

A much more advanced modification of a family is the `robust()` *wrapper* function. It takes a family as an argument, and returns a new family that has been robustified. Its definition begins

```
robust <-function(family, scale, k = 1.345, maxit = 10){
```

This allows for a rather elegant modification to the call to `glm()`, which now may have the form

```
> glm(formula, family = robust(binomial, scale = 1))
```

Instead of minimizing the usual sum of deviance contributions

$$D = \sum_{i=1}^{n} D(y_i, \mu_i),$$

the robust minimum-deviance estimate uses the tapered criterion

$$D_\omega = \sum_{i=1}^{n} \phi \omega_k \left(\frac{D(y_i, \mu_i)}{\phi} \right)$$

where ω_k dampens contributions larger than k^2

$$\omega_k(t) = \begin{cases} t, & \text{for } t \le k^2; \\ 2k\sqrt{t} - k^2 & \text{for } t > k^2. \end{cases}$$

The dispersion parameter ϕ is either supplied or else estimated if it is missing.

Two aspects of the IRLS iterations need to be modified to handle robust families:

- The iterative weights get multiplied by an additional weight factor w_r, which is 1 for suitably small deviance contributions, and decreasingly small for increasingly large contributions. The parameter k determines the point at which the weights get smaller than 1.

- The deviance function needs to be replaced by a tapered version.

The function `robust()` achieves these changes by modifying the `weight` and `deviance` components of its `family` argument. The details of how it does this are rather technical but interesting, since it involves augmenting the `weight` expression and `deviance` function with additional code, some of which depends on the arguments to `robust()`. We encourage readers to explore the details by reading the code of the `robust()` function.

6.3.4   Diagnostics

In Chapter 4, we describe some tools for generating diagnostic statistics for linear models. The function `lm.influence()` (page 129) takes a linear-model object and generates a list of three basic diagnostic elements that, together with the output of `summary.lm()`, can be used to construct all the currently popular diagnostic measures. The three diagnostics are the delete-one coefficient matrix, the delete-one standard-error estimate, and the diagonals of the hat matrix. The exact computation of a delete-one statistic, say a coefficient, is quite expensive for a nonlinear model. Essentially, each observation has to be removed one at a time and the fitting algorithm iterated until convergence. A cheaper but effective approximation is to remove the observation, and perform only one iteration of the IRLS algorithm toward the new solution, starting from the model fit to the complete data. It turns out that `lm.influence()` achieves exactly that when given a fitted `glm` object as its argument! This pattern is becoming familiar, and is once again a consequence of the fact that a GLM is fitted by iteratively-reweighted least squares, and the least-squares fit at the final iteration is part of the `glm` object.

We demonstrate these diagnostics on the `glm` object `kyph.glm4`, which is displayed in Figure 6.6. The Cook's distance diagnostic

$$C_i = \frac{(\hat{\boldsymbol{\beta}} - \hat{\boldsymbol{\beta}}_{(i)})^T \boldsymbol{X}^T \boldsymbol{W} \boldsymbol{X} (\hat{\boldsymbol{\beta}} - \hat{\boldsymbol{\beta}}_{(i)})}{p\hat{\phi}} \tag{6.8}$$

measures the effect of the ith observation on the coefficient vector. Writing $\hat{\boldsymbol{\eta}} = \boldsymbol{X}\hat{\boldsymbol{\beta}}$ and $\hat{\boldsymbol{\eta}}_{(i)} = \boldsymbol{X}\hat{\boldsymbol{\beta}}_{(i)}$ leads to

$$C_i = \frac{\left\| \hat{\boldsymbol{\eta}} - \hat{\boldsymbol{\eta}}_{(i)} \right\|_W^2}{p\hat{\phi}} \tag{6.9}$$

showing that it also measures an overall difference in the linear predictor when the observation is removed, suitably standardized. The W that subscripts the norm $\|\cdot\|_W$ reminds us that the squared norm is in fact a weighted sum-of-squares, or in other words we are computing the norm in the chi-squared metric. The term in the denominator is usefully viewed as the average variance of the fitted values, where once again we mean weighted average. Figure 6.8 is a plot of these distances against the sequence number for the fitted model `kyph.glm4`. There are three observations with large values for the distance—namely 11, 25, and 77. Possibly a more useful diagnostic in this scenario is the version of Cook's distances confined to a subset of the parameters, in particular those belonging to an individual term in the model. Figure 6.9 shows index plots of the Cook's distance for both the quadratic and

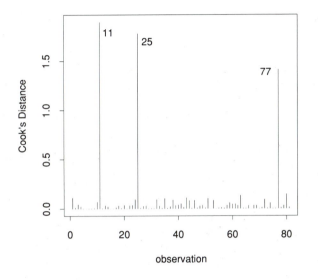

Figure 6.8: *An index plot of the Cook's distances for* `kyph.glm4`. *The numbers in the figure identify the observations with large distances.*

piecewise linear terms in `kyph.glm4`. The expression

$$C_i^j = \frac{\left\| \hat{\boldsymbol{f}}_j - \hat{\boldsymbol{f}}_{j(i)} \right\|_W^2}{\sum_{i=1}^n w_i \mathrm{var} \hat{f}_{ji}} \tag{6.10}$$

where $\hat{\boldsymbol{f}}_j = \boldsymbol{X}_j \hat{\boldsymbol{\beta}}_j$ denotes the subset of the model matrix and coefficient vector for the jth term in the model, and similarly $\hat{\boldsymbol{f}}_{j(i)} = \boldsymbol{X}_j \hat{\boldsymbol{\beta}}_{j(i)}$ is an approximation to the jth term fit with the ith observation removed. It now becomes apparent why some of the observations have large Cook's distances. In each case, the observation is a 1 in a region of nearly pure 0s. In logistic regression with binary response data, such points are highly influential. In a sense they are the most important observations in regions where the probability of a 1 is small. It is also clear that if we were to remove them, the fitted functions and hence coefficients could change quite dramatically, and in some cases diverge.

The overall Cook's distance in (6.8) can be computed most efficiently using the

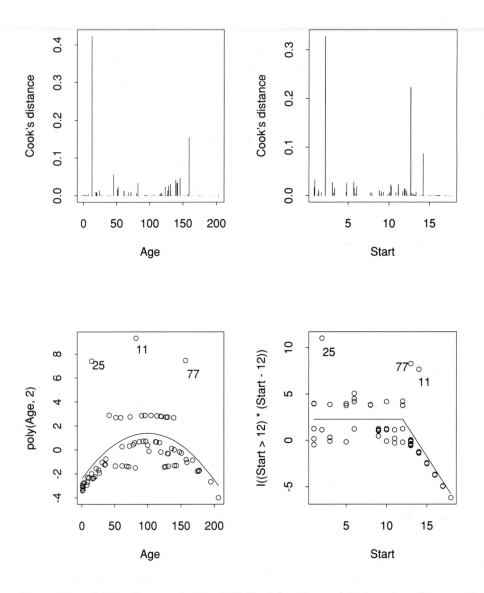

Figure 6.9: *Cook's distances for the individual functions, plotted against the respective variables. The lower figures reproduce the fitted functions in Figure 6.6 and identify the partial residuals corresponding to the points with largest distances.*

equivalent formula

$$C_i = \frac{r_i^2 h_{ii}}{p\hat{\phi}(1 - h_{ii})^2}$$

where the h_{ii} are the diagonal "hat" elements of the weighted least-squares projection matrix used in the final IRLS iteration. The r_i are the standardized residuals from this final regression, and $\hat{\phi}$ is the scaled chi-squared statistic. These are immediately available from the output of `lm.influence()` and `summary()` applied to `kyph.glm4`.

For the distances for the different terms, it is more convenient to use the delete-one coefficients themselves rather than a hat element equivalent. As an illustration of working with `glm` objects, we end this example by showing a function written to derive the Cook's distances for each of the terms in a `glm` or `lm` object:

```
Cook.terms <- function(fit)
{
    fit.s <- summary(fit)
    fit.infl <- lm.influence(fit)
    R <- fit$R
    I <- t(R) %*% R
    Iinv <- fits$cov.unscaled
    ass <- fit$assign
    D <- matrix(0, length(fit$residuals), length(ass))
    dimnames(D) <- list(names(fit$residuals), names(ass))
    Dcoef <- scale(fit.infl$coefficients,
                   center = fit$coefficients, scale= F)
    for(subname in names(ass)) {
        sub <- ass[[subname]]
        Dcoefi <- Dcoef[, sub, drop = F] %*% t(R[, sub, drop = F])
        denom <- I[sub, sub, drop = F] %*% Iinv[sub, sub, drop = F]
        denom <- sum(diag(denom)) * fit.s$dispersion
        D[, subname] <- apply(Dcoefi^2, 1, sum)/denom
    }
    D
}
```

6.3.5 Stepwise Model Selection

A typical GLM analysis proceeds in a stepwise fashion. We build models by adding in new terms and seeing how much they improve the fit, and by dropping terms that don't degrade the fit by a "significant" amount. This is usually a tedious task if performed manually, since many different models need to be tried, and the bookkeeping alone can get voluminous. Since each candidate model is fit iteratively, the computations can be time-consuming as well. Here we describe some ways to finesse these operations, and provide a function `step.glm()` for conducting a *stepwise* selection procedure. We illustrate the functions on the wave-soldering data.

The function `step.glm()` operates as follows:

1. It starts with an arbitrary `glm` object.

2. It takes a *step* by adding or removing that term from the current model that reduces the *AIC* selection criterion the most.

3. It stops when it hits a specified *model boundary* or when no step will decrease the criterion.

The second step is clearly the most work and would be quite time-consuming if carried out exactly. It assumes there is a current model from which to work. The *AIC* statistic

$$AIC = D + 2p\hat{\phi}$$

is used by `step.glm()` to evaluate different models, where D is the deviance, p the degrees of freedom in the fit, and $\hat{\phi}$ an estimate of the dispersion parameter. *AIC* is the likelihood version of the C_p statistic, and like C_p, changes in *AIC* due to augmenting or subsetting a model by a given term reflects both the change in deviance caused by the step, as well as the dimension of the term being changed (often terms involve more than 1 degree of freedom).

The idea is to fit all the models obtainable by deleting a single term from the current model, and computing the *AIC* statistic for each. Similarly, all models obtainable by adding a single term to the current model are fit, and the *AIC* statistic is computed. A step is taken toward the model having the smallest value for the *AIC* statistic; if none are smaller than the original model, the procedure terminates.

There is an inherent vagueness in the previous paragraph that is cleared up by the `scope=` argument to `step.glm()`. Although it may seem obvious what terms can be dropped at any stage, the terms available for inclusion have to be specified in some way. This information is supplied in the form of a list with two components, `"upper"` and `"lower"`. Each is a formula (for which only the right side is of relevance). Only those models are considered that include the terms in `scope$lower` and whose terms are included in `scope$upper`. Since the lower limit is often the null model, the `scope=` argument can be given simply as a formula, in which case it is interpreted as the upper formula, and the lower is taken to be the null model. Our example starts with the model `paov` fit to the wave-soldering data.

```
> formula(paov)
skips ~ Opening + Samt + Stype.th + PadType + Panel
> paov.step <- step(paov, scope = ~ . ^3)
```

This will potentially step through all models ranging from the full third-order interaction model down to the null model.

At each step, not all single terms in `scope$upper` are eligible for inclusion; similarly, not all terms in the current model (less those in `scope$lower`) are available for exclusion). The reason is that the model hierarchy has to be honored; for example, main-effect terms must be added before their interactions. The function `add.scope()` is useful for seeing what terms can be added in a hierarchical fashion:

```
> form1 <- formula(paov)
> add.scope(form1, update(form1, ~ .^3))
 [1] "Opening:Solder"  "Opening:Mask"    "Opening:PadType"
 [4] "Opening:Panel"   "Solder:Mask"     "Solder:PadType"
 [7] "Solder:Panel"    "Mask:PadType"    "Mask:Panel"
[10] "PadType:Panel"
```

So even though the upper formula included third-order interactions, only the second-order interactions could be added at this stage. Similarly `drop.scope()` determines what terms can be dropped at any stage:

```
> drop.scope(form1)
[1] "Opening" "Solder" "Mask"    "PadType" "Panel"
```

Both these functions are used before each step taken by `step.glm()`, and thus repeatedly throughout its execution.

Suppose `step.glm()` is considering as current model the initial model, `paov`. It has ten separate terms to consider for inclusion and five for deletion. A maximum-likelihood fit of each of these models requires iteration and would be time-consuming. We expect the fit of each of these subset or augmented models to be reasonably close to the parent model. This suggests that the quadratic approximation to the deviance can be used rather than the deviance itself in computing the selection criterion, and more importantly that we can use the one-step approach of the score test in computing all the subset fits. We use the Pearson chi-squared version of AIC, which is the C_p statistic for the local quadratic model, defined by $C_p = X^2 + 2p\hat{\phi}$.

The consequence of all this is that we can simply hand the `glm` object corresponding to the current model to `drop1.lm()` and `add1.lm()`, which compute all the subset and augmented models efficiently. This is another example of where we can exploit the inheritance properties of classes of models in a very natural way. We then select for deletion or addition the term corresponding to the smallest value of C_p, and complete the IRLS iterations for that model. The AIC statistic is then computed for this selected model; if it is lower than the AIC for the previous model, the new model becomes the current model, and the stepping continues, otherwise `step.glm()` terminates by returning the previous model.

Before we examine the output of `step.glm()`, let us make a slight diversion and take a closer look at `add1()` and `drop1()`:

```
> add1(paov, ~ . ^2)
Single Term Additions
```

```
Model: skips ~ Opening + Solder + Mask + PadType + Panel
                  Df Sum of Sq  RSS   Cp
          <none>                1045 1099
 Opening:Solder   2     23.07  1022 1082
   Opening:Mask   6     68.02   977 1049
Opening:PadType  18     46.26   999 1106
  Opening:Panel   4     10.93  1034 1100
    Solder:Mask   3     48.74   996 1059
 Solder:PadType   9     43.82  1001 1082
   Solder:Panel   2      6.48  1039 1098
   Mask:PadType  27     57.12   988 1122
     Mask:Panel   6     21.48  1024 1095
  PadType:Panel  18     14.63  1031 1138
```

```
> drop1(paov)
Single Term Deletions
```

```
Model: skips ~ Opening + Solder + Mask + PadType + Panel
          Df Sum of Sq  RSS   Cp
  <none>                1045 1099
 Opening   2    2101    3146 3194
  Solder   1     811    1856 1907
    Mask   3    1429    2474 2518
 PadType   9     473    1518 1545
   Panel   2      66    1111 1159
```

Although these are generic functions, the drop1.lm() and add1.lm() methods are
what actually get invoked since no particular methods exist for glm objects, which
inherit from the class "lm". They both return anova objects, and have arguments
scope= and scale=. The scale= argument is also used in step(), and allows the
user to specify the dispersion constant ϕ to be used in computing the C_p or AIC
statistic. If scale is missing, add1() and drop1() use the residual variance of the
original model. By default, step() uses the dispersion parameter for the original
glm object, which is 1 for binomial and Poisson models, and the scaled Pearson
chi-squared statistic in all other cases. Both these functions are described in some
detail in Section 4.3.2, where additional arguments such as keep= are described. See
also their detailed documentation for a precise description.

Typically, one calls step() using the trace=T argument, which then displays all
the drop1 and add1 anova tables along the way. For large models, such as the wave-
soldering example, the function can take a while to run so it is encouraging to see
the intermediate results:

```
> paov.step <- step(paov, ~.^3)
> paov.step$anova
Stepwise Model Path
Analysis of Deviance Table

Start: skips ~ Opening + Solder + Mask + PadType + Panel

Final: skips ~ Opening + Solder + Mask + PadType + Panel +
       Opening:Mask + Solder:Mask + Solder:PadType +
       Opening:Solder + Opening:PadType + Opening:Solder:Mask
```

	Step	Df	Deviance	Resid. Df	Resid. Dev	AIC
1				702	1130	1166
2	+ Opening:Mask	-6	-71.0	696	1059	1107
3	+ Solder:Mask	-3	-55.0	693	1004	1058
4	+ Solder:PadType	-9	-43.3	684	961	1033
5	+ Opening:Solder	-2	-32.2	682	929	1005
6	+ Opening:Solder:Mask	-6	-52.7	676	876	964
7	+ Opening:PadType	-18	-47.7	658	828	952

The result of `step.glm()` is a `glm` object corresponding to the final model selected. The object includes an `anova` object, printed above, which shows the path taken to the final model. No terms were deleted in this case, only added, including one third-order interaction.

There are other arguments to `step()`, namely `keep=`, `direction=`, `scale=`, and `steps=`; these are described in the detailed documentation, and also in some detail in Section 7.3.1.

As is pointed out in Chapter 4, we do not always want to drop or include entire terms. For example, we might wish to try different orders of a polynomial fit for one of the variables. It would be simple, for example, to modify `drop1.lm()` (and make it *less smart*) so that it dropped columns of the model matrix rather than subsets of the columns corresponding to terms. But this would not be sufficient, because we do not really want to drop these columns in any order.

In Chapter 7, a more general stepwise method `step.gam()` is described for additive models and can be used with GLMs and LMs as well. It allows a *regimen* of subterms to be specified for each term in the model, and performs stepwise backward and forward selection on subsets defined by these. For example, a particular regimen for polynomial regression may be

```
~1 + Age + poly(Age, 2) + poly(Age, 3)
```

There is an ordering in this sequence of subterms, ranging from no term at all to a third-degree polynomial. The price to be paid for this greater generality is speed, since each of the candidate models has to be fitted separately. Although this stepwise model works for both `glm` and `lm` objects as well, we postpone its full description until Chapter 7.

6.3.6 Prediction

Often we wish to evaluate the fitted model at some new values of the predictors, either for predictive purposes, or for validation. The method `predict.glm()` is used to make such evaluations. The expressions

```
predict(glmob)
fitted(glmob)
```

are simple ways of extracting the linear predictor and the fitted values from `glmob`. More generally, the syntax for `predict()` is

```
predict(glmob, newdata)
```

where `newdata` is a data frame consisting of the new data. This will once again produce values for the linear predictor evaluated at the new data. If the `newdata` argument is missing, predictions are made using the same data that were used to fit the model. Unless some of the options below are selected, the predictions in this case already exist on the fitted `glm` object, so no additional work is required.

Prediction for GLMs is not much different than that for LMs, which is discussed in Section 4.2.3. Our discussion here is meant to complement that section, as well as describe any features specific to GLMs. The `type=` argument allows a choice of either `"link"`, `"response"`, or `"terms"`; thus

```
fitted(glmob)
predict.glm(glmob, type = "response")
```

produce identical results. Choosing `type="terms"` results in a matrix of predictions, with a column for each term in the model. The construction of these columns is quite straightforward. Recall that a component of a `glm` or `lm` object is the `"assign"` list; it has an element for each term, and each element is a vector of numbers. For example,

```
> kyph.glm4$assign
$"(Intercept)":
[1] 1

$"poly(Age, 2)":
 1 2
 2 3

$"I((Start > 12) * (Start - 12))":
[1] 4
```

tells us that the columns 2 and 3 of the model matrix correspond to the quadratic term "poly(Age, 2)", and similarly elements 2 and 3 of the coefficients vector. The relevant components are extracted and multiplied together to form a single fitted term, and the same is done for all terms:

```
> predict(kyph.glm4, type = "terms")[1:5,]
  poly(Age, 2) I((Start > 12) * (Start - 12))
1    1.06519                     2.2648
2    0.06911                    -0.4133
3    1.07005                     2.2648
4   -2.28338                     2.2648
5   -2.35883                    -1.7523
```

These fitted terms are centered such that, when computed for the original data, they average zero. Fitted terms are typically used in plots—i.e., plotting a fitted polynomial term against its predictor(s). The terms are centered because the interesting features in such plots are typically the slope and shape, while the level is of no importance. The matrix of predicted terms returned has an attribute "constant", which is a single number; the sum of the terms plus this constant is then identical to the linear predictor. The centering is simply achieved by subtracting from each column of the model matrix the (weighted) mean of the corresponding column of the model matrix for the original data.

Standard errors can optionally be computed using the se.fit= argument in a call to predict(). The object returned is then a list with four components:

- fit: the usual output that would have been returned if se.fit=F; either a vector or a matrix if terms=T;

- se.fit: the standard errors of each element in the fit component, and therefore having the same shape as fit;

- residual.scale: the scale estimate used in constructing the standard errors;

- df: the degrees of freedom used in estimating the scale estimate.

The last two components are provided in case alternative scaling is required, and for computing pointwise confidence intervals.

When the type="terms" option is used, the "assign" component is once again used to extract the relevant sub-block of the estimated covariance matrix of the parameter estimates. Assume we have available the model matrix X, which has been centered in the fashion described above. Suppose we are constructing the fit and standard errors for the jth term, and we have extracted the relevant subset of the model matrix X_j, the coefficients β_j, and the covariance submatrix Σ_{jj}. Then the fitted term and its pointwise standard errors are given by

$$\eta_j = X_j \beta_j$$

$$\text{se}(\boldsymbol{\eta}_j) \;=\; \text{diag}\big(\boldsymbol{X}_j \boldsymbol{\Sigma}_{jj} \boldsymbol{X}_j^T\big)$$

Centering is even more important when standard errors are computed, once again in the context of plotting fitted terms. Since plots of this kind are produced by the `plot()` method for additive models, we defer further discussion of the centering issue until Section 7.3.5 (page 296).

When the `type="response"` option is used in combination with `se.fit=T`, the *delta-method* standard errors are computed:

$$\text{se}(\boldsymbol{\mu}) \approx \left|\frac{\partial \boldsymbol{\mu}}{\partial \boldsymbol{\eta}}\right| \text{diag}\big(\boldsymbol{X} \boldsymbol{\Sigma} \boldsymbol{X}^T\big) \tag{6.11}$$

Let's look at an example. The mean of the binary response `Pick` in the market survey data has a strong dependence on `Usage`, as we see in Chapters 7 and 9. We fit the simple model

```
survey.fit <- glm(pick ~ log(Usage + 2), binomial)
```

The sample is rather large (759 observations), and our model is simple; rather than plot all the fitted values, we choose to represent them over a grid of 50 evenly spaced values of `Usage`. The following computations produce the required fitted values and standard errors:

```
Usage.grid = seq(from = min(Usage), to = max(Usage), length = 50)
survey.pred <- predict(survey.fit, list(Usage = Usage.grid),
     type = "resp", se = T)
```

Notice that even though we have only one predictor, the `newdata=` argument of predict expects a data frame or a list. In fact, a matrix is acceptable as well, as long the number of columns coincides with the number of coefficients. Since an intercept is included in this model, we would have to provide a two-column matrix, so

```
predict(survey.fit, cbind(1, Usage.grid), type = "resp", se = T)
```

would give the same predictions as the previous expression. In general, it is at least as simple to provide a list or data frame of new data, and far safer when factors might be involved. Figure 6.10 shows the fitted proportions at the selected points, as well as line segments joining the twice-upper and -lower pointwise standard errors. The fitted values are not too extreme, and the delta method seems to have worked well. Had the fitted values been around zero or one, it is quite likely that the standard-error bands could have gone outside [0, 1], which is not acceptable for binomial data. An alternative approach in situations such as this is to instead compute the bands on the scale of the linear predictor and then *invert* the upper and lower bands using the inverse logit transform. We give an example in Section 7.3.3 (page 291).

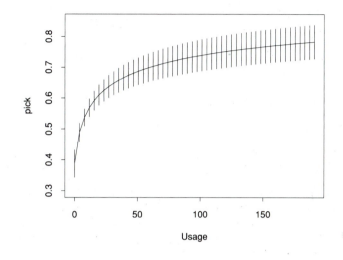

Figure 6.10: *The central curve is the fitted proportion for the binary response* Pick, *modeled linearly in the logarithm of* Usage *for the market survey data. The curve was evaluated at 50 points over the range of* Usage *using* predict(), *as were the pointwise standard errors. The vertical bars join the upper and lower twice-standard-error points, meant to represent approximate 95% confidence intervals for the mean response.*

GLMs also inherit from LMs the pitfalls in prediction arising from the general expressions allowed in model formulas, as described in Section 4.2.3. For example, $\sim$ log(x - min(x) + 1) is a perfectly valid formula expression; any coefficient for this term is likely to be meaningless when the same expression is applied to new data. The predict.gam() method is designed to avoid these pitfalls at a slight loss in computational efficiency; we defer discussion of this safer form of prediction to Section 7.3.3.

6.4 Statistical and Numerical Methods

Here we give a brief overview of the estimation of generalized linear models by maximum likelihood, and the associated iteratively-reweighted least-squares algorithm. We describe the inference tools available for this model of analysis, which are similar to those for linear models.

In Section 6.4.3, we give some additional algorithmic details about glm(), and discuss starting values in Section 6.4.4.

6.4.1 Likelihood Inference

If $f(y; \mu)$ is the density or probability mass function for the observation y given μ, then the log-likelihood, considered as a function of μ, is simply

$$l(\mu; y) = \log f(y; \mu).$$

Large values of $l(\mu; y)$ correspond to more likely values of the parameter μ, for a given value of y. Now suppose we have a sample of n independent observations $y_1, \ldots, y_n$ with $Ey_i = \mu_i$; then the log-likelihood for the entire sample is

$$l(\boldsymbol{\mu}; \boldsymbol{y}) = \sum_i \log f(y_i; \mu_i). \tag{6.12}$$

If $g(\mu_i) = \boldsymbol{x}_i^T \boldsymbol{\beta}$, then the parameters $\boldsymbol{\beta}$ can be estimated by maximizing (6.12). For example, for independent binary response data the log-likelihood of the sample is

$$l(\boldsymbol{\mu}; \boldsymbol{y}) = \sum_i \{y_i \log(\mu_i) + (1 - y_i) \log(1 - \mu_i)\}$$

If we use the logit link, then the log-likelihood can be written as

$$l(\alpha, \boldsymbol{\beta}; \boldsymbol{y}) = \sum_i \left[y_i \boldsymbol{x}_i^T \boldsymbol{\beta} - \log\{1 + \exp(\boldsymbol{x}_i^T \boldsymbol{\beta})\} \right] \tag{6.13}$$

The Bernoulli distribution is a member of the *exponential family* of distributions, whose members have densities of the form

$$f(y; \theta, \phi) = \exp\{ (y\theta - b(\theta))/a(\phi) + c(y, \phi) \} \tag{6.14}$$

Other familiar members are the binomial, Poisson, gamma, and Gaussian distributions. For our purposes, $a(\phi_i) = \phi/w_i$, where ϕ is referred to as the *dispersion parameter*, and the w_i are prior weights. If the *dispersion parameter* is known, then the distributions are one-parameter members; the binomial, Bernoulli, and Poisson all have $a(\phi) = 1$, while for the Gaussian distribution $\phi = \sigma^2$, the variance parameter. The parameter θ is known as the *natural parameter;* the links in the table on page 197 are the corresponding *natural links* that transform the mean of a family to the natural parameter. For this class of link functions, the regression parameters enter the log-likelihood as linear combinations

The *deviance* function is linearly related to the log-likelihood, and is often used as a goodness-of-fit criterion. The deviance $D(\boldsymbol{y}; \boldsymbol{\mu})$ is defined by

$$\frac{D(\boldsymbol{y}; \boldsymbol{\mu})}{\phi} = 2l(\boldsymbol{\mu}^*; \boldsymbol{y}) - 2l(\boldsymbol{\mu}; \boldsymbol{y}) \tag{6.15}$$

where μ^* maximizes the log-likelihood over μ unconstrained. Often it is the case that $\mu^* = y$, as it is in the exponential family models that we discuss here. For the Gaussian distribution, the deviance is

$$D(\boldsymbol{y}; \boldsymbol{\mu}) = \sum_i w_i(y_i - \mu_i)^2$$

and is simply the residual sum-of-squares. Since the first term in (6.15) does not depend on the parameters, maximum-likelihood estimation is identical to minimum-deviance estimation, with the latter being more natural since it has the interpretation of a distance.

To compute the maximum-likelihood estimates, we solve the score equations $\partial l(\boldsymbol{\beta}; \boldsymbol{y})/\partial \boldsymbol{\beta} = \mathbf{0}$. These score equations are nonlinear in $\boldsymbol{\beta}$, so iteration is required to solve them. The IRLS algorithm for exponential-family models consists of the following steps:

- Compute a *working response* with typical value

$$z_0 = \eta_0 + (y - \mu_0)\left(\frac{\partial \eta}{\partial \mu}\right)_0$$

  where $\eta = g(\mu)$ is the *linear predictor*, and the subscript 0 denotes evaluation at the current value of the parameters $\boldsymbol{\beta}_0$.

- Compute weights

$$W_0^{-1} = \left(\frac{\partial \eta}{\partial \mu}\right)_0^2 V_0.$$

- Regress z_0 on the predictors $x_1, \ldots, x_j$ with weights W_0 to obtain an updated $\boldsymbol{\beta}_1$.

These steps are iterated until the relative change in the coefficients is below some small threshold. For the binary response example, the working response and weights are

$$\begin{aligned} z_0 &= \eta_0 + \frac{(y - \mu_0)}{\mu_0(1 - \mu_0)} \\ W_0 &= \mu_0(1 - \mu_0) \end{aligned}$$

The IRLS algorithm can be justified in a variety of ways. When the natural link function is used, it is equivalent to the Newton-Raphson algorithm for iteratively solving the score equations. For other link functions it is equivalent to the Fisher-scoring algorithm, a close relative to the Newton-Raphson algorithm. The maximum-likelihood score equations can be written as

$$\partial l(\boldsymbol{\beta}; \boldsymbol{y})/\partial \beta_j = \sum_i \frac{(y_i - \mu_i)}{V_i} \frac{\partial \eta_i}{\partial \mu_i} x_{ij} = 0 \tag{6.16}$$

for each predictor x_j. These equations represent a form of orthogonality between the residual $(y - \mu)$ and the linearized version of the model at the solution, in the metric of V^{-1}, where V is the variance of the residual; in this sense they are the analogues of the usual least-squares normal equations.

Although these *estimating equations* are formally derived from maximum-likelihood principles, one could simply write them down knowing the link and variance functions. We may not know or be willing to postulate the distribution of a response, but may be happy to pin down its mean-variance relationship. These estimating equations behave like score equations in several important respects, and provide a basis for estimation in this case. This is known as quasi-likelihood estimation, and extends the class of generalized linear models considerably.

We have seen that fitting a generalized linear model is not much harder than fitting a linear model. The interpretation generalizes in a similar way. The steps in comparing two models are very similar to those for comparing two Gaussian-error linear models. The difference in deviance between two nested models measures the contribution of the parameters by which they differ, just as does the numerator of the F-test statistic. The distribution theory is asymptotic; under appropriate assumptions and the null hypothesis that the smaller model μ_1 is correct, the difference in deviance

$$D(\hat{\mu}_1, \hat{\mu}_2) = D(\hat{\mu}_1, y) - D(\hat{\mu}_2, y)$$

has an asymptotic $\phi \chi^2$ distribution with degrees of freedom $\nu = \nu_1 - \nu_2$ equal to the difference in the dimension of the linear spaces implicit in μ_1 and μ_2. These approximations can be poor in small sample situations, but the difference in deviance between two models can still be useful as a screening device.

Also of interest is the distribution of the parameter estimates $\hat{\beta}$. Under the same asymptotics and a correct model, the distribution of $\hat{\beta}$ approaches that of a $N(\beta, (X^T W X)^{-1} \phi)$ distribution, where X is the model matrix and W, the diagonal matrix of weights. This approximation is often used to perform hypothesis tests on subsets of the parameters. The form of the asymptotic-covariance matrix alerts us to a possible additional difference between these models and the usual linear models: even if the predictors are orthogonal, the nonlinearity of the model can induce correlations in their associated parameter estimates.

6.4.2   Quadratic Approximations

Each step of the IRLS algorithm minimizes the weighted least-squares criterion

$$\sum_{i=1}^{n} W_i(z_i - x_i^T \beta)^2 \tag{6.17}$$

where $z_i = \boldsymbol{x}_i^T \boldsymbol{\beta}_0 + (y_i - \mu_i)\left(\partial \eta_i / \partial \mu_i\right)$ and $W_i^{-1} = \left(\partial \eta_i / \partial \mu_i\right)^2 V_i$, and each of μ_i, η_i and V_i are evaluated at $\boldsymbol{\beta}_0$. At convergence, $\boldsymbol{\beta}_0 = \hat{\boldsymbol{\beta}}$, which means that the criterion (6.17) reduces to the Pearson chi-squared statistic

$$X^2 = \sum_{i=1}^{n} \frac{(y_i - \mu_i)^2}{V(\hat{\mu_i})}. \tag{6.18}$$

By a simple Taylor series expansion, (6.18) can be seen to be a quadratic approximation to the deviance $D(\boldsymbol{y}; \boldsymbol{\mu}) = 2\sum_i \{l(y_i; y_i) - l(\mu_i; y_i)\}$ at the minimum.

All the first-order asymptotic theory for maximum-likelihood estimates is based on this approximation. A simple way of viewing this theory is through the working response, evaluated at the MLE $\hat{\boldsymbol{\beta}}$. The asymptotics can be derived by assuming that the z_i are asymptotically independent with mean η_i and variance $\left(\partial \eta_i / \partial \mu_i\right)^2 V_i \phi$, where η_i and V_i now depend on the *true* mean μ_i.

As an example, consider the asymptotic covariance matrix of the parameter estimates:

$$\text{var}\hat{\boldsymbol{\beta}} = (\boldsymbol{X}^T \boldsymbol{W} \boldsymbol{X})^{-1}\phi \tag{6.19}$$

where $\boldsymbol{W}$ is a diagonal matrix with elements W_i. This is exactly the covariance matrix that would be obtained from the weighted linear regression of z_i on $\boldsymbol{x}_i$ with weights W_i. The weights W_i are estimated, based on the fitted values. The dispersion parameter is either assumed known (as is often assumed in the case of binomial or Poisson regression), or else is estimated by X^2/ν, where ν is the residual degrees of freedom. Both these quantities are computed automatically by the `summary.lm()` function when handed a weighted least-squares fit object; a `glm` object inherits from the class `"lm"`, and so has all the components of weighted least-squares object.

It is also clear that all the other functions designed for `lm` objects will work when handed `glm` objects. These include `print.lm()`, `lm.influence()`, `drop1.lm()`, and `add1.lm()`. They will produce appropriate weighted least-squares output, and so can be interpreted as the Pearson chi-squared equivalent of the usual output. In each case we need to be careful about interpreting the results. For example, the results of `drop1.lm()` and `add1.lm()` are conditioned on the weights and the working response of the current fit when computing the subset or augmented fits. This is equivalent to taking one IRLS step toward the new solution, and is the same philosophy that motivates Rao's score test.

6.4.3 Algorithms

The IRLS computations in `glm()` are performed by the function `glm.fit()`, which uses repeated QR decompositions. The only nonstandard aspect of `glm.fit()` is its `family=` argument, which is the same as in `glm()`, and the generality of the response `y`. Otherwise, it receives an X matrix and optional prior weights, offset and starting values, as well as some iteration constants. The implications of this are that the fitting mechanism of `glm()` could be replaced without much knowledge of the workings of `glm()`. The only constraints on such a replacement function would be:

- the `family` argument would have to be used in much the same way as it is by `glm.fit()`;

- the object returned should include the essential components of a `glm` object.

There are several reasons why one might embark on such a venture:

- For standard families, all the computations could be performed in C or FOR-TRAN. This would be faster than the current setup, where the iteration updates are computed in S, and only the weighted least-squares fit is computed in FORTRAN. Whether or not a family is standard could be identified by the `family` component of a family object, which is an identification tag.

- This modularity allows dramatic functionality changes to be made with relative ease. For example, one could create a function to fit the Cox's proportional hazards model with relative ease, perhaps by using a `cox` family. The `initialize` expression of the `family` object would be responsible for untangling the death times and censoring information, and `glm.fit.cox()` would fit the model using the usual Newton-Raphson algorithm for minimizing the partial likelihood.

Currently such modifications to `glm()` can be implemented via the `method=` argument, which defaults to `"glm.fit"`. Using the Cox model as an example, a typical calling sequence would look like:

```
glm(Times ~ Dose + Age, family = "cox", method = "glm.fit.cox")
```

Alternatively, users could write their own version of `glm.fit()` that identified what algorithm to use by examining the `family` object. Such a (more permanent) modification could simultaneously deal with both the generalizations suggested above.

6.4.4   Initial Values

As is always the case, the Newton-Raphson iterations are not guaranteed to converge unless step-length optimization is used. No such step-length calculations are performed in `glm()`, simply because they add to the computational burden and are rarely needed.

There is an art, however, in selecting starting values for the iterations. It might seem that an obvious starting point should be the sample mean for the fitted values. This causes problems, however, when the response is binary and the sample mean is too small or large. Similarly, it might seem even more natural to start the iterations of the stepwise algorithms from the previous fit, or even better, from the fit produced by the single step taken by `drop1()` or `add1()`. This strategy, too, is dangerous; while it may save one or perhaps two iterations most of the time, these starting values cause convergence problems far more often than those we describe below. It is also worthwhile to note that selection of a model is computationally far more intensive than iterating the selected model to convergence.

We use the data themselves as starting values for the fitted response. These need to be corrected in some cases. For binomial data, we use $r_i + (0.5 - r_i)/n_i$, where r_i is the proportion of 1s, and n_i is the number of trials for the ith response. This shrinks the response toward 0.5 to avoid a proportion of 0 or 1; for binary data the shrinking is dramatic (all initial values are 0.5). For Poisson counts and gamma data, a zero response is replaced by $1/6$.

There is a `start` argument to `glm()`, so in the rare situations when convergence problems are encountered, alternative starting values for the fitted response can be tried.

Bibliographic Notes

GLMs have become popular over the last 10 years, partly due to the computer package GLIM (Generalized Linear Interactive Modeling; Baker and Nelder, 1978). GLIM is a FORTRAN-based interactive environment that features a formula language for describing the components of the linear predictor, and easy specification of the different components of a GLM. Apart from the formula language, the core of GLIM is the iteratively-reweighted least-squares (IRLS) algorithm; the remainder is an environment for setting up the data structures and summarizing and plotting the fitted models. Our formula language, described in Chapter 2 and used throughout this book, was inspired by the the Wilkinson and Rogers (1973) formula language used in GLIM, and features many enhancements. Several recent books have appeared on the practice of using GLIM and on the GLM mode of modeling (Healy, 1988; Aitkin et al., 1989).

McCullagh and Nelder's (1989) research monograph is a comprehensive text about the theory of generalized linear models. They study many examples covering a wide spectrum of applications, and give an excellent overview of the recent advances in the field. They discuss in some detail the asymptotic results referred to in this chapter, and also summarize the recent work that has been done to improve them.

The `robust()` function follows the conventions for resistant GLMs established by Pregibon (1982), and described in Hastie and Tibshirani (1990).

Chapter 7

Generalized Additive Models

Trevor J. Hastie

In the previous chapters we introduced a number of new functions for fitting linear models. In particular, `glm()` fits linear models in a variety of settings such as ordinary regression or logistic regression for binary data. The `data` and `formula` arguments provide a flexible language for specifying the variables and their form in a model, and the `family` argument supplies information on the error structure and the link function. The output of `glm()` can be fed into a number of auxiliary functions for summarizing the estimated coefficients and evaluating and examining the fits. In particular, residual and partial residual plots are used to identify discrepant observations and to identify nonlinearities. This chapter describes some tools for identifying nonlinearities in a more direct way by incorporating them into the model. This practice is not new, of course, if by nonlinearities we mean polynomial terms and parametric transformations. We use a more adaptive approach; the techniques described here allow us to model the terms nonparametrically using a scatterplot smoother, and in so doing let the data suggest the nonlinearities. The `gam` functions share many of the features of `glm()` and `lm()`, with some added flexibility. The output of `gam()`, being graphical in nature, tends to complement rather than overlap with `glm()`.

7.1   Statistical Methods

An additive regression model has the general form

$$\eta(\boldsymbol{x}) = \alpha + f_1(x_1) + f_2(x_2) + \cdots + f_p(x_p) \tag{7.1}$$

where each of the x_i are predictors and the f_i are functions of the predictors or terms. The name *additive* refers to the multivariate assumption underlying the model, namely that the p-predictor function η has a low-dimensional additive structure. Such models are attractive if they fit the data, since they are far easier to interpret than a p-dimensional multivariate surface.

So far we have entertained additive models in which each of the terms is a parametric function of the predictor—in fact, a function *linear* in its parameters. Examples are simple transformations such as logarithm, polynomials, and sinusoids, as well as step functions introduced by transforming numeric variables into ordered factors. More elaborate functions can be used to generate piecewise linear or polynomial functions with breakpoints at specified values of the predictors.

The innovations in this chapter are additional flexible methods for modeling an individual term in an additive model. This relieves the user of the burden of fishing around for the correct transformation for each variable. The functions are fitted using *scatterplot smoothers*, nonparametric techniques for fitting a regression function in a flexible data-defined manner. Several smooth terms are fitted simultaneously in an additive model by using the scatterplot smoothers iteratively. Of course, nothing comes for free; for the nonparametric techniques to be successful and remain parsimonious, the underlying functions need to be reasonably smooth.

The *additive predictor* η can be used in all the situations where the *linear predictor* was used for generalized linear models. Lets look at a few examples:

- A univariate smoother estimates the unknown function in the simple additive model $y = f(x) + \varepsilon$, having only one term.

- A semiparametric model $y = \boldsymbol{x}^t \boldsymbol{\beta} + f(z) + \varepsilon$ is an additive model, where only one term is modeled nonparametrically. Of course, the purely linear model is additive as well! These semiparametric models have received attention in the analysis of agricultural field trials; the linear terms usually correspond to design effects and the nonlinear function models spatial ordering of the plots.

- The additive model $y = \boldsymbol{\beta}^t \boldsymbol{x} + f_1(z_1) + f_2(z_2) + \cdots + f_q(z_q) + \varepsilon$ is also semiparametric, but more complicated. There are several linear and several nonparametric terms.

- The nonparametric logistic regression model has the form

$$\mathrm{logit}\, P(\boldsymbol{x}) = \log\Big(\frac{P(\boldsymbol{x})}{1 - P(\boldsymbol{x})}\Big) = \eta(\boldsymbol{x}).$$

Once again the additive predictor $\eta(\boldsymbol{x})$ can be a single term, a semiparametric term, a full additive model as in (7.1), or a mixture.

Often the nature of the variables determines to a certain extent how we model them. For example, if a variable x_j is a factor, we would most likely fit it as a set of constants corresponding to each level, and so it would appear in the linear part of the semiparametric predictor. If a variable is quantitative, we can choose whether we want to model it in a parametric or nonparametric fashion—often we want to try both.

The examples above may give the impression that each of the terms is univariate. On the contrary, they may be parametric compound variables, or even multivariate nonparametric terms, which might be fitted using a surface smoother.

We have presented these models from a rather formal statistical point of view. In fact, each of the items above represents somewhat independent and rather large sections of the statistical modeling literature. The next step is to define algorithms for fitting these models, subject to suitable constraints on the nonparametric functions. The methods we use unify this large literature, and in fact allow us to discuss all these different models under the description *generalized additive models*. One approach is based on penalized likelihood and smoothing splines. Others are based on the idea of local estimation. Section 7.1.2 gives a brief overview of the steps involved; a more detailed account is given in Section 7.4.

7.1.1   Data Analysis and Additive Models

New and adventurous techniques abound in the applied statistics literature. In order for these new techniques to be used and become popular, they must blend with the existing technology. Additive models and the associated methodology naturally blend well, and the software described in this chapter emphasizes this.

For example, the formula

```
ozone ~ wind + s(temp)
```

specifies an additive model in which wind is to appear linearly and temp is to be modeled by a nonparametric smooth term. This model looks very similar to the model

```
ozone ~ wind + poly(temp, 3)
```

except that smoothers allow the term in temp to be modeled in a more flexible way. Readers familiar with smoothing will wonder about the amount of smoothing or *smoothing parameters* implicit in the term s(temp). Using the concept of the *equivalent degrees of freedom* of a smoother, we are able to prespecify the value of the smoothing parameter. So in the example above, s(temp) implies a smooth term fit using a smoothing spline with df=5 (the default), while the general form would be s(temp, df). Similarly, the term lo(temp) implies a smooth term using the loess() smoother, with a default amount of smoothing.

The fitted model is therefore very similar to a parametric model, and similar analyses are possible. For example, we can perform tests (albeit crude) for whether terms should be linear or simply smooth, by fitting the two separate models and analyzing the change in deviance relative to the change in df. Similarly, we can use stepwise model-building algorithms for automatically selecting terms. We can compute pointwise standard-error bands for the curves, and make predictions at new observations. The fitted functions have strong graphical appeal, and are almost always plotted.

Some analysts may not be comfortable with the more confirmatory type of analysis proposed above. Additive models can also be used in a diagnostic mode as a tool for suggesting parametric transformations or alternative forms for terms in the model. Once the transformations have been discovered, subsequent fitting and testing can then be based on these parametric transformations.

7.1.2 Fitting Generalized Additive Models

This section briefly outlines how we fit additive models, leaving the details to Section 7.4. A general and efficient algorithm for fitting a generalized additive model (GAM) consists of a hierarchy of three modules:

- *Scatterplot smoothers*, which are used to fit individual functions, can be thought of as a general regression tool for fitting a functional relationship between a response and, in our case, a one- (or typically low-) dimensional predictor variable. So locally weighted polynomials (loess() in Chapter 8), smoothing splines, kernel and near-neighbor smoothers fall into this category. So do linear parametric regression fitters such as simple and polynomial regression.

- The *backfitting algorithm* cycles through the individual terms in the additive model and updates each using an appropriate smoother. It does this by smoothing suitably defined partial residuals. Known as the Gauss-Seidel algorithm in numerical analysis, the cycles continue until none of the functions change from one iteration to the next. Typically, three or four smooths per variable are required.

- The *local-scoring algorithm* is similar to the Fisher-scoring algorithm or IRLS used to fit generalized linear models. Each iteration produces a new working response and weights, which are handed to a weighted backfitting algorithm, which produces a new *additive predictor*.

These three steps are a rather natural and intuitive generalization of the usual linear model algorithms, and that is how they were originally conceived. The algorithm can be placed on firmer theoretical ground if we are willing to commit ourselves to particular classes of smoothers. For example, if all the smooth terms in the additive model are *polynomial smoothing splines*, then the local scoring algorithm solves an appropriately penalized likelihood problem.

7.2 S Functions and Objects

This section presents the S functions for fitting and understanding generalized additive models (GAMs). Readers who visit Chapters 4 and 6 will soon discover that the tools are the same, and most of the new functions introduced here can be used with `lm` and `glm` objects as well.

7.2.1 Fitting the Models

Readers familiar with S may have used `lowess()`, an example of a scatterplot smoother. Figure 7.1 (left panel) shows the `lowess` smooth of `Mileage` against `Weight`

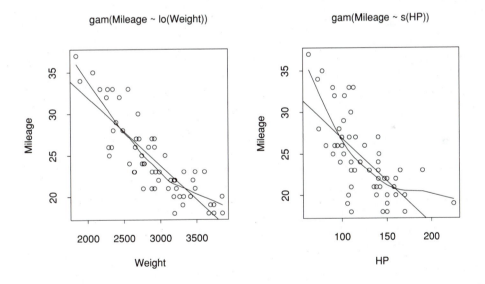

Figure 7.1: *Scatterplot smoothers summarize the relationship between the variable* `Mileage` *as a response and* `Weight` *and* `HP` *as predictors in the automobile data. The first panel uses the* `loess()` *smoother, while the second panel uses a smoothing spline. They can also be viewed as nonparametric estimates of the regression function for a simple additive model. The straight line in each case is the least-squares linear fit.*

for the automobile data, using the default smoothing parameter and no iterative reweighting. We can also view this smooth as an estimate of the regression function f in the simple additive model $y = f(x) + \varepsilon$ where y is `Mileage`, x is `Weight`, and the ε represents *iid* errors. We can go further, and estimate that there are 4.5 equivalent degrees of freedom (df) used in the smooth fit, versus 2 df for the linear fit. Comparing the residual sum-of-squares (*RSS*) for the two models, we can perform an approximate F-test for the hypothesis that the regression is linear:

$$F = \frac{(380.8 - 317.3)/(4.5 - 2)}{317.3/(60 - 4.5)} = 4.45$$

and compute the corresponding percentage point of the F distribution

```
> 1 - pf(4.45, 4.5-2, 60 - 4.5)
[1] 0.01063
```

which is significant. So it seems that, although visually undramatic, the nonlinearity exhibited by the smooth is real. This model was actually fit by the expression

```
gam(Mileage ~ lo(Weight))
```

This is a call to `gam()` with a model formula that specifies a single smooth term in `Weight`, using the smoother `lo()`, which is an abbreviation for `loess()` (the newer version of the S function `lowess()`, described in Chapter 8). The amount of smoothing is set to the default (`span=1/2`); otherwise this parameter could be passed as well, as in the expression

```
gam(Mileage ~ lo(Weight, span = 1/3))
```

The `span=` argument gives the fraction of data to be used in computing the local polynomial fit at each point in the range of the predictor. Since this model only involves a single term rather than a sum of terms, it could also have been fit using the `loess()` function described in Chapter 8.

The plot on the right in Figure 7.1 also displays a scatterplot smooth, using a different predictor `HP` (horsepower) and a different smoother. It was created by the call

```
gam(Mileage ~ s(HP))
```

where `s(Mileage)` requests a smooth term to be computed using a smoothing spline. The smoothing parameter is also set to the default, which in the case of `s()` is `df=4` for the smooth term, or `5` in all for the overall fit. The `df=` argument stands for *degrees of freedom*, and is a useful way of calibrating a smoother. Smoothing splines are discussed in a bit more detail in Section 7.4.1; there we mention the stand-alone smoothing-spline function `smooth.spline()`, which could also have been used to produce the spline curve in Figure 7.1.

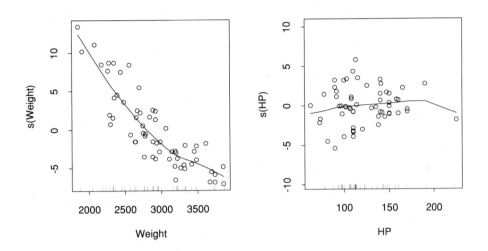

Figure 7.2: *An additive model relates* `Mileage` *to* `Weight` *and* `HP`. *Each plot is the contribution of a term to the additive predictor, and has as "y" label the expression used to specify it in the model formula. Each curve has been centered to have average 0. The effect of* `HP` *in this joint fit is greatly reduced from that in Figure 7.1.*

What smoother to use is a matter of taste, and the very question has given rise to a large research literature; visually the performance of the two used here seems comparable. In practice, both have complementary advantages and disadvantages. For example, it is almost as easy to fit two- or higher- dimensional surfaces with `loess()` as it is to fit one-dimensional curves; the computational complexity of smoothing splines increases dramatically as we move from curves to surfaces. Smoothing splines, on the other hand, minimize a data-defined convex criterion, while the `loess()` method is based on sensible heuristics; one consequence is that both the theoretical and numerical behavior of smoothing splines is cleaner than for `loess()`. We discuss the use of different smoothers in additive models in more detail in Section 7.3.4.

The variables `Mileage`, `Weight`, and `HP` in the data frame `car.test.frame` are available by name, because we attached the data frame for use in the entire session:

```
attach(car.test.frame)
```

This is a useful alternative to supplying the `data=` argument each time we fit a model.

We can model `Mileage` additively in `Weight` and `HP`:

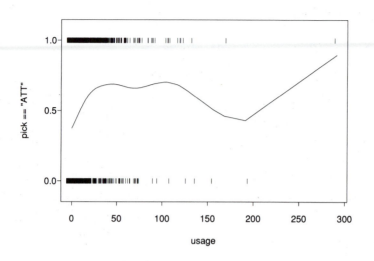

Figure 7.3: *A scatterplot smooth for binary data (jittered to break ties). The smooth estimates the proportion of* ATT *subscribers (1s) as a function of* usage.

```
auto.add <- gam(Mileage ~ s(Weight) + s(HP))
```

and plot the fitted model

```
plot(auto.add, residuals=T)
```

Figure 7.2 shows the result, which seems to indicate that the effect of HP is dramatically reduced in the presence of Weight. The curves in the plot are produced by the plot() method for gam objects, which joins up the fitted values for each term by straight line segments.

We can get a numerical summary of the fit by simply printing the gam object:

```
> auto.add
Call:
gam(formula = Mileage ~ s(Weight) + s(HP))

Degrees of Freedom: 60 total; 51 Residual
Residual Deviance: 306.4
```

Similarly the model fit to Weight alone prints as

```
> gam(Mileage ~ s(Weight))
Call:
gam(formula = Mileage ~ s(Weight))

Degrees of Freedom: 60 total; 55 Residual
Residual Deviance: 313.6
```

and we see that the residual deviance (or residual sum-of-squares in this case) has not increased much (relative to the average residual deviance of the bigger model). We should not be too surprised by this particular result; heavier cars tend to have higher horsepower, so on its own HP acts as a surrogate for the more important predictor Weight.

The overall predictor-response relationships are evident in Figure 7.1 without the smooth fits, although the finer details are not. Often the structure of interest is not at all evident for bivariate data. Figure 7.3 shows a plot of a binary variable

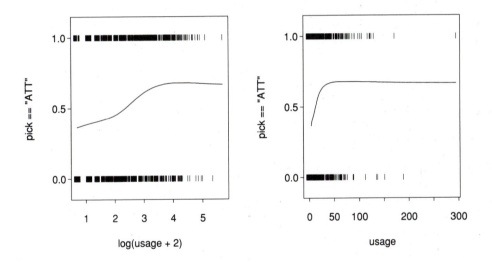

Figure 7.4: *The left figure smooths* pick *against the transformed* log(usage+2). *The right figure plots the same fit against the untransformed* usage.

pick against a numeric variable usage, two variables from the frame market.survey. The response pick indicates whether a household chose ATT or OCC (Other Common Carrier) as their long-distance carrier. These data are described in some detail in Chapter 3. Of interest in this particular plot is the proportion of ATT subscribers as a function of usage. Although we have randomly perturbed (*jittered*) usage to break ties, it is still difficult to detect the trend from the data alone. The scatterplot smooth is really needed here; it shows an initial increase which then flattens off. Although we could simply have smoothed the 0-1 data directly, that was not done here. The curve in Figure 7.3 was fit on the *logit* scale, which guarantees that the fitted proportions (which is what is plotted) lie in $[0, 1]$ (scatterplot smoothers do not in general guarantee this). Our fitting mechanism also takes the changing binomial variance into account. This model was fit by the call

```
gam(pick == "ATT" ~ s(usage), family = binomial)
```

which should look familiar to those readers who are reading this book serially. Notice the form of the response in the formula. The variable pick is a factor with levels "OCC" and "ATT", and we want to make sure that we are modeling the proportion of AT&T subscribers. We can express that preference directly in the formula, by creating the binary (logical) variable pick=="ATT". The fit in the right tail of usage appears to track the data rather closely. This is not surprising, since the data are very sparse in this region. Smoothers such as smoothing splines and loess give high leverage to outlying points such as these, and as a consequence the variance of the fit is high. In situations such as this, it is useful to transform the predictor prior to smoothing to bring in the long tails. In this case, the log transformation seems appropriate (a histogram of log(usage+2) appears symmetric with short tails). The transformation can be applied directly in the model formula, as in

```
mkt.fit1 <- gam(pick == "ATT" ~ s(log(usage + 2)), family = binomial)
```

The fitted values are shown in Figure 7.4, plotted against both log(usage+2) and usage. The fitted values are easily obtained using fitted(mkt.fit1), which produces the fitted mean of the response—in this case the fitted proportion of AT&T subscribers. So, for example, the right plot in Figure 7.4 was produced by the sequence of expressions

```
> plot(usage, pick == "ATT", type = "n", ylim = c(-.1, 1.1), yaxt = "n")
> axis(2, at = c(0, 1))
> points(jitter(usage), pick == "ATT", pch = "|")
> o <- order(usage)
> lines(usage[o], fitted(mkt.fit1)[o])
```

Along with fitted(), other generic functions such as residuals(), summary(), predict(), family(), deviance(), formula() produce appropriate results when applied to gam objects. Printing the fitted gam object

```
> mkt.fit1
Call:
gam(formula = pick == "ATT" ~ s(log(usage + 2)), family = binomial)

Degrees of Freedom: 1000 total; 995.01 Residual
Residual Deviance: 1326.9
```

we notice that the residual degrees of freedom is not an integral quantity. As we will see, the df of a nonparametric term is an intuitively defined quantity, and can take on fractional values. Our smoothing spline was requested to produce a fit with 4 df (the default), and it returned one with 3.99, which is certainly close enough for our purposes. One might compare such a fit to a parametric fit with four parameters, such as a quartic polynomial; only here the functional form is not nearly as rigid.

We now move on to some data used in Chapter 6 on spinal bending in children. Figure 7.5 shows the fitted functions for the additive logistic regression model:

$$\text{logit} P(Kyphosis) = \alpha + f_{Age}(Age) + f_{Number}(Number) + f_{Start}(Start)$$

relating the prevalence of Kyphosis, a spinal deformity in young children, to three possible predictors: Age, Number, and Start. The response indicates the presence or absence of this deformity, a forward flexion of the spine, after an operation to correct it. The last two predictors refer to the number of vertebrae involved in the operation, and the position of the first. These data are also used in Chapter 6 and are described there. The plot is produced by the call

```
plot(kyph.gam1, residuals = T, rug = F)
```

Each of the functions represented in the plot is the contribution of that variable to the fitted *additive predictor*, the analogue of the linear predictor in GLMs. The curves are drawn by connecting the points in plots of the fitted values for each function against its predictor. The vertical level of these plots is of no importance since there is an intercept in the model; the fitted values for each function are adjusted to average zero. We have included the partial residuals, which automatically put all the figures on the same scale. Notice that for Age and Start there are regions where the residuals are all below the fitted curve. These correspond to pure regions in the predictor space where only zeros occur in the sample, and tend to dramatize the fitted curves in those regions.

This model was fit by the call:

```
kyph.gam1 <- gam(Kyphosis ~ s(Age) + s(Number) + s(Start),
                 family = binomial, data = kyphosis)
```

The summary function produces

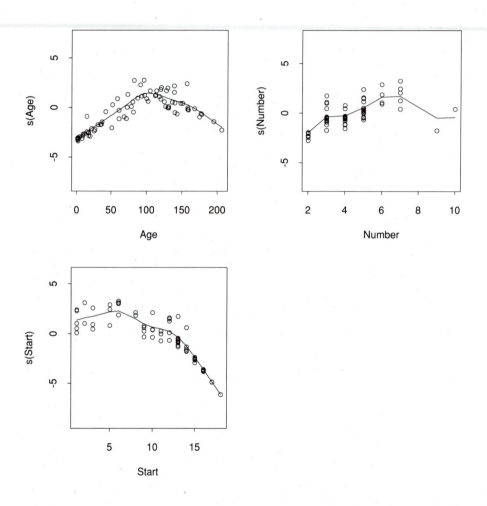

Figure 7.5: *A graphical description of the generalized additive model fit of the binary response* Kyphosis *to three predictors. The figures are plotted on the logit scale, and each plot represents the contribution of that variable to the fitted logit. Included in each of the plots are partial residuals for that variable.*

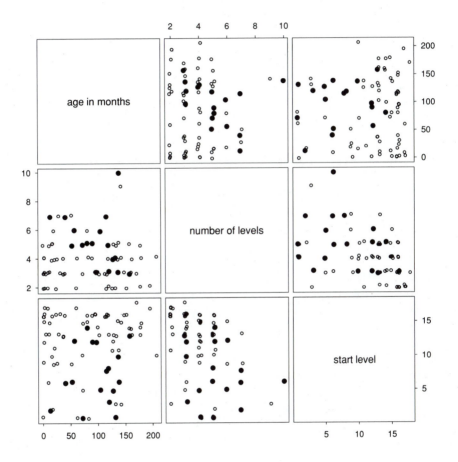

Figure 7.6: *A scatterplot matrix of the three predictors in the Kyphosis data. The presence (solid dots) and absence (hollow circles) of* Kyphosis *is indicated in the plots.*

```
> summary(kyph.gam1)

Call: gam(formula = Kyphosis ~ s(Age) + s(Number) + s(Start),
         family = binomial, data = kyphosis)
Deviance Residuals:
     Min       1Q   Median        3Q     Max
  -1.3603 -0.45752 -0.16406 -0.009855 2.0945


(Dispersion Parameter for Binomial family taken to be 1 )

    Null Deviance: 83.234 on 80 degrees of freedom

Residual Deviance: 40.53 on 68.086 degrees of freedom

Number of Local Scoring Iterations: 7

DF for Terms and Chi-squares for Nonparametric Effects

             Df Npar Df Npar Chisq P(Chi)
(Intercept)  1
    s(Age)   1      2.9        5.743 0.1162
 s(Number)   1      3.0        5.777 0.1263
  s(Start)   1      3.0        5.838 0.1181
```

The last part of the summary gives a crude breakdown of the degrees of freedom be-tween terms, and separates the parametric and nonparametric contributions within terms. This is represented in an anova table, which is a component of the output of summary.gam(). The column labeled "Npar Chisq" represents a type of score test to evaluate the nonlinear contribution of the nonparametric terms. In this case it seems to indicate that none of the nonlinear components are significant. Further details can be found in Section 7.4.5.

Figure 7.6 is a scatterplot matrix of the three predictors, with the response encoded. We see that Start and Number are negatively associated, and so it is possible that the fit would not suffer much by removing one of them. Let's remove the term in Number using the update() function:

```
> kyph.gam2 <- update(kyph.gam1, ~ . - s(Number))
> summary(kyph.gam2)

Call: gam(formula = Kyphosis ~ s(Age) + s(Start), family = binomial,
         data = kyphosis)
Deviance Residuals:
     Min       1Q   Median        3Q     Max
  -1.6917 -0.44878 -0.20098 -0.030184 2.0857
```

(Dispersion Parameter for Binomial family taken to be 1)

    Null Deviance: 83.234 on 80 degrees of freedom

Residual Deviance: 48.615 on 72.24 degrees of freedom

Number of Local Scoring Iterations: 6

DF for Terms and Chi-squares for Nonparametric Effects

```
              Df Npar Df Npar Chisq P(Chi)
(Intercept)   1
    s(Age)    1      2.9       6.122 0.1016
  s(Start)    1      2.8       7.639 0.0469
```

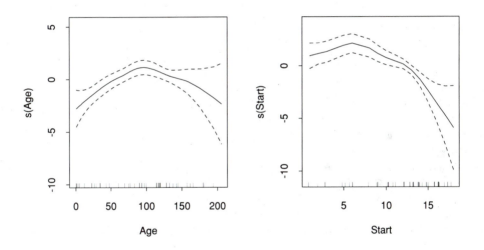

Figure 7.7: *The additive logistic fit of* Kyphosis *to* Age *and* Start. *The dashed curves are pointwise 2 × standard-error bands.*

We see that the term s(Start) has apparently gained in importance. The fitted functions for s(Start) look much the same for the two models, and are displayed in Figure 7.7. They were produced by the expression

```
plot(kyph.gam2, se = T)
```

We can use the anova() function to make the comparison of the two models for us:

```
> anova(kyph.gam1, kyph.gam2, test = "Chi")
Analysis of Deviance Table

Response: Kyphosis

                        Terms Resid. Df Resid. Dev      Test
1 s(Age) + s(Number) + s(Start)    68.09     40.53
2            s(Age) + s(Start)     72.24     48.61 -s(Number)

        Df Deviance Pr(Chi)
1
2 -4.154   -8.085   0.097
```

and find that the omitted term is not quite significant. These nonparametric curves suggested a quadratic term in `Age` and perhaps a low-order spline in `Start`; they led to the final parametric model selected for these data, as displayed in Figure 6.6 on page 221.

7.2.2   Plotting the Fitted Models

Since `gam()` fits an additive model consisting of a sum of flexible components, the emphasis in many of the functions in this chapter is on the individual terms in the formula. The `plot()` method for `gam` objects tries to produce a sensible plot for each term. To illustrate, suppose we fit the two-term model `mkt.fit2` to the market share data:

```
> summary(mkt.fit2)

Call: gam(formula = pick == "ATT" ~ s(log(usage + 2)) +
        income, family = binomial, data = market.survey,
        na.action = na.omit, trace = T)
Deviance Residuals:
     Min     1Q  Median    3Q    Max
  -1.6625 -1.1008 0.79345 1.1373 1.4811

(Dispersion Parameter for Binomial family taken to be 1 )

     Null Deviance: 1086.9 on 784 degrees of freedom

Residual Deviance: 1038.7 on 774.02 degrees of freedom

Number of Local Scoring Iterations: 2

DF for Terms and Chi-squares for Nonparametric Effects
```

```
                  Df Npar Df Npar Chisq P(Chi)
     (Intercept)  1
s(log(usage + 2)) 1        3      6.194 0.1014
          income  6
```

where `income` is a five-level factor. Notice that we have used the `na.action=` argument in creating `mkt.fit2`, since there are missing observations for the predictor `income`. The `na.omit()` action removes all observations missing any values in the model frame derived from `market.survey`. Section 7.3.2 gives more details on the use of `na.actions()`. The pair of plots in the top row of Figure 7.8 was created by the expression

```
plot(mkt.fit2, se = T, scale = 3)
```

A plot was produced for each term in the model, and since the `se=T` option was set, each includes upper and lower pointwise 2 × standard-error bands. Some features to notice in the pair of plots:

- The curve for `s(log(usage + 2))` is plotted against `usage` itself rather than against `log(usage+2)`. The *rugplot* at the base of the plot indicates the locations of the observed values of `usage`, randomly perturbed or *jittered* to break ties. The rugplot offers an explanation for the wide standard-error bands, since the observations are sparse in the right region of the plot.

- The variable `income` is an ordered factor with seven levels, and so an appropriate step function is produced. The jittering at the base of this plot results in solid bars because there are so many observations; the width of the bars is proportional to the level membership for the factor.

The `plot()` method is set up to produce a variety of different plots, depending on the nature of the term and the predictors involved. It will plot a curve for any term that can usefully be represented as a curve, a step function for a factor term, and two-dimensional surfaces for terms that are functions of two variables. In Section 7.3.5, we outline how users can add their own plot functions to the list.

For example, `plot.gam()` is used in Figure 6.6 in Chapter 6 to plot a polynomial term. The coefficients of polynomial terms and other similar terms that result in a curve, such as the B-splines produced by terms that are expressions in `bs()`, are kept together and the composed polynomial or spline is plotted as a function of the argument, `Age` in that case. Even straight line fits are usefully plotted in this fashion, for visual comparison with other terms in the model. The details and options given in this section are therefore pertinent for plotting fitted `lm`, `glm`, and even some `aov` models as well.

In Figure 7.8 we give an argument `scale=4` in the call to `plot.gam()`. The scale argument is a lower bound for the range of each vertical axis, in this case large

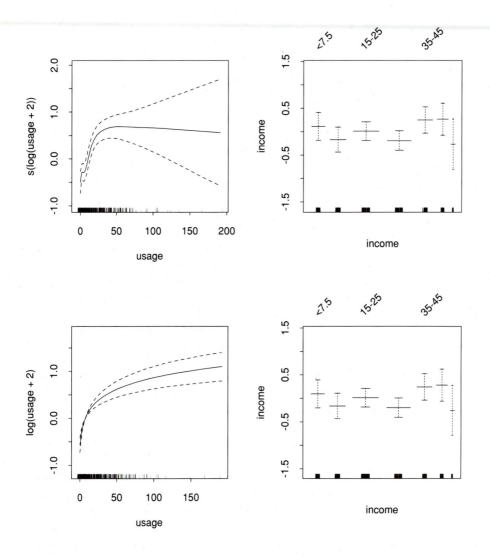

Figure 7.8: *Representations of some additive fits to the market share data. The top row was created by the expression* plot(mkt.fit2, se = T, scale = 3). *The x-axis in each plot is labeled according to the "inner" predictor in the term, such as* usage *in the term* s(log(usage+2)); *the y-axis is labeled by the term label itself. The bottom row is a plot of the* GLM *model* glm(pick=="ATT" ∼ log(usage+2) + income, binomial), *using a similar call to the function* plot.gam().

enough to ensure that both the plots are on the same vertical scale. This allows us to make visual judgments of how important different functions are relative to each other. Setting a common scale is essential when plotting purely linear terms; otherwise the lines would all be plotted as 45-degree diagonals.

The argument `se=F` is a logical flag, and is a request for pointwise standard-error curves. These are constructed at each of the fitted values by adding and subtracting two (pointwise) standard errors, resulting in an upper and lower curve that define an *envelope* around the fitted term. Under additional assumptions of no bias, these can be viewed as approximate 95% pointwise confidence intervals for the "true" curve.

The `rug=T` flag causes a frequency plot of the *x*-values to be represented at the base of the plot. Ties are broken by randomly jittering the values, and each value is represented by a vertical bar. Usually this information is available if residuals are included in the plot, which is also an option. In some cases, especially for generalized linear or additive models, adding residuals to a plot is unhelpful because they can distort the scale dramatically. Any interesting features in the functions get lost because of a few large residuals, even though they may carry a very small weight. In cases such as these, where residuals are omitted, the rugplot is useful since it warns us about influential predictor values. For example, in the top left panel in Figure 7.8, we see that three values of `usage` occupy about half its range! No wonder the standard-error bands are so wide in that region.

The `residuals=` argument can either be a logical flag, or a residual vector. If `residuals=T`, the plot for each term includes partial deviance residuals; these are simply the fitted term plus the deviance residuals. The `residuals=` argument can also be supplied with a residual vector rather than a logical flag, thus allowing the user to choose what partial residuals to use in the plots. For example

```
> plot(gamob, residuals = residuals(gamob, type = "working"))
```

would construct partial residuals by adding the working residuals to the fitted values for each term. The points resulting from this choice have the property that if the term in question is refit to the points using the final iterative weights, in a univariate fashion, then the same fitted term values would be obtained. We use the deviance residuals as the default because they have the variance information built in.

We have seen that a variety of choices have to be made in plotting the terms of an additive model. Not all terms are suitable for plotting; for example, plots are not available for interaction terms. Other choices have to be made for each plot, regarding standard-error curves, partial residuals, the rugplot, and the scale. For these reasons, `plot.gam()` has an interactive mode, initiated by the plotting parameter `ask=T`. When used, a menu is displayed on the screen:

```
> plot(mkt.fit2, ask = T)
Make a plotter selection:
```

```
1: plot: s(log(usage + 2))
2: plot: income
3: plot all terms
4: residuals on
5: rug off
6: se on
7: scale (0)
8: browser
9: exit
Selection:
```

Some of the menu options, such as "`se on`" in item 6, are flags that can be switched on or off; for example, the display "`se on`" means that currently the standard-error flag is off, but choosing this item will flip the switch. Subsequent plots would include standard-error bands.

Options 4, 5, 6, and 8 in this example are similar flags. The "`browser`" option allows the user access to a local frame within the function call, and therefore access to all the variables, the fitted values and the component functions. This is useful for augmenting the standard plots, or creating a new plot. It is up to the user to define the plotting region before invoking `plot()`. By using the "`browser`" option, users can reset the plotting parameters, or even change devices without leaving `plot.gam()`.

When the "`scale`" item (7 above) is selected, the user is prompted to enter a new vertical scale; thereafter, each plot will be produced using the new scale. After each plot is completed, the scale actually used is printed on the screen; this helps in selecting an appropriate scale for all the plots. We give further details of the plot function in Section 7.3.5.

Getting back to the examples, it is interesting to note that on page 264 the `summary()` function flags the nonlinear part of the nonparametric term for `usage` in `mkt.fit2` as nonsignificant. This claim is supported by fitting the model

```
pick == "ATT" ~ log(usage + 2) + income
```

So although we are using logarithms to improve the behavior of the smoother, it appears that the nonparametric term is well approximated by a term linear in `log(usage + 2)`. The bottom row of Figure 7.8 displays this model. The standard-error bands for the log-linear term are much narrower for large values of `usage` since three less parameters are being used in the fit.

7.2.3   Further Details on gam()

The functions for fitting additive models look very similar to those for fitting linear and generalized linear models; the only change is in the `formula=` argument. To fit the functions in Figure 7.5 on page 260, we used the formula

```
Kyphosis ~ s(Age) + s(Number) + s(Start)
```

The s() function indicates that a smooth term is to be fitted as a function of its argument, using a smoothing spline as the smoother and the default amount of smoothing. We are able to mix smooth terms with linear terms or factors as in

    pick ~ s(usage) + income

and mix smoothers within an additive model

    Kyphosis ~ poly(Age, 2) + lo(Number) + s(Start)

All the other possibilities available for GLMs and LMs can be mixed in with smooth terms; so, for example, both

    ozone ~ log(ibt) + poly(dpg, 3) + s(ibh)
    ozone ~ ibt*ibh + s(dpg)

are also accommodated. The former indicates a linear term in the log of ibt, a cubic polynomial in dpg, and a smoothing spline term in ibh. The latter indicates an interaction term between the quantitative predictors ibt and ibh, consisting of main effects and tensor product interaction, and a smoothing spline term in dpg.

The function s() can have more than one argument. A second argument is df=, which determines how much smoothing is done (default df=4). The units are in degrees of freedom, a convenient but approximate way of calibrating a nonparametric smoother. A third possible argument is the more customary but less intuitive smoothing parameter that we call spar. These parameters are described in more detail in Section 7.4.

The term lo() can be used to specify functions of more than one variable, using the loess() smoother. For example, lo(wind, rad, 1/2) implies a smooth nonparametric loess surface as a function of wind and rad, using a span or neighborhood size of 50%. There is an optional degree= argument to lo() that can be 1 (default) or 2 for local quadratic fits.

To avoid any possible confusion, it is probably worth noting up front an important detail about the implementation of gam(), which is described at greater length in Sections 7.3.4 and 7.4.5. A term using lo() or s(), such as s(Age), does not itself evaluate to a smooth term. It evaluates to an object that conveniently packages up the information needed for gam() to jointly model that term with the others in the model. In the case of s(Age), the evaluation results in Age itself, with some attributes containing the other arguments to s() and more. This behavior is consistent with a linear parametric term such as poly(Age, 3), which does not evaluate to a cubic polynomial itself, but rather to a basis for polynomial regression.

The software does not currently fully accommodate interaction terms of the sort a:s(b) or a:lo(b) (anything interacting with smooth term), although in principle this is possible. Even more plausible would be terms of the type a/s(b), or separate nonparametric curves within each level of the factor a. We say "not fully", since using such terms is not illegal; they simply result in the usual linear interaction between a and the term that results from the evaluation of s(b) or lo(b). Of course, parametric versions of these interactions could be used instead; this issue is resumed in the next two sections.

In principle, any number of different smoothers can be used; see Section 7.3.4 for details on adding personalized smoothers to the list.

The object returned by gam() has the following components:

```
> names(kyph.gam1)
 [1] "coefficients"         "residuals"      "fitted.values"
 [4] "R"                    "rank"           "smooth"
 [7] "nl.df"                "df.residual"    "var"
[10] "assign"               "terms"          "call"
[13] "formula"              "family"         "nl.chisq"
[16] "y"                    "weights"        "iter"
[19] "additive.predictors"  "deviance"       "null.deviance"
```

A gam object inherits from class "glm":

```
> class(kyph.gam1)
[1] "gam" "glm" "lm"
```

and has a few additional components:

- $smooth is a matrix of fitted smooth functions with as many columns as there are s() or lo() terms in the formula;

- $var is a matrix, like $smooth, of pointwise variances;

- $nl.df is a vector of the effective degrees of freedom for the nonlinear part of each smooth term; and

- $nl.chisq is a vector of chi-squared statistics that approximate the effect of replacing each nonparametric curve by its parametric component.

Once again functions like residuals(), fitted, and predict() are useful for extracting particular components from the fitted object.

7.2.4   Parametric Additive Models: bs() and ns()

The focus of this chapter is on flexible methods for fitting terms in an additive model. The greatest flexibility for any single term is achieved by using a non-parametric regression smoother, with user control over the smoothing parameter. In some applications we might prefer an intermediate level of control. Flexible parametric methods exist that fill this gap. These typically construct the fit for a term using a parametric fit to a set of basis functions. Some simple examples that we have already seen include step functions—e.g., `cut(Age, 4)`—and polynomial terms—e.g., `poly(Age, 3)`. Although somewhat less flexible than the nonparametric techniques, these models (albeit large at times), are fitted using weighted least squares rather than the iterative algorithms required to fit nonparametric GAMs. Although this implies a slightly faster fitting method, the more important consequence is that the fit is a least-squares projection, while the nonparametric fits are not. Consequently, issues such as degrees of freedom, standard-error bands, and tests of significance are straightforward, while for nonparametric GAMs we rely on approximations and heuristics.

A special class of parametric linear functions with a flexibility approaching that of smoothing splines and `loess()` are the piecewise polynomials and splines as specified by `bs()` (B-splines) and `ns()` (natural splines). We give a very brief description here, and refer the reader to the literature for more details.

A piecewise polynomial requires the placement of interior *knots* or breakpoints at prechosen places in the range of the predictor. These knots separate disjoint regions in the data, and the regression function is modeled as a separate polynomial piece in each region. It is common to require the pieces to join smoothly; a *polynomial spline* requires that the $d - 1$th derivatives be continuous at the knots when the pieces are dth-degree polynomials. Cubic splines are very popular. The space of functions that are cubic splines with a given set of k interior knots is a linear space of dimension $k + 4$, and so $k + 4$ basis functions are needed to represent them:

$$f(x) = \alpha + \sum_{j=1}^{k+d} \alpha_j B_j(x)$$

where $d = 3$ for cubic splines. B-splines are one particular class of basis functions that represent piecewise polynomials, popular in the numerical analysis literature. The function `bs()` computes the values of the $k + d$ B-spline basis functions at the n values of its argument, and returns the $n \times (k + d)$ matrix of evaluations. So, for example, a cubic spline term in `Start` with one interior knot at `12` can be specified by `bs(Start, knots = 12)`, and will result in a term with 4 df.

Other arguments to `bs()` besides the variable itself are

- `knots=`: a vector of (interior) knot locations. The degree of continuity at a given knot can be dropped by duplicating the knot. So j copies of a knot for a d-degree spline results in continuity of order $d - j - 1$.

- `df=`: rather than specify knots, one can simply give the degrees of freedom, and have `bs` place $df - d$ knots uniformly along the range of the predictor.

- `degree=`: the degree d of the spline, with a default of 3 for cubic splines.

- `intercept=`: by default, `intercept=F` and `bs()` evaluates to a matrix whose columns are all orthogonal to the column of 1s.

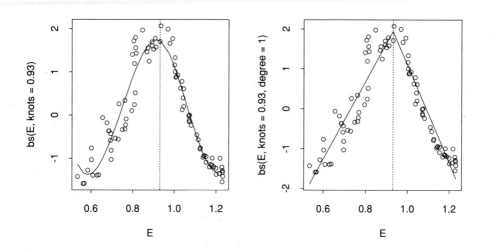

Figure 7.9: *A demonstration of B-spline functions using* bs()*. The y-labels show the term used in a call to* lm()*; the functions were plotted using* plot.gam()*. The dotted vertical lines are included to show the placement of the single knot.*

Figure 7.9 shows two splines fitted to the scatterplot of NOx versus E, two variables from the ethanol data frame. The response NOx is a measure of nitric oxide concentration in exhaust emissions from automobiles, and the predictor E is the equivalence ratio, a measure of fuel/air mixture. These data are described in more detail in Chapter 8. The function on the left, a cubic spline with a single interior knot at 0.93, was created by a call to lm():

```
gas.bs1 <- lm(NOx~ bs(E, knots = 0.93))
```

and then plotted using plot.gam(). The function on the right was created by the expression

```
gas.bs2 <- lm(NOx~ bs(E, knots = 0.93, degree = 1))
```

The function ns() is similar to bs(); it produces a basis for a *natural cubic spline* function in its first argument. A natural cubic spline is a cubic spline with the additional constraint that the function is linear beyond the boundary knots, which we take to be the endpoints of the data. Natural cubic splines tend to have better behaved tails than cubic splines, and the 2 degrees of freedom saved by the endpoint constraints can be spent on additional knots in the interior. As an illustration, we refit the model above using ns() rather than bs():

```
gas.ns <- lm(NOx~ ns(E, knots = c(0.7, 0.93, 1.1)))
```

Here we used two additional interior knots, and end up with the same degrees of freedom as `gas.bs1`:

```
> anova(gas.ns, gas.bs1, gas.bs2)
Analysis of Variance Table

Response: NOx

                              Terms Resid. Df   RSS    Test Df
1 ns(E, knots = c(0.7, 0.93, 1.1))         83  9.09
2               bs(E, knots = 0.93)        83  9.83 1 vs. 2  0
3  bs(E, knots = 0.93, degree = 1)         85 12.98 2 vs. 3 -2

  Sum of Sq F Value      Pr(F)
1
2    -0.743
3    -3.149   14.38 4.343e-06
```

Notice that the anova table does not report F values for model 1 versus 2 since they have the same degrees of freedom; the difference is still large though, relative to the residual variance of about 0.1. Figure 7.10 graphs the fitted natural spline, showing the knot placement as well as pointwise twice standard-error curves. Although these models were all fit by `lm()`, the plots were all produced by `plot.gam()`.

Further details can be found in the detailed documentation of `bs()` and `ns()`.

7.2.5   An Example in Detail

The example in the previous section is based on one of the two predictors in the `ethanol` data frame. In this section, we explore the `ethanol` data further using additive models; these data also receive considerable attention in Chapter 8. We perform many fits, summaries, and plots, and in so doing demonstrate the ease with which quite complex analyses can be simply performed using the hierarchy of additive modeling software.

Figure 7.11 (top row) shows `NOx` plotted against both `E` and `C`. `C` stands for compression ratio, and the other two variables are described in the previous section. The right plot seems to suggest that an additive model would be ideal for modeling these data; in fact `C` does not seem to play a role. Indeed, we fit the model

```
> attach(ethanol)
> eth1 <- gam(NOx ~ C + lo(E, degree = 2))
> eth1
Call:
gam(formula = NOx ~ C + lo(E, degree = 2))
```

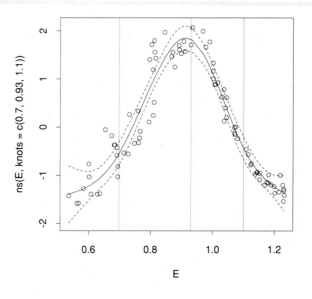

Figure 7.10: *A demonstration of a natural cubic B-spline using* `ns()`. *The vertical lines indicate the placement of knots. The upper and lower curves are pointwise twice standard-error bands.*

```
Degrees of Freedom: 88 total; 80.118 Residual
Residual Deviance: 5.1675
```

and show the plotted functions in the bottom row of Figure 7.11. We have included the partial residuals and the pointwise twice standard-error curves. The fit seems acceptable at face value. The figure suggests that perhaps the linear term in C is not needed; we can check this by simply dropping it:

```
> eth2 <- update(eth1, ~ . - C)
> eth2
Call:
gam(formula = NOx ~ lo(E, degree = 2))

Degrees of Freedom: 88 total; 81.118 Residual
Residual Deviance: 9.1378
```

and we see that the residual deviance has increased dramatically relative to the error variance estimate 5.2/80.2=0.06 for 1 `df`.

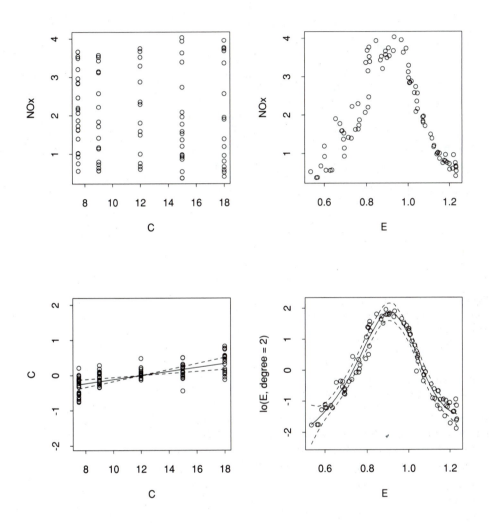

Figure 7.11: *The top row shows* NOx *plotted against* C, *the combustion level, and* E, *the equivalence ratio for the* ethanol *data. The bottom row shows an additive model fit, where* C *is modeled linearly and* E *is modeled by a locally quadratic smooth term, specified by* lo(E, degree = 2).

The next question is whether an additive surface is sufficient. We fit a two-dimensional smooth surface, again using loess():

```
> eth3 <- gam(NOx ~ lo(C, E, 1/4, degree = 2))
> eth3
Call:
gam(formula = NOx ~ lo(C, E, 1/4, degree = 2))

Degrees of Freedom: 88 total; 62.19 Residual
Residual Deviance: 1.74
```

We use a span of 1/4, the square of the span used for the loess curve in the additive model fit. This ensures (approximately) that the marginal span for E is the same for both fits, and as a consequence that the models are approximately nested. The anova() function can be used to compare the models:

```
> aov1<- anova(eth2, eth1, eth3, test = "F")
> aov1
Analysis of Deviance Table

Response: NOx
```

	Terms	Resid. Df	Resid. Dev	Test	Df
1	lo(E, degree = 2)	81.12	9.138		
2	C + lo(E, degree = 2)	80.12	5.168	+C	1.00
3	lo(C, E, 1/4, degree = 2)	62.19	1.740	2 vs. 3	17.92

	Deviance	F Value	Pr(F)
1			
2	3.970	141.9	0.000e+00
3	3.427	6.8	5.143e-09

and the effect looks rather small. However, the residual deviance is also small, and the F-test shows that the interaction surface is strongly significant.

Figure 7.12 is a perspective plot of the the bivariate smooth term in eth3, produced by the expression plot(eth3). The surface in fact looks additive, and a similar plot of the additive surface defined by eth2 shows no perceptible difference. Perspective plots have their limitations; in this particular case, the structure is dominated by the strong quadratic effect of E, and departures from additivity are not evident.

There is a noticeable roughness in the perspective plot, an artifact due to the technique used to construct it. The default style for plotting gam objects is to plot the fitted values for each term against the corresponding predictors, using the data that were used to fit the object. In the case of surfaces such as the one here, these fitted values occur at irregularly placed (E, C) pairs in the plane. The S function interp() is used within gplot.matrix() (Section 7.3.5) to approximate the

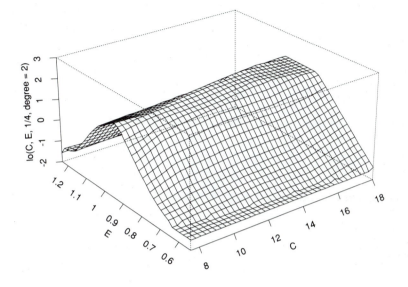

Figure 7.12: *A perspective plot of the bivariate surface smooth term in the model* `eth3`, *produced by the* `plot()` *method for* `gam` *objects.*

fitted values on a grid, the required form of input for `persp()`. The interpolation algorithm used by `interp()` introduces the roughness evident in the surface. A more sophisticated method for extracting fitted curves and surfaces on a grid of values, described in Section 7.3.3, does not produce these anomalies.

Next we use the condition plot routine `coplot()` described in Chapter 8 to show the interaction structure remaining in the residuals from the additive model fit. The separate least-squares fits in each plot show that the slope of `C` changes with the level of `E`. Such behavior cannot be modeled simply using an additive model. The following sequence of expressions produces the `coplot` displayed in Figure 7.13:

```
> to.do <- function(x, y){
        points(x, y)
        abline(lsfit(x, y))
    }
```

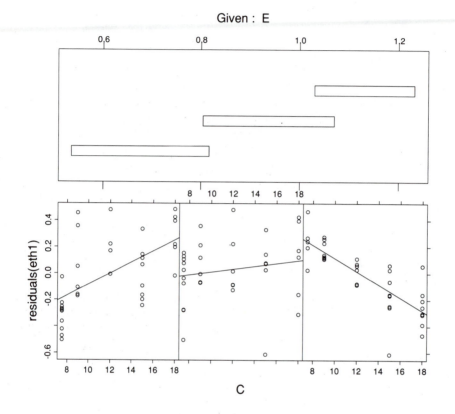

Figure 7.13: *A* `coplot()` *shows the residuals from the additive model fit plotted against* C, *given three different overlapping intervals of the values of* E. *The interaction structure is evident.*

```
> E.int <- co.intervals(E, number = 3, overlap = 0.1)
> coplot(residuals(eth1) ~ C|E, given = E.int, panel = to.do)
```

These data are pursued in more detail in Chapter 8, where a conditionally parametric model seems to fit the bill. The `coplot()` function is described there as well.

Finally, to wrap up this example for the moment, let's proceed along more conventional lines. The following `anova()` display compares two parametric linear models which address the interaction question:

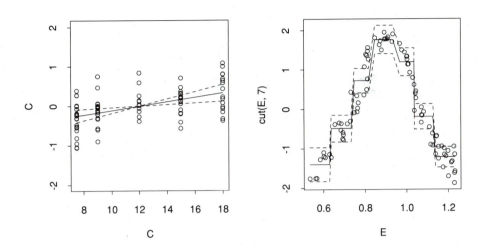

Figure 7.14: *A plot of the terms in the parametric model* glm(NOx ~ C + cut(E, 7)), *with pointwise twice standard-error bands and partial residuals. The* plot.gam() *method represents a categorical transformation of a quantitative predictor as a piecewise constant function.*

```
> anova(eth4, eth5, test = "F")
Analysis of Variance Table

Response: NOx

           Terms Resid. Df   RSS        Test Df Sum of Sq
1 C + cut(E, 7)     80     12.68
2 C * cut(E, 7)     74      8.21  +C:cut(E, 7)  6     4.472

   F Value      Pr(F)
1
2   6.717 1.046e-05
```

Here we have approximated the quadratic shape for the E effect by making seven cuts along the range of E. Once again the F-test rejects the no-interaction hypothesis, but we had to sacrifice a good deal of fit in the process. It is interesting to plot the additive fit eth4:

```
> plot.gam(eth4, residuals = T, scale = 4, se = T, rug = F)
```

The results are shown in Figure 7.14, and we see the piecewise constant approximation to the quadratic curve.

A better approximation could have been achieved using piecewise linear or cubic functions rather than piecewise constant:

```
> eth6 <- gam(formula = NOx ~ C * bs(E, df = 6))
> eth6
Call:
gam(formula = NOx ~ C * bs(E, df = 6))

Degrees of Freedom: 88 total; 74 Residual
Residual Deviance: 2.3963
```

Here we used the same degrees of freedom as in the model eth5 above, but achieved a much better fit. These last two models, eth5 and eth6, can be viewed as conditionally linear models; they serve as parametric counterparts for the nonparametric conditionally linear model proposed for these data in Chapter 8. Notice that we used gam() to fit the model eth6, although it is entirely parametric. No extra cost is incurred in doing this, and it facilitates making comparisons with the two preceding models:

```
> anova(eth3,eth6,eth5,test="Cp")
Analysis of Deviance Table

Response: NOx


                          Terms Resid. Df Resid. Dev   Test    Df Deviance
1 lo(C, E, 1/4, degree = 2)      62.2        1.74
2           C * bs(E, df = 6)    74.0        2.40 1 vs. 2 -11.8   -0.66
3           C * cut(E, 7)        74.0        8.21 2 vs. 3   0.0   -5.81

    Cp
1 3.18
2 3.18
3 8.99
```

We compare the models using the C_p statistic since they are not nested. The first two models are roughly equivalent in terms of C_p; the better fit of the nonparametric model is offset by the extra degrees of freedom required to fit it.

7.3   Specializing and Extending the Computations

7.3.1   Stepwise Model Selection

In Section 6.3.5, we describe an efficient stepwise model selection method step.glm() for selecting a GLM model. Here we describe a much more general but less computationally efficient version that will also work on `glm` and `lm` objects.

The function `step.gam()` allows one to step through arbitrary models along a prespecified path. The syntax of the function is

```
step.gam(object, scope,  scale, direction, keep, ...)
```

The argument `scope=` is a list, with each element corresponding to a term in the model. This is different from the `scope=` argument of `step.glm()`. Each of the elements of this list is a `formula` object that specifies an *ordered regimen* of candidate forms for the term. Each candidate model is constructed by pasting together a formula consisting of a single element from each term formula, in conjunction with the formula used in the initial model `object`. Here are some examples of term formulas, the elements of the `scope` list:

- `~1 + income`: selecting the 1 for this term is the natural way for `step.gam()` to remove a term in `income` from the model. Otherwise `income` enters linearly or as a factor, depending on its class.

- `~1 + log(usage + 2) + s(log(usage + 2))`: this allows the choice between no term, a log-linear term, or a smooth term in `usage`.

- `~1 + age + poly(age, 2) + s(age) + s(age, 7)`: select the form of the non-linear effect from a class ordered in richness.

- `~1 + education + income:education`: this makes most sense if `income` is in the model by default. It checks whether `education` should enter, enter additively, or as an interaction with `income`.

Suppose the four examples above were actually the four elements of the scope argument to a call to `step()`; then an example of a valid formula within the scope of `step()` is

```
 . ~ 1 + log(usage + 2) + s(age) + education
```

where the first term is represented by the 1 (and is effectively left out), the second term by `log(usage + 2)`, and so on. The "." gets replaced by the response in `object`.

We restrict the term formulas in `scope` to be ordered; this means that `step.gam()` will only look ahead or back one step each time from the current version of a term. If there are p terms with k choices each, this reduces the number of models that need to be tried for each term change from $(k-1)p$ to at most $2p$. The function ensures that it never visits a model more than once.

The first argument `object` to `step()` is a `gam` object, which the function uses as its starting model. Each of the formulas in the `scope` argument must be represented in the formula of `object`; otherwise an error is reported.

Starting with `object` as the *current model*, a series of models is constructed and fitted by successively moving each term up or down in its scope formula. The argument `direction=` to `step.gam()` determines whether steps are made in a `"backward"` or `"forward"` direction, or in the default direction, which is `"both"`. Thus, if the example above was the starting model and `direction="both"`, the first few trial models considered by `step()` would have formulas

```
. ~ income + log(usage + 2) + s(age) + education
. ~ 1 + s(log(usage + 2)) + s(age) + education
. ~ 1 + 1 + s(age) + education
. ~ income + log(usage + 2) + s(age) + education
. ~ 1 + log(usage + 2) + s(age, 7) + education
```

and so on.

The model that results in the biggest decrease in *AIC*:

$$AIC = D + 2df\phi$$

is then selected as the current model, and the updating is repeated. Here df is the effective degrees of freedom used in fitting `object`, and ϕ is the dispersion parameter. The argument `scale=` to `step.gam()` is the *cost* (divided by 2) per `df` incurred by adding or dropping a term. It defaults to the dispersion parameter for the initial model, which is 1 for binomial or Poisson models, else the scaled Pearson chi-squared statistic. If all the modified models cause `AIC` to increase, the function stops and returns the `gam` object of the best-fitting model visited (in the *AIC* sense).

The object returned by `step()` has two additional components:

- `$anova` is an `anova` object that summarizes the models selected along the path to the final model;

- `$keep` is a list of the items created by the `keep()` function, an optional argument to `step.gam()`. This list is constructed by applying `keep()` to every model visited during the model search, and then repacking the results in a convenient form. By default nothing is kept.

We illustrate the `step()` method on the marketing data described in Sections 2.1.3 and 8.2.1. There are missing observations in these data, so we use the 759 observations in `market.clean` for which complete data are available:

```
> market.clean <- na.omit(market.survey)
```

We deliberately do not use any of the alternative missing data strategies, since with model selection going on, it seems important to fit all the models to the same data. The response is the binary `pick` (`ATT` or `OCC`), and all the predictors are either factors or ordered factors, with the exception of `usage`, which is quantitative (Section 7.2.1).

Our initial model is simply additive in all the factors, with a linear term in `usage`:

```
> mkt.start <- gam(pick ~ ., data = market.clean,
+       family = binomial)
> mkt.start
Call:
gam(pick ~ ., data = market.clean, family = binomial)

Degrees of freedom: 759 total; 723 residual
Residual Deviance: 942.2
```

We need to construct the `scope` argument. For this example, we will simply allow each of the factors to be in or out, and allow the term in `usage` to be in, out, or a smooth term `s(usage)`. Although this could easily be done manually, it seems such a useful default scope argument that we provide the function `gam.scope()` for producing it as a default:

```
> mkt.scope <- gam.scope(market.clean)
> mkt.scope
$income:
 ~ 1 + income
$moves:
 ~ 1 + moves
$age:
 ~ 1 + age
$education:
 ~ 1 + education
$employment:
 ~ 1 + employment
$usage:
 ~ 1 + usage + s(usage)
$nonpub:
 ~ 1 + nonpub
$reach.out:
 ~ 1 + reach.out
$card:
 ~ 1 + card
```

We also write a function `mkt.keep()` to save particular components of the models visited:

```
> mkt.keep <- function(object, AIC)
+     list(df.resid = object$df.resid, deviance = object$deviance,
+     term = as.character(object$formula)[3], AIC = AIC)
```

The component `"term"` is a character version of the right side of the formula. We now execute `step.gam()` (and probably go for a cup of coffee; 50 models are visited in this example!)

```
> mkt.step <- step(mkt.start, mkt.scope, keep = mkt.keep)
> mkt.step
Call:
gam(formula = pick ~ moves + s(usage) + nonpub + reach.out +
        card, family = binomial, data = market.clean)

Degrees of Freedom: 759 total; 742.06 Residual
Residual Deviance: 954.32
```

The final model consists of five of the nine predictors, with the smooth term for usage selected. The deviance has increased by about 12 for an increase of 19 residual degrees of freedom.

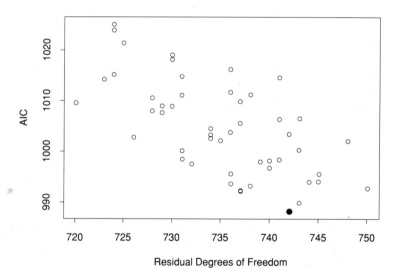

Figure 7.15: *An AIC plot of the models visited during the stepwise search by* step.gam() *for the (stepwise) optimal* mkt.step. *The best-fitting model is plotted using a black dot.*

The anova component of mkt.step summarizes the path taken to the final model:

```
> mkt.step$anova
Stepwise Model Path
Analysis of Deviance Table

Initial Model:
```

```
pick ~ income + moves + age + education + employment + usage +
       nonpub + reach.out + card


Final Model:
pick ~ moves + s(usage) + nonpub + reach.out + card


Scale: 1
```

	From	To	Df	Deviance	Resid. Df	Resid. Dev	AIC
1					723.0	942.2	1014
2	income		6.000	5.45	729.0	947.6	1008
3	employment		6.000	6.53	735.0	954.2	1002
4	age		5.000	4.58	740.0	958.8	997
5	usage	s(usage)	-2.941	-10.32	737.1	948.4	992
6	education		5.001	5.88	742.1	954.3	988

Starting from the initial model, the table reports the changes in the order that they occurred.

The `keep` component of `mkt.step` is a list with elements

```
> names(mkt.step$keep)
[1] "df.resid" "deviance" "term"      "AIC"
```

Each is a vector of length 50, the number of models visited by `step.gam()`. Figure 7.15 plots the AIC statistic against the residual deviance for each of the models visited, with the final model indicated. One might also use an S function such as `identify()` in conjunction with this plot to see what models are close to the best model.

We end this section with a brief discussion of further differences between the model selection methods for GLMs and GAMs. The method `step.glm()` is based on the primitives `drop1()` and `add1()`, which in turn are based on a quadratic approximation to the likelihood for GLM models. With the more flexible `scope=` argument to `step.gam()`, `drop1()` and `add1()` are no longer applicable, because terms get *changed* rather than added or dropped. There are in fact no `drop1()` and `add1()` methods for `gam` objects, but since `gam` objects inherit from class `"lm"`, these functions would produce results. We have deliberately not blocked this usage since the results can be interpreted with caution. Parametric terms in the model can be dropped or added, with the effect of freezing the nonlinear components of any nonparametric terms present. The results provided for the nonparametric terms should be ignored. However, the `summary()` method for `gam` objects fills the gap, and gives approximate "drop" information about the nonlinear parts of the nonparametric terms; Section 7.4.5 has more details.

7.3.2 Missing Data

All the modeling functions have an `na.action=` argument. An `na.action()` is a
filter function that takes a data frame as input and returns a "clean" data frame.
For example, in Section 7.2.2 we pass the function `na.omit()` as the `na.action=`
argument in creating `mkt.fit2`. The action of `na.omit()` is to omit all rows from
the data frame missing values on *any* of the variables. There are two ways of using
`na.omit()`:

- When `na.omit()` is passed as an argument to the fitting function, it gets
  applied to the model frame. This is the data frame created internally that
  consists of only those observations required to compute the fit. Consequently,
  only those variables will be checked for missing observations; in some cases,
  as in the construction of `mkt.sm` for Figure 7.4, no observations are omitted.
  In the case of `mkt.fit2`, 215 observations were omitted.

- One can create a new data frame at the onset by applying `na.omit()` to the
  original data frame. For example, in Section 7.3.1 we create the new frame

  ```
  market.clean <- na.omit(market.survey)
  ```

  which we then use as the data in the stepwise model selection procedure. This
  is important in such applications, since one would like model comparisons to
  be based on the same sample sizes.

The `gam()` function has a way of dealing with missing data in a reasonably
natural way in conjunction with *replacement* `na.action()` functions. A replacement
`na.action()` is one that replaces missing observations rather then removing them.
The simplest of these treat each of the variables in a data frame separately, and apply
some replacement rule depending on the class of the variable. More complicated
methods would treat the data frame as a whole in order to impute values for the
missing data. Which particular methods are best, if any, is an ongoing debatable
subject, and consequently we have not recommended any particular `na.replace()`
function for general use with this software. Rather we outline a particular example,
`na.gam.replace()`, that blends nicely with additive models; another, similar choice
suitable for tree-based models, `na.tree.replace()`, is described in Chapter 9.

The function `na.gam.replace()` operates on each of the variables in its argument
`frame` separately, and replaces them in the following fashion:

- Quantitative predictors have their missing observations replaced by the mean
  of the nonmissing observations.

- Quantitative matrices are treated similarly, where a row is regarded as missing
  if any observation in any of the columns is missing for that row.

- Factors or ordered factors get endowed with a new level, labeled `NA`, that records all the missing observations.

- If the data frame is a model frame, and in particular if the response can be identified, then all rows having missing response values are removed.

So `na.gam.replace()` returns a *clean* data or model frame. This `na.action()` is not specific to GAMs, and can be used in any other context, such as with GLMs or LMs in particular.

Let's understand what happens when a linear model is fit to the variables in this filtered data frame. Factors have an extra effect estimated that isolates the missing data. The coefficient of a term linear in a quantitative predictor with missing observations is not directly influenced by them since they have all been replaced by the mean of the nonmissing values for that predictor. With a constant in the model, we can regard the predictors as centered (zero mean), in which case those recorded at exactly the mean have zero leverage. The fitted term values for those observations that are missing data for that particular term are therefore zero.

The backfitting algorithm in `gam()` cycles around the smooth terms, updating them in an iterative fashion. At each stage the fitting operation involves a single term. The natural thing to do in `gam()`, when missing predictor values are encountered, is mimic the behavior described above for linear models:

- ignore the missing predictor values when computing the smooth term for a particular predictor, and

- return zero as the fitted smooth values corresponding to the missing observations for that term.

This makes sense since fitted nonparametric terms in an additive model are centered to have average value zero, so the missing predictors get assigned the mean value of the curve.

This all works because `gam()` and its associated functions are set up to anticipate the missing data. During the construction of the model frame, the functions `lo()` and `s()` always detect missing observations, and if present, attach an `"na"` attribute to their output, recording the rows that are missing. After replacement functions such as `na.gam.replace()` have made their changes, these `"na"` attributes remain on the "smooth" terms, even though the values of the terms have been modified. The consequence of this is that the model matrix can be constructed on the replaced data frame, just as it would be for linear models; the model matrix is used to compute the parametric component of the fit. These `"na"` attributes are picked up again when `gam()` comes to compute a particular term for the nonparametric component; the missing observations are omitted from the smooth, and their fitted values are returned as zero.

This strategy will work for any replacement `na.action()`, as long as it does not inadvertently strip off the `"na"` attribute of the variables during its replacement operation. If it were to do that, the function would still produce results, but they would not be as interpretable.

7.3.3   Prediction

Often we wish to evaluate the fitted model at some new values of the predictors: for predictive purposes, for plotting, or for validation. There is a `predict()` method for `gam` objects, just as there is for other classes of objects. However, we will see that the method `predict.gam()` also serves as the "safe" method for predicting from new data for both `glm` and `lm` model objects.

Just as for `glm` objects, the expression `predict(gamob)` is a simple but clean way of extracting the additive predictor from `gamob`—in other words, making predictions at the same predictor values that were used to fit `gamob`. Similarly, `fitted(gamob)` extracts the fitted values. Once again

```
predict.gam(gamob, type = "response")
```

is identical to

```
fitted(gamob)
```

The other choice is `type="terms"`, in which case a matrix of fitted terms is returned. The argument `terms=` is used in conjunction with the choice `type="terms"`, and specifies a character vector of term labels for which predictions are desired; by default, all terms are predicted. The `se.fit=T` argument causes a list to be returned that contains both fitted values and standard errors. All these arguments are exactly the same as for `predict.glm()`, and are discussed in some detail in Section 6.3.6. The difference occurs when the `new.data=` argument is used.

In Sections 4.2.3 and 6.3.6 we draw attention to problems that can occur in using coefficients from a fitted `lm` or `glm` object on new data. Both `predict.lm()` and `predict.glm()` compute a model matrix using the new data, which is then multiplied by the coefficients extracted from the original object. This will work as long as the expressions defining the terms in the formula do not depend on the entire data vector for their evaluation. The functions `poly()` and `lo()`, for example, normalize their arguments to have length 1, an operation that depends on the entire vector of values. A term like `bs(E, df = 5)` is an elegant way of specifying a B-spline with 5 degrees of freedom (excluding the intercept). This corresponds to 2 interior knots, which `bs()` places at the 1/3 and 2/3 quantiles of `E`. This is a highly data-dependent operation, and as a consequence `bs(E, df = 5)` applied to a new version of `E` will result in different knots, and hence different basis functions. Any coefficients estimated for the former will make no sense when applied to the latter. In situations such as these and many others, the predictions are incoherent. Since

GAM objects tend to be made up of either nonparametric smooth terms or often somewhat complicated parametric terms, they are likely to face these prediction problems more frequently.

As a consequence, the function `predict.gam()` operates quite differently from the other `predict()` methods, when presented with new data. On the negative side, it is slower since essentially it has to refit the model. On the brighter side, it is a "safe" method of prediction that overcomes the problems described above. It can also handle `glm` and `lm` objects (with less work than `gam` objects), and so is the safe method of prediction for them as well.

Here follows a brief outline of the steps that are taken in the execution of `predict(gamob, new.data, ...)`:

1. A new data frame, `both.data`, is constructed by combining the data used to fit `gamob`, say, `old.data` with the data in `new.data`, retaining only the relevant predictor variables.

2. The model frame and model matrix are constructed from the combined data frame `both.data`. The model matrix is then separated into the two pieces X^o and X^n corresponding to the old and new data.

3. The parametric part of the object `gamob` is refit using X^o. In most situations, the fitted values should be identical to those in `gamob`. In some, such as in the case of `bs()` above, the fit will not be identical; the percentage difference between the old fit and the new fit is reported as a warning in cases such as this.

4. The coefficients from this new fit are then applied to X^n to obtain the new predictions.

5. For `gam` objects with both parametric and nonparametric components, an additional step is taken to evaluate the fitted nonlinear functions at the new data values. In principle, most smoothers produce a fitted function that can be evaluated anywhere (at least in the domain of the original data).

Details of steps 3 and 5 are discussed in Section 7.4.

To illustrate the use of `predict()`, let's first return to the model `eth3` fitted to the `ethanol` data:

```
> formula(eth3)
NOx ~ lo(C, E, 1/4, degree = 2)
```

The perspective plot in Figure 7.12 on page 277 has some irregularities, introduced by the function `interp()`. We can predict the values of the surface exactly on a grid:

```
> attach(ethanol)
> new.eth <- expand.grid(
+     C = seq(from = min(C), to = max(C), len = 40),
+     E = seq(from = min(E), to = max(E), len = 40)
+     )
> eth.grid <- predict(eth3, new.eth)
```

The function `expand.grid()` produces a data frame of points from a grid, with marginal values supplied as arguments. So `new.eth` has 1600 rows and two columns. Figure 7.16 shows the fitted surface, which is much smoother than that of Fig-

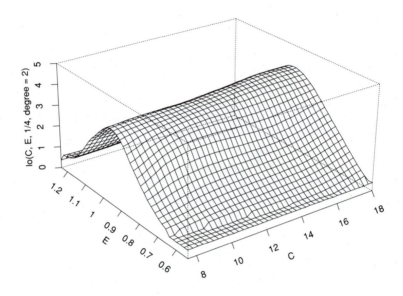

Figure 7.16: *The fitted surface corresponding to* `eth3`, *evaluated on a* 40×40 *grid of values of* C *and* E *using* `predict()`. *Compare with Figure 7.12.*

ure 7.12. A slight additional difference between the two is that Figure 7.12 is a display of the fitted term (excluding the intercept), while Figure 7.16 includes the intercept. Of course, we could have removed the intercept here as well. The model `eth3` is not really additive, even though we are able to accommodate it within the

GAM framework. Rather, it is an example of a general multivariate smooth surface, and models such as this are covered in much more detail in Chapter 8. In particular, a variety of plotting methods are described there for displaying such surfaces.

Our next example *is* particular to additive models. By the nature of additive models, we need only evaluate the individual terms on a reasonable grid of predictor values to be able to reconstruct the entire multivariate additive surface *everywhere*. So suppose we construct a data frame `new.data`, where each component has 40 evenly spaced observations over a suitable range. Then

```
predict.gam(gamob, new.data, type = "terms")
```

will return a 40×`length(terms)` matrix of fitted terms, from which we can construct the additive surface.

We return to the `kyphosis` example to illustrate this feature, in particular to the model `kyph.gam2`:

```
> formula(kyph.gam2)
Kyphosis ~ s(Age) + s(Start)
> attach(kyphosis)
> kyph.margin <- data.frame(
    Age = seq(from = min(Age), to = max(Age), len = 40),
    Start = seq(from = min(Start), to = max(Start), len = 40)
    )
> margin.fit <- predict(kyph.gam2, kyph.margin, type = "terms")
```

The matrix `margin.fit` has two columns labeled "s(Age)" and "s(Start)." If we were to plot these two columns of `margin.fit` against the two columns of `kyph.margin`, we would see exactly the same function as in Figure 7.7 on page 263, except the abscissae would be at 40 evenly spaced values.

The following sequence of expressions constructs the additive surface on a bivariate grid defined by the margins in `kyph.margin`:

```
> kyph.surf <- outer(margin.fit[, 1], margin.fit[, 2], "+")
> kyph.surf <- kyph.surf + attr(margin.fit, "constant")
> kyph.surf <- binomial()$inverse(kyph.surf)
```

The first line computes an *outer sum*, adding together the additive components at each of the 1600 elements of the 40 × 40 grid. The second line adds in the `"constant"` attribute of `margin.fit`, since the terms produced by `predict()` are centered to have weighted mean zero (Section 6.3.6). The third line converts the surface from the logit scale to the probability scale using the inverse link function. Of course, for a two predictor problem such as this, we could have simply produced the data frame with the 1600 points on the grid, as in the previous example, and predicted the fitted probabilities directly using `type="response"`. Often there are more terms, and then a procedure like the one described above is far more efficient for constructing the additive surface.

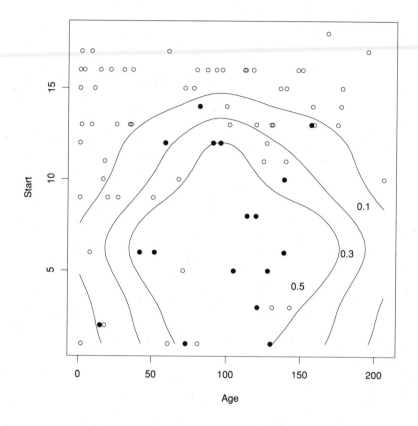

Figure 7.17: *A contour plot of the fitted probability surface derived from the fitted additive model* kyph.gam2. *The black dots indicate cases with* Kyphosis *present, the circles, absent.*

Finally, in Figure 7.17 we produce a contour plot from the fitted prevalence surface:

```
> plot(Age, Start, type = "n")
> points(Age[Kyphosis == "absent"], Start[Kyphosis == "absent"])
> points(Age[Kyphosis == "present"], Start[Kyphosis == "present"],
+     pch = 183)
> contour(kyph.margin$Age, kyph.margin$Start, kyph.surf,
+     add = T, v = c(0.1, 0.3, 0.5))
```

The contours appear to enclose the cases reasonably well (black dots), and the plot is a confirmation of the fit.

7.3.4   Smoothers in gam()

In all of our examples so far we have specified a smooth term in a formula using one of two smoothers:

- s(): as in s(Age, df = 5). This implies that a smoothing spline will be used in the backfitting algorithm for fitting that term in the model, using 5 df as the smoothing parameter.

- lo(): as in lo(Age, f = 0.5), which uses the loess() smoother described in Chapter 8. The lo() smoother additionally allows smooth surfaces to be included in the model, as in lo(wind, temp, 0.4).

Remember that the functions s() and lo() do not actually smooth the data, but rather set things up for the backfitting algorithm used by gam() to fit the model. These are both *shrinking* smoothers, in that their operation is not a projection but rather a shrunken fit onto a rich basis set.

In this section, we outline the steps needed to add a smoother to the existing set, and in so doing give an idea how it all works.

Adding a new smoother to gam() is extremely simple if it is of the *projection* type, such as regression splines or polynomials. All that is required is that functions such as bs() or poly() be written that create a basis matrix for representing the projection space. These can then be used by any of the regression fitters, not only gam().

Suppose someone wishes to fit additive models using their own favorite shrinking smoother rather than using smoothing splines or locally weighted regression (prejudices do exist in the smoothing field!). They would need an S function (typically interfaced with FORTRAN or C) that computes the smooth. Let's suppose their smoother is called kernel(). Three simple steps need to be taken in order to incorporate kernel() as a smoother to be used in fitting additive models in gam():

- An interface function to kernel() is needed that takes as input arguments x, y, and weights (the names are not important), as well as any parameters that control the amount and type of smoothing. The function should return a residual vector called residuals, and optionally a vector of pointwise variances called var, and the nonlinear degrees of freedom called nl.df. We discuss these latter two items in more detail in Section 7.4. Let's assume that kernel() itself has all these features. The function gam.s(), for example, is the interface function for the smoothing spline s().

- A shorter named function such as k() is needed for use in the model formula. Typically, k(x) will return its argument x, along with one or more attributes. One of the attributes is named "call", and gives the expression needed in the backfitting algorithm to update the residuals by smoothing against x. In the case of both lo() and s(), the data returned are exactly what is required for the parametric part of the fit. For example, the expression lo(x, degree = 2) evaluates to a two-column matrix consisting of basis vectors for quadratic regression. This would be incorporated into the model matrix. The "call" attribute, on the other hand, would reference the first column as the smoothing variable.

- The character vector gam.slist should be augmented with the character "k". This is used as an argument to the terms() function, which identifies all the *special* smooth terms in the formula.

To illustrate, let's look at the output of s():

```
> pred <- 1:4
> s(pred)
[1] 1 2 3 4
attr(, "spar"):
[1] 0
attr(, "df"):
[1] 4
attr(, "call"):
gam.s(data[["s(pred)"]], z, w, spar = 0, df = 4)
attr(, "class"):
[1] "smooth"
```

The "call" component is an S expression (of mode "call"); it gets evaluated repeatedly inside the backfitting algorithm all.wam() when fitting the terms in the model. The name "wam" stands for *weighted additive model*, while "all" refers to the fact that it can mix in smoothers of all types. This is in contrast to s.wam(), which is a specialized backfitting algorithm only used if all the smooth terms are splines; similarly, lo.wam() is used if all the smooth terms are to be fitted using loess(). So the only knowledge all.wam() has of the smoother to be used for a particular term is this call it is given to evaluate. The arguments z, data, and w in the call are local to all.wam(), and in fact data[["s(pred)"]] refers to exactly that component of the model.frame created by the expression s(pred). The other attributes of s(pred) are needed for the more efficient backfitting algorithm s.wam() described in Section 7.4.

Apart from allowing other regression smoothers, this modularity opens the door to other interesting generalizations. For example, included in this software is a function random() that is aimed at fitting a random effect factor term in an additive model. All it does is fit each level by a constant, but then shrinks all the constants toward the overall mean—another shrinking smoother. The function random() is used in the formula, and evaluates to its argument, which is a factor. It might be instructive for the reader to print the definition of random() or look at the detailed documentation (`?random`), as well as its workhorse `gam.random()`, for further illustration. Ridge regression can be accommodated similarly; one could write a simple function `ridge()` that would perform a similar shrinking.

In Section 7.4, we show how users can provide their own backfitting algorithms for efficiency. This opens the door to even more adventurous generalizations.

7.3.5 More on Plotting

The `plot.gam()` method has a modular construction, and consequently users can tailor-make their own plotting functions very straightforwardly. Before we describe the method `plot.gam()` itself, we first describe the `preplot()` method for `gam` objects.

In order to produce a plot, a certain amount of computation is required. Typically, there are many different choices to be made in representing a function or surface; in the case of GAMs, these choices involve rugplots, standard-error bands, residuals, vertical scales, and which terms to plot. With not much loss in efficiency, all the extra "data" needed to produce these plots can be computed once and for all, and stored away for future plotting. This is the idea behind the `preplot()` functions.

In its standard usage,

```
preplot(gamob)
```

produces a list with class `"preplot.gam"`, with an element for each term in the model. Each of these elements is named according to the particular term label. Let's examine the contents of one of these term elements in the context of a specific example, say the term `"s(Age)"` in `kyph.gam2`. The expression

```
kyph.preplot <- preplot(kyph.gam2)
```

evaluates to a two-element list, with names `"s(Age)"` and `"s(Start)"`, and we shall dissect the first of these. It has components:

- `$x`, the values of the *inner* variable in the expression `s(Age)`, in this case `Age` itself, to be used as abscissa in the plot. Had the expression been `s(log(Age))`, the x component would still have been `Age`. The values in `Age` are exactly those used to fit the model.

- `$xlab`, which is the character name `"Age"` in this case.

- `$y`, the fitted term to be used as the ordinate in the plot. It is exactly the term returned by `predict(object, type = "terms")`.

- `$ylab`, which is simply `"s(Age)"`.

- `$se.y` is the vector of pointwise standard errors corresponding to the term, as returned when the `se=T` option is used with `predict()`.

It should be clear that a major part of this work is done by `predict.gam()` in producing the fitted terms and their standard errors. This element `"s(Age)"` of `kyph.preplot` is also an object of class `"preplot.gam"`, and it might seem to be ready for plotting as it is. Although true in this case, in general the `x` component can be more complex. It may be a factor or a category, or it may be a list of two or more variables. The `preplot()` method goes to some effort to identify these inner predictor(s), and different plotting methods are appropriate depending on their data class.

The function that actually performs the plot is named `gplot()`, and it currently has four methods, corresponding to the data classes `numeric`, `category`, `matrix`, and `list`. The last of these can deal only with two-element lists, which it reshapes into a matrix and then calls `gplot.matrix()`.

The expression `plot(kyph.preplot)` produces two plots, one after the other, for each of the two terms in `kyph.preplot`. One can use `plot()` on either the entire `preplot.gam` object, or else on any of the elements separately. Ultimately, the appropriate `gplot()` method is invoked.

Some readers may wonder why we bothered to invent the generic `gplot()` instead of simply using `plot()`. The reason is that we did not wish to *claim* the name `plot.matrix`, for example, for the very specific type of plots we have in mind here.

The `gplot()` methods are quite straightforward, and produce the style of plots we have seen in this chapter and the last. It would be easy for users to create their own `gplot()` functions, either to modify the styles we have chosen, or else to accommodate other classes. Currently we do not provide plots for interaction terms, or for terms involving more than two variables. There will also be data classes for which we have no methods. Related to this, and as a caveat, we note that the concept of extracting the `x` variable is somewhat fragile. Although most of the time it should produce the desired results, there is no guarantee that what is extracted as `x` will be suitable as an abscissa for plotting. Indeed, it may not even have the correct length, or correspond in any way to an abscissa. Expressions in formulas are governed only by rules dictating the data class of the object they evaluate to, and can be built up in general from objects of any size, shape, etc.

The `plot()` method for `gam` objects is built up in a straightforward way from the plot method for `preplot.gam` objects. In its simplest usage, it computes the `preplot.gam` object and plots it! If its `x` argument has a `preplot` component, it uses it instead. This suggests a convenient place to stash `preplot` objects, and in fact the usage

```
kyph.gam2$preplot <- preplot(kyph.gam2)
```

is quite standard.

Further details on `plot.gam()` can be found in the detailed documentation in the appendix, as well as online (`?plot.gam.`)

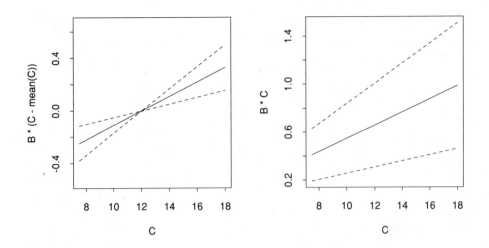

Figure 7.18: *The plot on the left shows the centered, fitted term for* C *in the model* eth7, *together with pointwise twice standard-error bands. The plot on the right is the uncentered version.*

We end this section with an explanation of why centering is important when computing fitted terms and their standard errors, as is done by the predict() methods (Section 6.3.6, page 239). We require the fitted terms most often for plotting purposes, and wish to represent the composed polynomial, B-spline, smooth term, etc. as a function of their argument. The slope and shape of each term is important for this kind of examination, and not the level; rather, an overall level in the model is usually only interpretable if the nonconstant terms average zero. These are not compelling reasons, however, to go to the extra expense of centering since we could simply ignore the level when examining the plots. Centering becomes essential when we compute pointwise standard errors. We illustrate on the simplest of terms, namely a linear term involving a single coefficient, why this is the case. The pointwise standard errors for such a term, if it is centered, are given by the expression $|x_j - \bar{x}_j| \, \sigma_{jj}$; these are zero at $\bar{x}_j$ and increase linearly for values away from $\bar{x}_j$. If the term is not centered, the pointwise standard errors are simply $|x_j| \, \sigma_{jj}$. To illustrate the difference, consider the linear model

```
> eth7 <- lm(NOx ~ C + ns(E, 5), data = ethanol)
```

Figure 7.18 shows the centered and uncentered versions of a plot for the fitted linear term in C, together with pointwise twice standard-error terms. The centered plot is informative, since it shows us both the range of the effect of C, as well as an approximate 95% t interval for the variation in the slope (in units of this effect). The uncentered plot also shows us the same range of effect, but there is no such interpretation for the standard-error bands.

7.4   Numerical and Computational Details

This section gives some background to additive models, and further insight into this particular implementation of the algorithms needed to fit them. Generalized additive models are a fairly recent innovation, and the methodology will likely be unfamiliar. Fortunately, the way smooth terms appear additively in formulas is quite intuitive, and the fitted models look and feel just like GLM models, only more flexible. All we can hope to do in the following few pages is shed some insight into the technical background that supports the algorithms we use, and guide the reader to the references for a more detailed explanation.

7.4.1   Scatterplot Smoothing

A basic element of our additive model routines is the scatterplot smoother. We have favored two approaches to smoothing in constructing the GAM software:

- lo: short for loess or locally weighted regression, is a direct method of smoothing, and extends naturally to dimensions higher than 1.

- s: short for smoothing splines, is an indirect methods of smoothing, driven by penalized least squares. Although smoothing splines are also defined in higher dimensions, the computational complexity increases dramatically.

There are other types of smoothers, such as kernel smoothers, nonlinear smoothers (for example running medians), and trigonometric series smoothers, to name a few. All of these, and in fact any smoother, can be used as a building block for fitting additive models.

For our purposes, it is easiest to motivate a smoother in terms of the simple model $y_i = f(x_i) + \varepsilon_i$, where f is some unknown and arbitrary function of the *predictor* x and the ε_i represent zero mean independently distributed errors. Since $E(y_i \mid x_i) = f(x_i)$, any estimate of f can be viewed as an estimate of this conditional expectation. For simplicity, we assume x is univariate. However, we seldom have more than one observation at any given point x_i, so we have to relax the definition in order to get a reasonable estimate. The typical assumption that is made toward this end is to assume that f is smooth in some sense, and then exploit this smoothness in defining the estimate. Smoothers differ chiefly in the way they exploit the smoothness assumption.

A locally weighted regression smoother estimates $f(t)$ at an arbitrary point t by computing a weighted average of all those values y_j in the sample that have predictors x_j *close* to t, and the weights depend smoothly on this closeness.

We can represent it analytically as

$$\hat{f}(t) = \sum_{j=1}^{n} S_\lambda(t, x_j) y_j, \tag{7.2}$$

where $\{x_j, y_j\}$, $j = 1, \ldots, n$ is the series of n data points, t is the target point, and S_λ is a weight function parametrized by λ. The $n \times n$ matrix $\{S_\lambda(x_i, x_j)\}$ is often called the *smoother matrix*. All *linear* smoothers can be represented in the form (7.2); they differ chiefly in the construction of the weights:

- The `loess()` smoother imposes a tricube weight function on the $\lambda = span$ nearest neighbors in x to the target point t, and then computes the fit at t by a weighted linear (or optionally quadratic) regression. The *span* controls how much smoothing is performed: a large span results in smoother but less local fitted functions whereas a small span results in rougher (higher variance), more local fits. Locally weighted regression smoothers are discussed in depth in Chapter 8, so we do not dwell further on them here.

- Smoothing splines exploit the smoothness from a different more explicit angle. A cubic smoothing spline fit to our data is that function $\hat{f}$ that minimizes

$$PRSS = \sum_{i=1}^{n} \left(y_i - f(x_i) \right)^2 + \lambda \int (f''(t))^2 dt \tag{7.3}$$

over all functions with continuous first and integrable second derivatives. The solution is a *function*; it is a natural cubic spline with interior and boundary knots at the unique values of the x_i. The smoothing parameter λ controls the tradeoff between fidelity to the data and smoothness. Smoothing splines are also *linear* smoothers, and so can be represented as in (7.2). The function `smooth.spline()`, which is the workhorse underlying `s.wam()` and `gam.s()`, fits a smoothing spline; see the detailed documentation for a description of its arguments (`?smooth.spline.`)

Both smoothing methods have a smoothing parameter that needs to be specified. In practice, it is common to use automatic techniques such as cross-validation for selecting the smoothing parameters. We prefer using fixed or user-specified smoothing parameters, since concepts such as cross-validation are very expensive to implement for generalized additive models.

The *span* for `loess()` is rather intuitive and can be selected subjectively, while the λ parameter for smoothing splines is not easy to prespecify. Another parameter useful for calibrating a smooth is the equivalent degrees of freedom, related, of course, to the *span* or λ. The simplest definition is df $= \mathrm{tr}(S)$, where S is the smoother matrix that produces the fit $\hat{\boldsymbol{y}} = S\boldsymbol{y}$ at each of the n data points x_i. For smoothing splines, a convenient smoothing parameter is df $= \mathrm{tr}(S)$ itself, which implies a value of λ.

7.4.2 Fitting Simple Additive Models

We now use the scatterplot smoother as a building block for fitting simple nonparametric additive models of the form $y_i = \sum_{j=1}^{p} f_j(x_{ij}) + \varepsilon_i$. Consider the system of p (vector) equations:

$$
\begin{aligned}
\boldsymbol{f}_1 &= S_1(\boldsymbol{y} \;-\; \cdot \;-\; \boldsymbol{f}_2 \;-\; \boldsymbol{f}_3 \;-\; \cdots \;-\; \boldsymbol{f}_p) \\
\boldsymbol{f}_2 &= S_2(\boldsymbol{y} \;-\; \boldsymbol{f}_1 \;-\; \cdot \;-\; \boldsymbol{f}_3 \;-\; \cdots \;-\; \boldsymbol{f}_p) \\
\boldsymbol{f}_3 &= S_3(\boldsymbol{y} \;-\; \boldsymbol{f}_1 \;-\; \boldsymbol{f}_2 \;-\; \cdot \;-\; \cdots \;-\; \boldsymbol{f}_p) \\
&\;\;\vdots \\
\boldsymbol{f}_p &= S_p(\boldsymbol{y} \;-\; \boldsymbol{f}_1 \;-\; \boldsymbol{f}_2 \;-\; \boldsymbol{f}_3 \;-\; \cdots \;-\; \cdot)
\end{aligned}
\tag{7.4}
$$

where the dots in the equation are placeholders showing the term that is missing in each row. Here a vector of the form $\boldsymbol{f}_j$ represents the function f_j evaluated at the n observed values of x_j in the sample; S_j represents the smoother operator matrix for smoothing against predictor x_j. The jth equation is reasonable for fitting f_j if we pretend we know all the other functions appearing on the right side; since we don't, we plan to solve all the equations simultaneously. There are several more rigorous motivations for these equations than that just given. For example, if all the smoothers are smoothing splines, then this system solves the penalized least-squares problem

$$
PRSS = \sum_{i=1}^{n} \left(y_i - \sum_{j=1}^{p} f_j(x_i) \right)^2 + \sum_{j=1}^{p} \lambda_j \int (f_j''(t))^2 dt
\tag{7.5}
$$

The S_j can represent any smoother; in particular, the choice $S_j = H_j$, the least-squares projection matrix, produces a linear fit for the term x_j. If all the smoothers are projections, system (7.4) reduces to the usual normal equations for least squares.

One method for solving (7.4) uses the Gauss-Seidel iterative method, also known as *backfitting*. The algorithm cycles through the equations, each time substituting the most current versions of the functions in the right side. Let's look at the algorithm applied to some air-pollution data contained in the frame `air`. The response is `ozone`, and there are three predictors: radiation, wind speed, and temperature. We fit the model

```
> gam(ozone ~ s(radiation) + s(wind) + s(temperature), trace=T)

WAM   iter   rss/n      term

        1    0.251   Parametric -- lm.wfit
        1    0.242   Nonparametric -- s(radiation)
        1    0.208   Nonparametric -- s(wind)
        1    0.183   Nonparametric -- s(temperature)
Relative change in functions: 0.089

        2    0.181   Parametric -- lm.wfit
        2    0.18    Nonparametric -- s(radiation)
        2    0.181   Nonparametric -- s(wind)
        2    0.18    Nonparametric -- s(temperature)
Relative change in functions: 0.015

        3    0.18    Parametric -- lm.wfit
        3    0.18    Nonparametric -- s(radiation)
        3    0.18    Nonparametric -- s(wind)
        3    0.18    Nonparametric -- s(temperature)
Relative change in functions: 0.003

        4    0.18    Parametric -- lm.wfit
        4    0.18    Nonparametric -- s(radiation)
        4    0.18    Nonparametric -- s(wind)
        4    0.18    Nonparametric -- s(temperature)
Relative change in functions: 0.001

GAM all.wam loop 1: deviance = 19.944
```

Figure 7.19 shows the additive model fitted above. Included in the figure are the univariate scatterplot smooths of ozone against each of the predictors separately. Due to correlations in the three predictors, we can see how backfitting had to adjust the smooth terms to achieve a joint fit.

Although backfitting is an efficient method for solving (7.4), convergence can be slow if variables are correlated, as in this example. By arranging the iterations sensibly, we can eliminate most of the problems. Two important strategies are:

- All the linear terms in a semiparametric fit are lumped together and treated as one term in the iterations.

- Even the terms to be smoothed are separated into a parametric and nonparametric part: $f_j(x_j) = \beta_j x_j + g(x_j)$. The linear coefficient is fitted together with the linear parts of all other terms in the model.

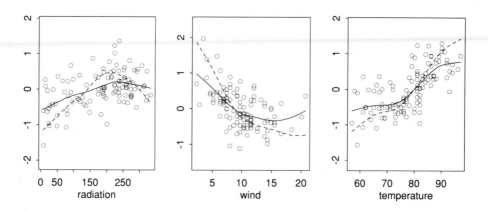

Figure 7.19: *The solid curves represent the additive model fit to* ozone *using three atmospheric variables. The points in the figures represent the partial residuals (fitted function + overall residuals). The broken curves show the functions obtained by smoothing the variables separately against* ozone.

The reason for this latter strategy is simple; a smoothing spline fit, for example, can be exactly decomposed into a component that is a projection onto the space of fits linear in its predictor, and a nonprojection component. The effect is that all the terms have one or more linear components, fitted jointly by least squares; some also have a nonparametric component. Splitting nonparametric terms up in this fashion avoids extended iterations in situations such as that above, where the overall slope of a function can change when fitted jointly. This strategy is generalized when fitting locally quadratic fits specified by lo(); both the linear and quadratic part is a projection component and is fit parametrically. Similarly, generalizations occur for surfaces.

7.4.3 Fitting Generalized Additive Models

The algorithm for fitting a GAM is exactly analogous to the algorithm for GLMs. For simplicity, we use the binary logistic regression model as an example.

Suppose the current estimate of the additive predictor is η_i^{old}, and, via the inverse of the *logit* link, we get $\mu_i^{old} = \exp(\eta_i^{old})/(1 + \exp(\eta_i^{old}))$. Then we

- compute the working response:

$$z_i = \eta_i^{old} + \frac{\left(y_i - \mu_i^{old}\right)}{\mu_i^{old}\left(1 - \mu_i^{old}\right)};$$

- compute weights $w_i = \mu_i^{old}\left(1 - \mu_i^{old}\right)$;

- obtain η_i^{new} by fitting a weighted additive model to z_i. This simply means that the smoothers in the backfitting algorithm incorporate the additional weights, and weighted least squares is used for the linear parts.

These steps are repeated until the relative change in the fitted coefficients and functions is below a tolerance threshold (say 0.001).

Apart from having intuitive appeal, this algorithm can also be justified on more rigorous grounds. For example, if an appropriate additive penalized likelihood is used as the criterion, the Newton-Raphson step for updating all the unknown functions simultaneously requires a system identical to 7.4 to be solved, with z instead of y, and weighted cubic spline smoothers for the S_j. For other error models and link functions, all that changes is the formula for constructing the working response and the weights, just as in the GLM case.

For additive and generalized additive models we can also compute approximate df_j for each of the terms, and hence perform crude likelihood-ratio tests in an informal way. Alternatively, we can use the fitted functions to suggest parametric transformations, and then use the linear model for inferences.

7.4.4 Standard Errors and Degrees of Freedom

An entirely parametric GLM is computed by weighted least squares, and the usual weighted least-squares covariance matrix is the inverse of the estimated Fisher information matrix for GLMs. This is readily available from the output of `summary.glm()`. The standard-error curves for composite functions are constructed in the obvious way, since they are linear combinations of fitted coefficients that have a covariance matrix. Even for GLMs that have no scale parameter, we use the scaled chi-squared statistic to estimate a scale parameter and use it in the calculations. This gives protection against overdispersion, and typically results in conservative standard errors.

When smooth terms are present in the model, the procedure is far more complicated. An exact analysis requires the computation of an *operator* matrix G_j for each smooth term s_j, such that $s_j = G_j z$. Here z is the working response from the last IRLS fit, which one can argue has an asymptotic Gaussian distribution. Then the covariance matrix of the fitted term is given by $G_j cov(z)G_j^t$, and is estimated by $\hat{\phi}G_j W^{-1} G_j^t$ where W is diagonal in the final IRLS weights. Since all of this is extremely expensive to obtain in general, and is asymptotic anyway, we have resorted to some even cruder approximations.

First, we approximate $\hat{\phi}G_j W^{-1} G_j^t$ by $\hat{\phi}G_j W^{-1}$. This is exact for weighted projections, is usually conservative for nonprojection smoothers in that it is larger, and can also be justified on Bayesian grounds for smoothing splines. One can orthogonally decompose G_j further into $G_j = H_j + N_j$, where H_j produces the parametric part of s_j, and N_j the nonparametric part. Although we do have H_j, we do not have the latter, and so approximate it by S_j^n, the operator for the nonprojection part of the smoother itself. The diagonal of S_j is all we need to compute the diagonal of S_j^n, and the former is usually available as part of the output of the smoothing operation.

Thus, in summary, our standard-error curves for nonparametric curve estimates are derived from the sum of 2 variance curves. The variance curve for the parametric part of the function reflects the joint covariance behavior, whereas the variance for the nonparametric part reflects only marginal information.

The procedure outlined above is admittedly *ad hoc*. Exact methods for computing the operators G_j exist, but the least expensive version we know takes $O(n^2)$ operations (with a large constant) to compute. The approximations described here have shown empirical success on a number of examples, as long as the pairwise correlations among the predictors are not too high. In practice one can always approximate the nonparametric term parametrically (and even conservatively) using functions such as `bs()` or `ns()`, and use the inexpensive parametric standard-error curves.

The df$_j$ in the case of nonparametric terms are computed as $\mathrm{tr}(S_j) - 1$; see the references for more details.

7.4.5 Implementation Details

In this section, we describe some of the implementational details. The style will tend to be somewhat narrative since the details can get complicated, and interested programmers will want to have a listing of some of the examples alongside.

First a general overview. The beginning of `gam()` is almost identical to that of `glm()` or even `lm()`. A slight difference is that when the `terms()` function is invoked, the vector `gam.slist` is passed as a `specials=` argument. At present, `gam.slist` consists of the three character strings `"s"`, `"lo"`, and `"random"`, and all that happens is that `terms()` makes a note of which terms are *special* in this way.

As we have seen, terms in `s()` and `lo()` evaluate to vectors or matrices in the model frame, with certain attributes. In fact, whatever they evaluate to is included in the model matrix. So smoothing splines evaluate to their argument vector since this will comprise its projection part. A locally quadratic bivariate term specified by `lo(C, E, 1/4, degree = 2)`, on the other hand, will evaluate to a five-column matrix consisting of terms of degree 2 or less. This is its projection part. All in all, a model matrix is constructed that one way or another represents all the terms in the model. The backfitting algorithms then cycle around, performing one large least-squares projection step, then one smooth for each of the nonparametric terms (to update the nonprojection parts). This is repeated until the relative change in the functions is below the threshold `bf.eps` supplied by default by the function `gam.control()`.

The most general backfitting function is called `all.wam()`. It performs the iterations described above, and computes each smooth by simply evaluating the `"call"` attribute of the corresponding term in the model frame. Recall the example from Section 7.3.4:

```
gam.s(data[["s(pred)"]], z, w, spar = 0, df = 4)
```

This works because `data` is the local name in `all.wam()` for the model frame, and `z` is the local name for the partial residual to be smoothed, and so on. By insisting that the result of the evaluation of this call have a component labeled `residual`, the term can be updated. In addition, the smoothers return a component called `var`. This is optional; if users provide their own smoothers without `var`, they will simply see less accurate standard-error curves. As described earlier, `var` is approximated by $\text{diag}(S_j^n W^{-1})$, where S_j^n is the (weighted) smoother operator with the (weighted) projection part removed. The component `nl.df` is the approximate degrees of freedom used in computing this nonprojection part of the smoother; `nl.df` is given by $\text{tr}(S_j^n)$.

This general backfitter `all.wam()` is quite modular but computationally rather inefficient. Part of the reason why is that the smoothers `gam.s()` and `gam.lo()` are both S functions (interfaced to FORTRAN), and in order to retain their simplicity and modularity we cannot take advantage of the fact that all that is changing during backfitting is the response.

If all the smooth terms in the model use the same smoother, say `s()`, a vector `gam.wlist` is consulted (currently the same as `gam.slist`). Since `"s"` is present, the implication is that a specialized, more efficient backfitter `s.wam()` exists and should be used. Some of the speedups are achieved in this case by:

- precomputing all the information needed for sorting and condensing the predictors prior to smoothing;

- performing the least-squares part and all the smoothing iterations within one FORTRAN subroutine

In fact, little of the work done by `s.wam()` is done in S! Users wishing to *hard wire* their own backfitting algorithms in this fashion will have to print out `s.wam()` or `lo.wam()` for further details.

In addition to the (inner) backfitting iterations, `gam()` performs the (outer) local scoring iterations; the backfitter is thus called repeatedly.

The local-scoring update step and test for convergence is organized in a single expression that is evaluated at each iteration. This expression can even be evaluated from within the FORTRAN subroutine itself, so in fact `s.wam()` and `lo.wam()` are invoked only once, but initiate a *back-chat* to `gam()` to get their updated response and weights.

The anova table at the end of the summary produced by `summary.gam()` reports a type of score test for the effect of each nonparametric function. For each nonparametric term in the model, the nonlinear component is set to zero and the parametric part of the model is refit by weighted least squares, holding the other nonlinear components fixed. There are two levels of approximation here:

- To refit the model completely, one should also adjust the nonlinear components of the other smooth terms; we only adjust their linear components, and hold the nonlinear parts fixed.

- For generalized additive models, we are making the typical score test approximation by using the weights and working response from the final local scoring iteration.

The change in the Pearson chi-squared statistic is recorded for each term so dropped. These computations can be done simultaneously for all the smooth terms in an efficient way, using the QR decomposition of the final IRLS iteration. Readers interested in further details can print out the short function `gam.nlchisq()` to see the details.

Finally, a few additional comments on `predict.gam()`. In order to produce fitted values at new observations, `predict.gam()` needs to produce both the parametric components and the nonparametric components of the predictions. For the parametric components, it simply needs to refit the coefficients using the derived model matrix and the final IRLS information in the fitted `gam` object, as outlined in Section 7.3.3. Rather than using the working response derivable from the `gam` object, the values in the linear predictor are used instead. For `gam` objects, this is the additive predictor less the nonparametric smooth terms. For `lm` objects the linear predictor is the vector of fitted values. When the model is refit to this response (using the final working weights), the residuals are expected to be zero. The situations where this is not the case are a subset of the cases where "safe" prediction is necessary, and a warning message is issued reporting the percentage difference.

To get the nonparametric components, a bit of trickery is used. We modify the `"call"` attribute of each smooth term in the model frame to include the argument `"xeval=xnew"`. The smoothers `gam.lo()` and `gam.s()` both respond differently if they have an `xeval` argument; instead of performing the smooth, and returning the residuals, variances, etc., they return the fit evaluated at the new predictor points `xnew`. The `predict()` method simply cycles through each smooth term and sets up the local variables appropriately.

Bibliographic Notes

The topic of nonparametric smoothing and additive models has a long history, although most of the material treated here has appeared in the last 15 years. Two popular smoothers are implemented in this chapter. One is the locally weighted polynomial smoother of Cleveland (1979) and Cleveland and Devlin (1988). This is the `loess()` smoother of Chapter 8, referred to as `lo()` in GAM formulas. The `lowess()` smoother in ⬚S is the one-dimensional predecessor of `loess()`. The other smoother implemented here is the cubic smoothing spline, first introduced by Whittaker (1923). The monograph by Wahba (1990) is a comprehensive account of the theory and applications of smoothing splines. The `s()` function in GAM formulas refers to a cubic smoothing spline term, and the stand-alone function `smooth.spline()` is also provided. Our underlying FORTRAN code is a modified version of the subroutine BART written by Finbarr O'Sullivan, known to some as the S function `bart()`.

The ACE algorithm of Breiman and Friedman (1985) was the first fully nonparametric proposal for fitting additive models, allowing a nonparametric transformation of the response as well as the predictors. The backfitting algorithm was proposed by Friedman and Stuetzle (1981) in the context of projection pursuit regression, and its convergence properties were studied by Breiman and Friedman (1985) and Buja et al. (1989).

A full historic account of generalized additive models with ample references can be found in the research monograph by Hastie and Tibshirani (1990). The style of working with additive models as extension of linear models, fixing smoothing parameters via degrees of freedom, and using approximate chi-squared tests to evaluate smooth terms was developed in this last reference.

Chapter 8

Local Regression Models

William S. Cleveland
Eric Grosse
William M. Shyu

Local regression models provide methods for fitting *regression functions*, or *regression surfaces*, to data. Two examples are shown in Figures 8.1 and 8.2 In the first figure, there is one predictor, and the fitted function is the curve. In the second figure, there are two predictors, and the fitted surface is shown by a contour plot. These two examples will be explained in detail later.

Consider any point x in the space of the predictors. One basic specification in a local regression model is that there is a neighborhood containing x in which the regression surface is well approximated by a function from a specific parametric class; for the S implementation described in this chapter, there will be two classes— polynomials of degree 1 or 2. The specifications of local regression models lead to methods of fitting that consist of smoothing the response as a function of the predictors; thus the fitting methods are nonparametric regression procedures.

Recall that in Chapters 4 to 6, responses are modeled as parametric functions of the predictors. Then, in Chapter 7, generalized additive models are introduced that lead to an element of nonparametric fitting. For such an additive model, a regression surface of two or more predictors is specified to be well approximated by additive functions of the predictors. In other words, the specification rules out certain interactions or rules out interactions altogether. But for local regression models, there is no explicit specification that rules out interactions. If a regression surface is additive, then the methods of Chapter 7 are appropriate since, in such a

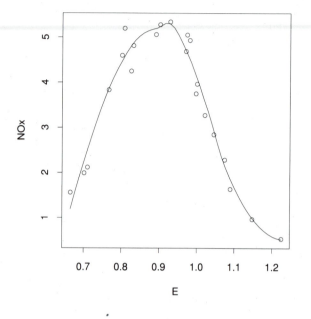

Figure 8.1: *Local regression model with one predictor—fitted curve.*

case, they provide more parsimonious descriptions of the surface and have better estimation properties. However, if additive fits are unlikely to result in a good approximation of the surface, the methods in this chapter are appropriate; the surface in Figure 8.2 is one example.

In Section 8.1.1, local regression models are defined; that is, the various specifications are described. The specifications of a particular model determine the details of the method used to fit the model; the fitting method, which is called *loess*, is described in Section 8.1.2.

In Section 8.2, we discuss the S functions and objects for local regression models by working through a number of examples. Our goal is to show how the data are analyzed in practice using S. This means we must discuss diagnostic methods. The specifications of a local regression model are impositions on the data, and these impositions need to be thoroughly checked if we are to have estimates and inferences with a demonstrated validity. Thus, diagnostic checking is an essential part of the practice of fitting local regression models, and, as with all model building, omitting it results in demonstrated validity being replaced simply by hope.

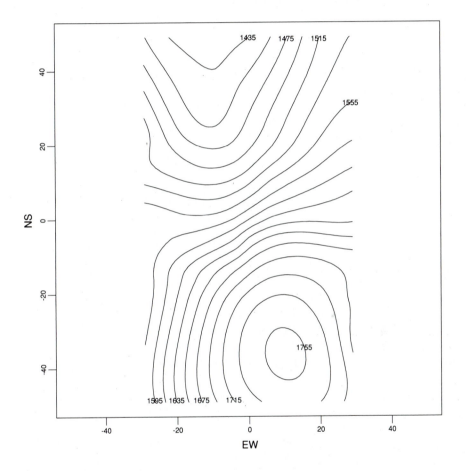

Figure 8.2: *Local regression model with two predictors—contours of fitted surface.*

8.1   Statistical Models and Fitting

8.1.1   Definition of Local Regression Models

Suppose, for each i from 1 to n, that y_i is a measurement of the response and x_i is a corresponding vector of measurements of p predictors. In a local regression model the response and predictors are related by

$$y_i = g(x_i) + \varepsilon_i,$$

where g is the regression surface and the ε_i are random errors. If x is any point in the space of the predictors, $g(x)$ is the value of the surface at x; for example, $g(x_i)$ is the expected value of y_i. In the fitting of local regression models we specify properties of the regression surface and the errors; that is, we make assumptions about them. We will now discuss the specifications that are allowable using the S functions and objects that are described in Section 8.2.

Specification of the Errors

In all cases, we suppose that the ε_i are independent random variables with mean 0. One of two families of probability distributions can be specified. The first is the Gaussian. The second is symmetric distributions, which allow for the common situation where the errors have a distribution with tails that are stretched out compared with the normal (leptokurtosis), and which lead us to robust methods of estimation.

We can specify properties of the variances of the ε_i in one of two ways. The first is simply that they are a constant, σ^2. The second is that $a_i \varepsilon_i$ has constant variance σ^2, where the *a priori weights*, a_i, are positive and known.

Specification of the Surface

Suppose, first, that all predictors are numeric; that is, none are factor variables. For each x in the space of the predictors, we suppose that in a certain neighborhood of x, the regression surface is well approximated by a function from a parametric class. The overall sizes of the neighborhoods are specified by a parameter, α, that is defined in Section 8.1.2. Size, of course, implies a metric, and we will use Euclidean distance. For two or more numeric predictors, the shapes of the neighborhoods are specified by deciding whether to normalize the scales of the numeric predictors. We will elaborate on this later.

We will allow the specification of one of two general classes of parametric functions: linear and quadratic polynomials. For example, suppose there are two predictors, u and v. If we specify linear, the class consists of three monomials: a constant, u, and v. If we specify quadratic, the class is made up of five monomials:

a constant, u, v, uv, u^2, and v^2. We will let λ be a parameter that describes the specification; if $\lambda = 1$, the specification is linear, and if $\lambda = 2$, the specification is quadratic.

Suppose $\lambda = 2$ and there are two or more numeric predictors. We can specify that any of the monomials that is a square be dropped from the class. For example, suppose again that the predictors are u and v. If we drop the square for u, then the class has four monomials: a constant, u, v, uv, and v^2.

If there are two or more numeric predictors we can specify that the surface be *conditionally parametric* in any proper subset of the numeric predictors; this means that given the values of the predictors not in the subset, the surface is a member of a parametric class as a function of the subset. If we change the conditioning, or given values, the surface is still a function in the same class, although the parameters might change. For example, suppose the predictors are u and v. Suppose $\lambda = 1$, and we specify the surface to be conditionally parametric in u. Then given v, the surface is linear in u; this means the general form of the surface is $\beta_0(v) + \beta_1(v)u$. Suppose $\lambda = 2$, and we specify the surface to be conditionally parametric in u. Then given v, the surface is quadratic in u; the general form of the surface in this case is $\beta_0(v) + \beta_1(v)u + \beta_2(v)u^2$. It makes sense to specify a regression surface to be conditionally parametric in one or more numeric variables if exploration of the data or a priori information suggests that the surface is globally a very smooth function of the variables. Making such a specification when it is valid can result in a more parsimonious description of the surface.

Suppose now that there are factor predictors. A combined factor is formed by taking all combinations of levels of the predictors. For example, suppose there are two factor predictors with levels *male* and *female* for the first and *black* and *white* for the second. Then the combined factor has four levels: *black female*, *black male*, *white female*, and *white male*. In such a case, the above specifications for numeric predictors apply separately for each level of the combined factor; that is, we divide the data up into subsets according to the levels of the combined factor, and the specifications of the surface as a function of the numeric variables hold separately for each subset.

Summary of the Choices

Thus, the fitting of local regression models involves making the following choices about the specification of properties of the errors and the regression surface:

- Gaussian or symmetric distribution;

- constant variance or a priori weights;

- locally linear or locally quadratic in numeric predictors;

- neighborhood size;

- normalization of the scales;

- dropping squares;

- conditionally parametric subset.

8.1.2 Loess: Fitting Local Regression Models

The method we will use to fit local regression models is called *loess*, which is short for *local regression*, and was chosen as the name since a loess is a deposit of fine clay or silt along a river valley, and thus is a surface of sorts. The word comes from the German *löss*, and is pronounced *lōís*.

Identically Distributed, Gaussian Errors: One Numeric Predictor

Let's begin with the classical case of Gaussian errors with constant variance σ^2. Suppose there is just one numeric predictor. Let x be any value along the scale of measurement of the variable. The loess fitting procedure is a numerical algorithm that prescribes how $\hat{g}(x)$, the estimate of g at a specific value of x, is computed.

Let $\Delta_i(x) = |x - x_i|$, let $\Delta_{(i)}(x)$ be the values of these distances ordered from smallest to largest, and let

$$T(u;t) = \begin{cases} (1 - (u/t)^3)^3, & \text{for } 0 \le u < t \\ 0 & \text{for } u \ge t \end{cases}$$

be the *tricube weight function*.

The smoothness of the loess fit depends on the specification of the neighborhood parameter, $\alpha > 0$. As α increases, $\hat{g}$ becomes smoother. Suppose $\alpha \le 1$. Let q be equal to αn truncated to an integer. We define a weight for (x_i, y_i) by

$$w_i(x) = T(\Delta_i(x); \Delta_{(q)}(x)).$$

For $\alpha > 1$, the $w_i(x)$ are defined in the same manner, but $\Delta_{(q)}(x)$ is replaced by $\Delta_{(n)}(x)\alpha$. The $w_i(x)$, which we will call the *neighborhood weights*, decrease or stay constant as x_i increases in distance from x.

If we have specified the surface to be locally well approximated by a linear polynomial—that is, if λ is 1—then a linear polynomial is fitted to y_i using weighted least squares with the weights $w_i(x)$; the value of this fitted polynomial at x is $\hat{g}(x)$. If λ is 2, a quadratic is fitted. Note that as $\alpha \to \infty$, $\hat{g}(x)$ tends to a linear surface for locally linear fitting or a quadratic surface for locally quadratic fitting.

Identically Distributed, Gaussian Errors: Two or More Numeric Predictors

We continue to suppose the errors are identically distributed and Gaussian. The one additional issue that needs to be addressed for p numeric predictors with $p > 1$ is the notion of distance in the space of the predictors. Suppose x is a value in the space. To define neighborhood weights we need to define the distance, $\Delta_i(x)$, from x to x_i, the ith observation of the predictors. We will use Euclidean distance, but the x_i do not have to be the raw measurements. Typically, it makes sense to take x_i to be the raw measurements normalized in some way. We will normalize the predictors by dividing them by their 10% trimmed sample standard deviation, and call this the *standard normalization*. There are, however, situations where we might choose not to normalize—for example, if the predictors represent position in space.

Armed with the $\Delta_i(x)$, the loess fitting method for $p > 1$ is just an obvious generalization of the one-predictor method. For $\alpha < 1$, neighborhood weights, $w_i(x)$, are defined using the same formulas used for one predictor; thus, if $\lambda = 1$, we fit a linear polynomial in the predictors using weighted least squares, or, if $\lambda = 2$, we fit a quadratic. For $\alpha > 1$, the $w_i(x)$ are defined by the same formula except that $\Delta_{(q)}(x)$ is replaced by $\Delta_{(n)}(x)\alpha^{1/p}$.

Dropping Squares and Conditionally Parametric Fitting for Two or More Predictors

Suppose λ has been specified to be 2. Suppose, in addition, that we have specified the squares of certain predictors to be dropped. Then those monomials are not used in the local fitting.

Suppose a proper subset of the predictors has been specified to be conditionally parametric. Then we simply ignore these predictors in computing the Euclidean distances that are used in the definition of the neighborhood weights, $w_i(x)$. It is an easy exercise to show that this results in a conditionally parametric fit.

Symmetric Errors and Robust Fitting

Suppose the ε_i have been specified to have a symmetric distribution. Then we modify the loess fitting procedures to produce a robust estimate; the estimate is not adversely affected if the errors have a long-tailed distribution, but it has high efficiency in the Gaussian case.

The loess robust estimate begins with the Gaussian-error estimate, $\hat{g}(x)$. Then the residuals

$$\hat{\varepsilon}_i = y_i - \hat{g}(x_i)$$

are computed. Let

$$B(u; b) = \begin{cases} (1 - (u/b)^2)^2 & \text{for } 0 \leq |u| < b \\ 0 & \text{for } |u| \geq b \end{cases}$$

be the *bisquare weight function*. Let

$$m = \text{median}(|\,\hat{\varepsilon}_i\,|)$$

be the median absolute residual. The *robustness weights* are

$$r_i = B(\hat{\varepsilon}_i; 6m).$$

An updated estimate, $\hat{g}(x)$, is computed using the local fitting method, but with the neighborhood weights, $w_i(x)$, replaced by $r_i w_i(x)$; thus, points (x_i, y_i) with large residuals receive reduced weight. Then new residuals are computed and the procedure is repeated. The final robust estimate is the result of updating the initial estimate several times.

Factor Predictors

We can include one or more factor predictors in the fitting by dividing the data into subsets, one for each combination of levels of the factor predictors, and then fitting loess surfaces to y_i as a function of the numeric predictors for each subset. This allows for very general interactions between the numeric and factor predictors but, of course, requires that there be a sufficient number of measurements of the numeric predictors for each combination of the levels of the factor predictors. If we have specified the errors to be Gaussian, the fits for the subsets are not related in any way; for example, neighborhoods are determined separately for each subset. However, if the error distribution has been specified to be symmetric, the various fits are pooled in forming the median absolute residual, m.

Errors with Unequal Scales

Suppose we specify that $a_i \varepsilon_i$ have constant variance σ^2. Then, for the Gaussian-error estimate, the neighborhood weight, $w_i(x)$, is replaced by $a_i w_i(x)$, and for the robust estimate, the weight $r_i w_i(x)$ is replaced by $a_i r_i w_i(x)$.

8.2   S Functions and Objects

This section describes the S functions for local regression modeling. In each subsection we analyze a dataset, illustrating how S functions are used to explore the data, fit models, and then carry out graphical diagnostics to check the specifications of the fitted models. Our goal is to show how the data are analyzed in practice using S, and how each dataset presents a different challenge. We begin, however, by rapidly running through the S functions for fitting and inference to give an overview; the reader need not understand details at this point.

The basic modeling function is loess(), which returns an object of class "loess". Let's apply it to some concocted data in the data frame madeup, which has two numeric predictors:

```
> names(madeup)
[1] "response" "one"      "two"
> attach(madeup)
```

We will fit a Gaussian model with the smoothing parameter, α, equal to 0.8 and the degree, λ, of the locally-fitted polynomial equal to 1:

```
> madeup.m <- loess(response ~ one * two, span = 0.5, degree = 2)
> madeup.m
Call:
loess(formula = response ~ one * two, span = 0.5, degree = 2)

Number of Observations:         100
Equivalent Number of Parameters: 14.9
Residual Standard Error:        0.9693
Multiple R-squared:             0.76
Residuals:
   min   1st Q median  3rd Q    max
 -2.289 -0.5064 0.1243 0.7359 2.357
```

Notice that the printing shows the *equivalent number of parameters*, μ; this measure of the amount of smoothing, which is defined in Section 8.4, is analogous to the number of parameters in a parametric fit. Also shown is an estimate of σ, the standard error of the residuals. Let's update the fit by dropping the square of the first predictor and making it conditionally parametric:

```
> madeup.new <- update(madeup.m, drop.square = "one",
+ parametric = "one")
> madeup.new
Call:
loess(formula = response ~ one * two, span = 0.8, degree = 2,
    parametric = "one", drop.square = "one")

Number of Observations:         100
Equivalent Number of Parameters: 6.9
Residual Standard Error:        1.48
Multiple R-squared:             0.34
Residuals:
```

```
    min   1st Q  median  3rd Q    max
 -4.758 -0.6449 0.03682 0.9094 2.589
```

Until now we have been fitting Gaussian models because the argument that controls this, `family`, defaults to `"gaussian"`. Now let us fit a model with the error distribution specified to be symmetric:

```
> madeup.new <- update(madeup.new, family = "symmetric")
> madeup.new
Call:
loess(formula = response ~ one * two, span = 0.8, degree = 2,
    parametric = "one", drop.square = "one", family = "symmetric")

Number of Observations:          100
Equivalent Number of Parameters: 6.9
Residual Scale Estimate:         1.089
Residuals:
    min  1st Q  median  3rd Q    max
 -7.472 -0.726 -0.1287 0.6342 2.594
```

Also, we have been using the standard normalization to normalize the scales of the two predictors; this is controlled by the argument `normalize`, whose default is `TRUE`. Let's now remove the normalization:

```
> madeup.new <- update(madeup.new, normalize = FALSE)
```

The function `specs` shows all of the aspects of the fit, both the specifications of the local regression model and the computational options:

```
> specs(madeup.m)

DATA
     formula:      response ~ one * two
     model:        FALSE
ERRORS
     family:       gaussian
     weights:
SURFACE
     span:         0.8
     degree:       2
     normalize:    TRUE
     parametric:
     drop.square:
     enp:          9.7
COMPUTING
     surface:      interpolate
     statistics:   approximate
```

```
cell:          0.2
iterations:    4
method:        loess
```

In the above S expressions, we utilized the generic functions `print()` and `update()`. The generic function `predict()` can be used to evaluate a fitted surface at a set of points in the space of the predictors:

```
> range(one)
[1] -2.809549  3.451000
> range(two)
[1] -1.885139  1.859246
> newdata <- data.frame(one = c(-2.5, 0, 2.5,), two = rep(0, 3))
> newdata
    one two
1 -2.5   0
2  0.0   0
3  2.5   0
> predict(madeup.m, newdata)
[1]  8.15678 14.49359 14.85414
```

In this case, the second argument is a data frame, each of whose rows is a point in the space of the predictors, and the result is a vector of length equal to the number of rows. Its ith element is the evaluation at the ith row of `newdata`. Suppose, however, that the points over which we want to do the evaluation form a rectangular grid in the space of the predictors. For example, let us create the following:

```
> marginal.grid <- list(one = c(-2.5, 0, 2.5), two = c(-1.5, 0, 1.5))
> newdata <- expand.grid(marginal.grid)
> newdata
    one  two
1 -2.5 -1.5
2  0.0 -1.5
3  2.5 -1.5
4 -2.5  0.0
5  0.0  0.0
6  2.5  0.0
7 -2.5  1.5
8  0.0  1.5
9  2.5  1.5
```

The two components of `marginal.grid` are *marginal grid points*. The function `expand.grid()` expands this marginal information into a data frame whose rows are the coordinates of the grid points. Let's see what happens when this data frame is given to `predict()`:

```
> predict(madeup.m, newdata)
           two=-1.5 two= 0.0   two= 1.5
one=-2.5   5.072129  8.15678  -1.207997
one= 0.0  14.111210 14.49359  14.112857
one= 2.5   1.951178 14.85414   3.042429
```

Thus, in this case, predict() produces an array shaped according to the marginal grid values.

The function predict() can also be used to compute information about standard errors:

```
> newdata <- data.frame(one = c(-.5, .5), two = rep(0,2))
> newdata
   one two
1 -0.5   0
2  0.5   0
> madeup.se <- predict(madeup.m, newdata, se.fit = TRUE)
> madeup.se
$fit:
[1] 14.49181 14.38973

$se.fit:
[1] 0.2767463 0.2780086

$residual.scale:
[1] 0.9693021

$df:
[1] 81.23189
```

The components are fit, the evaluated surface at newdata; residual.scale, the estimate of the residual scale; df, the degrees of freedom of the t distribution upon which the confidence intervals are based; and se.fit, estimates of the standard errors of the fit. Now we can use pointwise() to compute upper and lower confidence intervals:

```
> madeup.ci <- pointwise(madeup.se, coverage = .99)
> madeup.ci
$upper:
[1] 15.22179 15.12303

$fit:
[1] 14.49181 14.38973

$lower:
[1] 13.76183 13.65642
```

The computations of `predict()` that produce the coefficients in the component `se.fit` are much more costly than those that produce `fit`, so the number of points at which standard errors are computed should be modest compared to those at which we do evaluations; this is not a limitation for the practice of local regression modeling since it makes statistical and graphical sense to compute intervals at a limited set of points.

In our first model, `madeup.m`, we took `span` to be 1/2. Can we increase it and still get a good fit? The best way to check is to use graphical diagnostics, but the analysis of variance can also provide some guidance:

```
> anova(update(madeup.m, span = .75), madeup.m)
Model  1:
loess(formula = response ~ one * two, span = 0.75, degree = 2)
Model  2:
loess(formula = response ~ one * two, span = 0.5, degree = 2)
Analysis of Variance Table
      ENP    RSS    Test    F Value    Pr(F)
1    10.1  93.219  1 vs 2     2.86   0.012145
2    14.9  74.583
```

The results suggest that the increase in `span` has led to a distortion.

The equivalent number of parameters, μ, is related, albeit somewhat roughly, to the smoothing parameter, α, by the following formula:

$$\alpha \approx 1.2\tau/\mu$$

where τ is the number of monomials used in the local fitting. (If factors are present in the model, then we must multiply the right side of the above approximation by the number of levels of the combined factor.) The function `loess` has an argument `enp` that can be used to specify a target value for μ. Then α is computed from this approximation. The actual equivalent number of parameters, which is what appears in the printing, will typically be somewhat different, as the following example shows:

```
> loess(response ~ one * two, enp.target = 15, degree = 2)
Call:
loess(formula = response ~ one * two, enp.target = 15, degree = 2)

Number of Observations:             100
Equivalent Number of Parameters: 15.4
Residual Standard Error:          0.968
Multiple R-squared:               0.76
Residuals:
    min  1st Q  median  3rd Q    max
 -2.292 -0.512 0.09987 0.7253 2.355
```

For exploratory data analysis and diagnostic checking, we will employ S graphics functions extensively, including `pairs()`, `panel.smooth()`, `scatter.smooth()`, and `coplot()`, which are discussed in Chapter 3. In addition, the generic function `plot()` takes `loess` objects and displays the fitted curve or surface. For the remainder of this chapter, we will set a graphics parameter and stick with it until we exit from the chapter:

```
par(pty= "s")
```

8.2.1   Gas Data

The data frame `gas` has 22 observations of two variables from an industrial experiment that studied exhaust from an experimental one-cylinder engine (Brinkman, 1981). The dependent variable, which will be denoted by NO_x, is the concentration of nitric oxide, NO, plus the concentration of nitrogen dioxide, NO_2, normalized by the amount of work of the engine. The units are μg of NO_x per joule. The predictor is the equivalence ratio, E, at which the engine was run. E is a measure of the richness of the air and fuel mixture. Here is a summary:

```
> summary(gas)
               NOx          E
    Mean 3.546591 0.9249545
  Median 3.899500 0.9490000
    Min. 0.537000 0.6650000
1st Qu. 2.118000 0.8070000
3rd Qu. 4.937000 1.0210000
    Max. 5.344000 1.2240000
    NA's 0            0
```

Data Exploration

We begin our analysis with an exploration of the data by the scatterplot of NO_x against E in Figure 8.3:

```
attach(gas)
plot(E, NOx)
```

The plot shows that there is substantial curvature as a function of E and that the errors have a small variance compared with the change in the level of NO_x.

Fitting a First Model

Because of the substantial curvature in the overall pattern of the data, we will fit a local regression model using locally quadratic fitting. A reasonable starting point for the smoothing parameter is $\alpha = 2/3$. Also, because variation about the overall

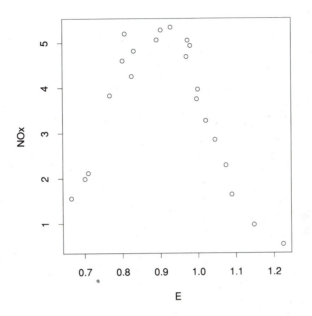

Figure 8.3: *Gas data—NO$_x$ against E.*

pattern shows no unusual behavior, we begin with the hope that an assumption of Gaussian errors is reasonable:

```
> gas.m <- loess(NOx ~ E, span = 2/3, degree = 2)
> gas.m
Call:
loess(formula = NOx ~ E, span = 2/3, degree = 2)

Number of Observations:          22
Equivalent Number of Parameters: 5.5
Residual Standard Error:         0.3404
Multiple R-squared:              0.96
Residuals:
    min   1st Q   median  3rd Q    max
 -0.5604 -0.213 -0.02511 0.1271 0.6234
```

The equivalent number of parameters of the fit is 5.5. The estimate of the residual variance is 0.3404, but we should not take this estimate seriously before carrying out the diagnostic procedures to come.

Evaluation and Plotting the Curve

Having fitted a model to gas, we can compute $\hat{g}(x)$ at values of the predictor, E:

```
> gas.fit.x <- c(min(E), median(E), max(E))
> gas.fit.x
[1] 1.1964144 5.0687470 0.5236823
> predict(gas.m, gas.fit.x)
[1] 1.1964144 5.0687470 0.5236823
```

We could compute the fitted values, $\hat{y}_i = \hat{g}(x_i)$, by:

```
predict(gas.m, E)
```

However they, as well as the residuals, $y_i - \hat{y}_i$, are stored on the loess object and can be accessed by the expressions:

```
fitted(gas.m)
residuals(gas.m)
```

If our goal is to evaluate the curve just to plot it, we can use plot() to both evaluate and graph:

```
plot(gas.m)
```

The result is shown in Figure 8.4. For one predictor, the plot() method for "loess" objects carries out an evaluation at equally spaced points from the minimum to the maximum value of the predictor, and makes the plot. An argument, evaluation, specifies the number of points at which the evaluation is carried out; the default is 50, so in Figure 8.4 the curve is evaluated at 50 equally spaced points and graphed by connecting successive plotting locations by line segments.

Diagnostic Checking

We turn now to diagnostic checking to accept or reject the specifications of the model we have fitted. To check the properties of $g(x)$ that are specified by the choice of $\alpha = 2/3$ and $\lambda = 2$, we plot the residuals, $\hat{\varepsilon}_i$, against E to look for lack of fit:

```
scatter.smooth(E, residuals(gas.m), span = 1, degree = 1)
abline(h=0)
```

The result is shown in Figure 8.5. The function scatter.smooth() makes a scatterplot and adds a smooth curve using loess fitting. No effect appears to be present in the diagnostic plot, so $\alpha = 2/3$ appears to have introduced no lack of fit. But is there surplus of fit, that is, can we get away with a larger α? To check this, we fit a new loess model with $\alpha = 1$:

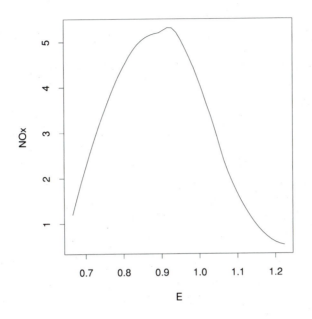

Figure 8.4: *Gas data—local regression fit.*

```
gas.m.null <- update(gas.m, span = 1)
```

The residual plot is shown in Figure 8.6. There is a strong signal in the residuals—a dependence of the level of the $\hat{\varepsilon}_i$ on E, so $\alpha = 1$ is too large, which suggests that $\alpha = 2/3$ is about as large as we can get away with. Thus, we have verified our specification of the form of $g(x)$ since there appears to be no surplus or lack of fit.

Next, we check the distributional specifications for the error terms. To see if the scale of the residuals depends on the level of the surface, we plot $\sqrt{|\hat{\varepsilon}_i|}$ against the fitted values, $\hat{y}_i$. Taking the square root tends to symmetrize the distribution of the absolute residuals. For our current example, with its small sample size of 22, we would not expect this method to reliably detect anything but a radical change in scale, but for illustrative purposes we show the plot in Figure 8.7:

```
scatter.smooth(fitted(gas.m), sqrt(abs(residuals(gas.m))), span = 1,
    degree = 1)
```

The graph does not show any convincing dependence. To check for dependence of the scale on E, a similar graph was made—but against E instead of the fitted

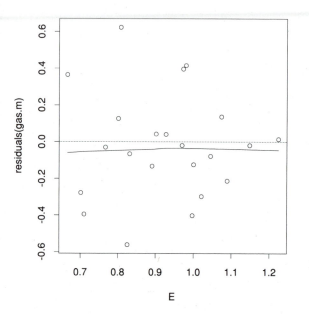

Figure 8.5: *Residuals against E with a scatterplot smoothing—first fit to* gas.

values—and, again, no convincing dependence was found. To check the assumption
of a Gaussian distribution of the errors, we will make a Gaussian probability plot
of the residuals. In order to judge the straightness of the points on such plots, we
will write a little function that draws a line through the lower and upper quartiles:

```
> qqline
function(x)
{
    data.quartiles <- quantile(x, c(0.25, 0.75))
    norm.quartiles <- qnorm(c(0.25, 0.75))
    b <- (data.quartiles[2] - data.quartiles[1])/
        (norm.quartiles[2] - norm.quartiles[1])
    a <- data.quartiles[1] - norm.quartiles[1] * b
    abline(a, b)
}
```

Now we make the plot:

```
qqnorm(residuals(gas.m))
```

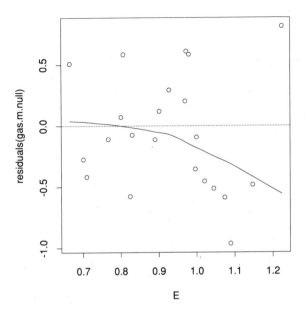

Figure 8.6: *Residuals against E with a scatterplot smoothing—second fit to* gas.

```
qqline(residuals(gas.m))
```

The result, shown in Figure 8.8, suggests that the Gaussian specification is justified.

Inference

gas.m has passed the diagnostic tests, which allows us to carry out statistical inferences with an assurance of validity. First, we compute 99% pointwise confidence intervals for $g(x)$ at seven values of E:

```
> gas.limits.x <- seq(min(E), max(E), length = 7)
> gas.se <- predict(gas.m, gas.limits.x, se.fit = TRUE)
> pointwise(gas.se)
$upper:
[1] 1.985621 4.109807 5.480230 5.566510 3.527610 1.710617
[7] 1.472049
```

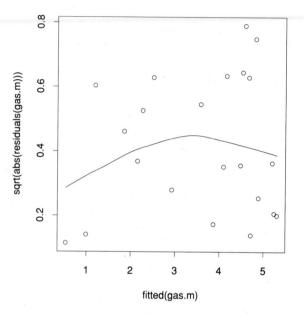

Figure 8.7: *Square-root absolute residuals against fitted values with a scatterplot smoothing.*

```
$fit:
[1] 1.1964144 3.6794968 5.0557086 5.1352603 3.1436568 1.1969317
[7] 0.5236823

$lower:
[1]  0.4072080  3.2491865  4.6311876  4.7040105  2.7597037  0.6832464
[7] -0.4246841
```

The function `plot()`, which was earlier used to plot the curve, will compute and add confidence limits to the plot, as shown in Figure 8.9:

```
plot(gas.m, confidence = 7)
```

The limits are added at `confidence` equally spaced points from the minimum to the maximum of the values of the predictor. Thus, the limits that are plotted in Figure 8.9 are the same as those we just computed.

We know from the diagnostic checking that `gas.m.null` does not fit the data. But for purposes of illustration we will carry out a statistical comparison of the two models:

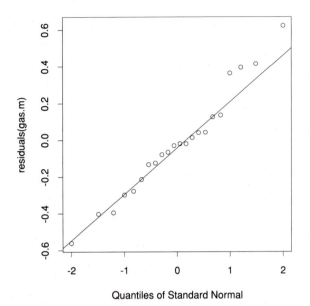

Figure 8.8: *Gaussian quantile plot of residuals with line passing through lower and upper quartiles.*

```
> gas.m
Call:
loess(formula = NOx ~ E, span = 2/3, degree = 2)

  Number of Observations:          22
  Equivalent Number of Parameters: 5.5
  Residual Standard Error:         0.3404
  Multiple R-squared:              0.96
  Residuals:
      min  1st Q  median   3rd Q    max
   -0.5604 -0.213 -0.02511 0.1271 0.6234
> gas.m.null
Call:
loess(formula = NOx ~ E, span = 1, degree = 2)

  Number of Observations:          22
```

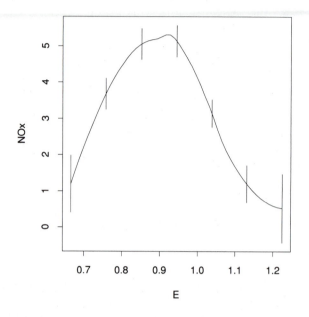

Figure 8.9: *Gas data—local regression fit with 99% pointwise confidence intervals.*

```
Equivalent Number of Parameters: 3.5
Residual Standard Error:         0.5197
Multiple R-squared:              0.9
Residuals:
      min   1st Q  median   3rd Q     max
  -0.9644 -0.4536 -0.1019  0.2914  0.8133
```

We can see that the increase in α for `gas.m.null` results in a drop in the equivalent number of parameters, but s, the estimate of σ, increases by a factor of about 1.5. This is to be expected in view of the lack of fit. We can test `gas.m` against `gas.m.null` by an analysis of variance:

```
> anova(gas.m.null, gas.m)
Model  1:
loess(formula = NOx ~ E, span = 1, degree = 2)
Model  2:
loess(formula = NOx ~ E, span = 2/3, degree = 2)
Analysis of Variance Table
```

```
       ENP    RSS     Test    F Value      Pr(F)
  1    3.5  4.8489   1 vs 2      10.14 0.0008601
  2    5.5  1.7769
```

The result, as expected, is highly significant.

8.2.2 Ethanol Data

The experiment that produced the `gas` data that we just analyzed was also run with gasoline replaced by ethanol. There were 88 runs and two predictors: E, as before, and C, the compression ratio of the engine. The data are in `ethanol`:

```
> summary(ethanol)
              NOx          C         E
    Mean 1.957375 12.03409 0.9264773
  Median 1.754500 12.00000 0.9320000
    Min. 0.370000  7.50000 0.5350000
 1st Qu. 0.944000  8.25000 0.7615000
 3rd Qu. 3.042000 15.00000 1.1115000
    Max. 4.028000 18.00000 1.2320000
    NA's 0        0         0
```

These data were analyzed previously in Chapter 7. It was discovered that an additive fit did not approximate the surface sufficiently well because of an interaction between C and E. Thus, we will try fitting a local regression model. To make typing easier we will attach `ethanol`:

```
attach(ethanol)
```

Exploratory Data Display

An exploratory plot useful for starting an analysis with two or more predictors is the scatterplot matrix, shown in Figure 8.10:

```
pairs(ethanol)
```

We will refer to panels in this and other multipanel displays by column and row, numbering as we would on a graph; thus, the lower left panel is (1,1) and the one to the right of it is (2,1). The (3,3) panel of the matrix, a scatterplot of NO_x against E, shows a strong nonlinear dependence with a peak between 0.8 and 1.0. This makes it immediately clear that we need to use locally quadratic fitting. The (2,3) panel of the scatterplot matrix shows no apparent dependence of NO_x on C; however, we should not at this point draw any firm conclusion since it is possible that a dependence is being masked by the strong effect of E. The (1,2) panel, which graphs the configuration of points in the space of the predictors, shows that

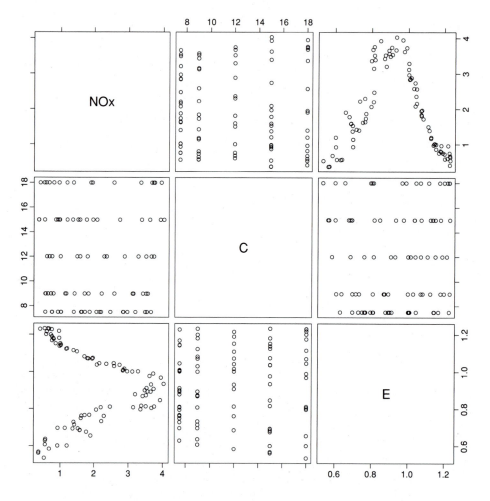

Figure 8.10: *Ethanol data—scatterplot matrix of* NO_x, *C, and E.*

the values of the two variables are nearly uncorrelated and that C takes on one of five values.

Coplots, introduced in Chapter 3, are an essential tool in fitting local regression models. Figure 8.11 is a coplot of the ethanol data. Thus, the dependence panels are the 3×3 array, and the given panel is at the top. On each dependence panel, NO_x is graphed against C for those observations whose values of E lie in an interval; on the panel, we are seeing how NO_x depends on C for E held fixed to the interval. The intervals are shown on the given panel; as we move from left to right through these intervals, we move from left to right and then bottom to top through the dependence panels.

To produce Figure 8.11, we begin by selecting the intervals:

```
E.intervals <- co.intervals(E, number = 9, overlap = 1/4)
```

The result is a 9×2 matrix that gives the left endpoints of the intervals in the left column and the right endpoints in the right column:

```
> E.intervals
        [,1]  [,2]
[1,]  0.535 0.686
[2,]  0.655 0.761
[3,]  0.733 0.811
[4,]  0.808 0.899
[5,]  0.892 1.002
[6,]  0.990 1.045
[7,]  1.042 1.125
[8,]  1.115 1.189
[9,]  1.175 1.232
```

The intervals produced by `co.intervals` have two properties: approximately the same number of observations lie in each interval and approximately the same number of observations lie in two successive intervals. The shared number is specified by the argument `overlap` as the fraction of points shared by the successive intervals. For example, if there are approximately 20 points in each interval and `overlap` is $1/2$, then successive intervals share about 10 points. Now we make the coplot:

```
coplot(NOx ~ C | E, given.values = E.intervals,
    panel = function(x, y) panel.smooth(x, y, degree = 1, span = 1))
```

The first argument specifies the response, the predictor, and the given variable by a formula; in the above expression, the formula is read, "Plot NO_x against C, given E." The argument `given.values` specifies the conditioning values. For a numeric given predictor, the values can be a two-column matrix as in the example, or they can be a vector, in which case each element is both the left and right endpoint of an interval, so the intervals have length 0. We can also condition on the levels of a

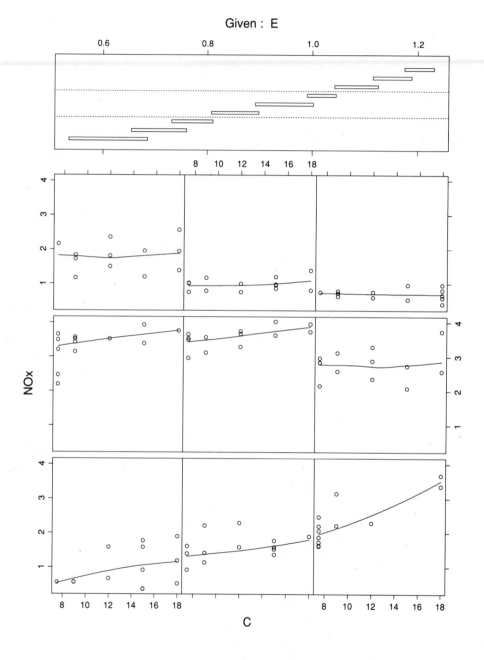

Figure 8.11: *Ethanol data—coplot of* NO_x *against C given E with scatterplot smoothings.*

factor; in this case the argument is a character vector. The argument `panel` takes a function with two arguments `x` and `y` that determine the method of plotting on each dependence panel; `x` refers to the abscissae on a panel and `y` refers to the ordinates. The default function is `points()`. In the above expression, the method of plotting is to create a scatterplot of the data for each panel and a scatterplot smoothing of the points. To do this, we used the function `panel.smooth`, which adds curves using loess smoothing. We defined a function on the fly that involves `panel.smooth`; this trick allows us to pass the arguments `span` and `degree` into `coplot()`.

Figure 8.12 is a coplot of NO_x against E given C. Since C takes on five values, we have simply conditioned on each of these five values:

```
> C.points <- sort(unique(C))
> coplot(NOx ~ E | C, given.values = C.points, columns = 3, rows = 2,
+    panel = function(x,y) panel.smooth(x, y, degree = 2, span = 2/3))
```

The arguments `columns` and `rows` have been used to specify the dependence panels to be arranged in an array with three columns and two rows.

What have we learned from these coplots? First, NO_x does in fact depend on C; for low values of E, NO_x increases with C, and for medium and high values of E, NO_x is constant as a function of C. Thus, there is an interaction between C and E. Second, over the range of values of E and C in the dataset, NO_x undergoes more rapid change as a function of E for C held fixed than as a function of C for E held fixed. Finally, the plots show that the amount of noise—that is, the variance, σ^2, of the ε_i—is small compared with the effect due to E, and is moderate compared with the effect due to C.

Modeling the Ethanol Data

It is quite clear from the exploratory plots that we must specify a locally-quadratic surface—that is, take λ to be 2—because of the substantial curvature as a function of E. Also, we will specify $\alpha = 0.5$ for the first fit:

```
> ethanol.first <- loess(NOx ~ C * E, span = 1/2, degree = 2)
> ethanol.first
Call:
loess(formula = NOx ~ C * E, span = 1/2, degree = 2)

Number of Observations:            88
Equivalent Number of Parameters: 13
Residual Standard Error:           0.2599
Multiple R-squared:                0.96
Residuals:
     min  1st Q  median  3rd Q   max
  -0.5017 -0.253 -0.06219 0.1333 0.43
```

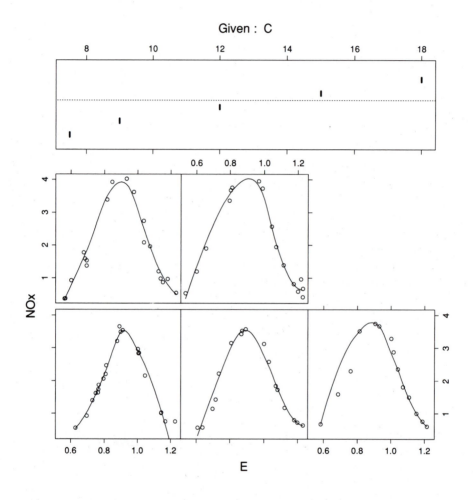

Figure 8.12: *Ethanol data—coplot of* NO_x *against* E *given* C *with scatterplot smoothings.*

We begin a search for lack of fit by plotting the residuals against each of the predictors:

```
> scatter.smooth(C, residuals(ethanol.first), span = 1, degree = 2,
+    ylab = "Residuals", main = "First Fit")
> abline(h = 0)
> scatter.smooth(E, residuals(ethanol.first), span = 1, degree = 2,
+    ylab = "Residuals", main = "First Fit")
> abline(h = 0)
```

The result is shown in the top two panels of Figure 8.13. Clearly there is lack of fit in the right panel. Thus, we drop span to $1/4$:

```
> ethanol.m <- update(ethanol.first, span = 1/4)
> ethanol.m
Call:
loess(formula = NOx ~ C * E, span = 1/4, degree = 2)

Number of Observations:          88
Equivalent Number of Parameters: 21.6
Residual Standard Error:         0.1761
Multiple R-squared:              0.98
Residuals:
      min    1st Q  median    3rd Q     max
   -0.3975 -0.09133 0.00862 0.06417 0.3382
```

The residual plots, shown in the bottom two panels of Figure 8.13, look much better. To enhance the comparison of the two sets of residual plots, the values of the arguments span and degree of scatter.smooth() are the same for all four panels, as are the vertical scales.

But we must check further; these marginal residual plots can, of course, hide local lack of fit in the (C, E) plane. We check this by the coplots in Figures 8.14 and 8.15:

```
> coplot(residuals(ethanol.m) ~ C | E, given.values = E.intervals,
+    panel = function(x, y)
+        panel.smooth(x, y, degree = 1, span = 1, zero.line = TRUE))
> coplot(residuals(ethanol.m) ~ E | C, given.values = C.points,
+    panel = function(x, y)
+        panel.smooth(x, y, columns = 3, rows = 2, degree = 1, span = 1,
+        zero.line = TRUE))
```

There is some suspicious behavior on the (1,2) and (2,2) dependence panels of Figure 8.14; almost all of the residuals are positive. The detected effect is, however, quite minor, so we will ignore it.

We can check the specifications of the error distribution by the same diagnostic methods used for the gas data—graph $\sqrt{|\hat{\varepsilon}_i|}$ against $\hat{y}_i$, graph $\sqrt{|\hat{\varepsilon}_i|}$ against C and

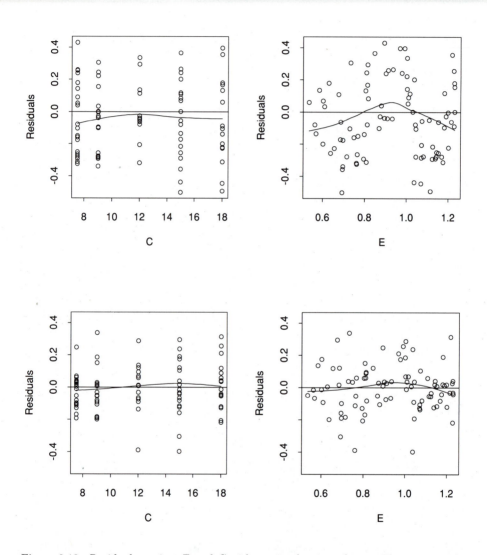

Figure 8.13: *Residuals against E and C with scatterplot smoothings. The top row of plots corresponds to the first fit, the bottom row to the second fit.*

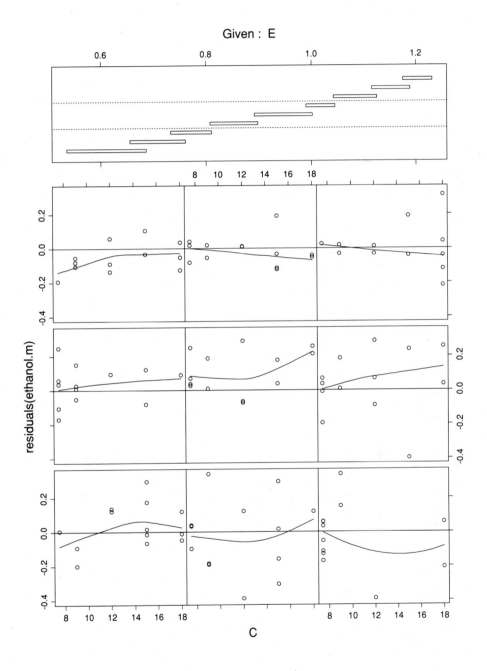

Figure 8.14: *Coplot of residuals against C given E with scatterplot smoothings.*

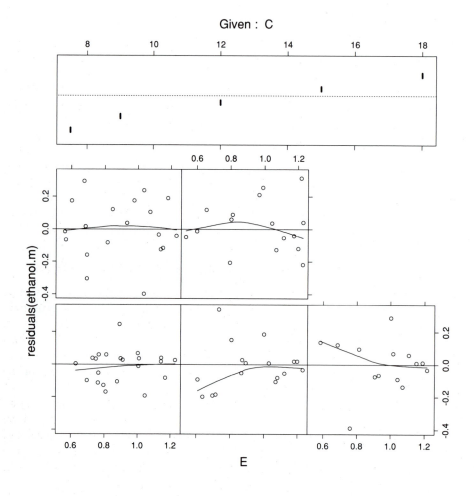

Figure 8.15: *Coplot of residuals against E given C with scatterplot smoothings.*

E, and make a Gaussian probability plot of $\hat{\varepsilon}_i$. This was done, and `ethanol.m` passed the tests.

Plotting the Surface

For `loess` objects with two predictors, `plot()` displays the fitted surface by coplots:

```
plot(ethanol.m, given = 16, evaluation = 50, confidence = 7,
    coverage = .99)
```

The result is shown in Figures 8.16 and 8.17. Let $\hat{g}(C, E)$ be the fitted surface. Consider a single panel of Figure 8.16. E has been set to a specific conditioning value, $E = E^*$; then $\hat{g}(C, E^*)$ has been evaluated for 50 equally spaced values of C ranging from the minimum value of C in the data to the maximum, and the surface values have been graphed on the panel against the equally spaced values of C. Also, 99% confidence intervals are drawn at seven equally spaced points from the minimum value of C in the data to the maximum. There are 16 equally spaced conditioning values of E ranging from the minimum value of E in the data to the maximum; the given panel in Figure 8.16 shows the 16 values. Similarly, Figure 8.17 shows the dependence of the fitted surface on E for 16 conditioning values of C.

Dropping Squares and Conditionally Parametric Surfaces

The coplot in Figure 8.16 show that the ethanol fit has an undesirable property: the surface as a function of C for fixed E has unconvincing undulations, especially in the (1,1) dependence panel. Our skepticism comes from two sources. First, in the coplot of the data in Figure 8.11, NO_x appears to be a very smooth function of C; in fact, the coplot suggests that given E, the dependence is actually linear in C. Second, the undulations in Figure 8.16 are small compared with the sizes of the confidence intervals.

As we saw from the diagnostic checking, if we increase α and thereby get more smoothness as a function of C, we introduce lack of fit. Instead, we will cut back on the variation of the fit as a function of C by dropping C^2 from the fitting variables; this leaves us with a constant, E, C, EC, and E^2. In addition, we will specify the surface to be conditionally parametric in C; this will result in a fit that is linear in C given E:

```
ethanol.cp <- update(ethanol.m, drop.square = "C", parametric = "C")
```

Let's compare the old fit and the new:

```
> ethanol.m
Call:
loess(formula = NOx ~ C * E, span = 1/4, degree = 2)
```

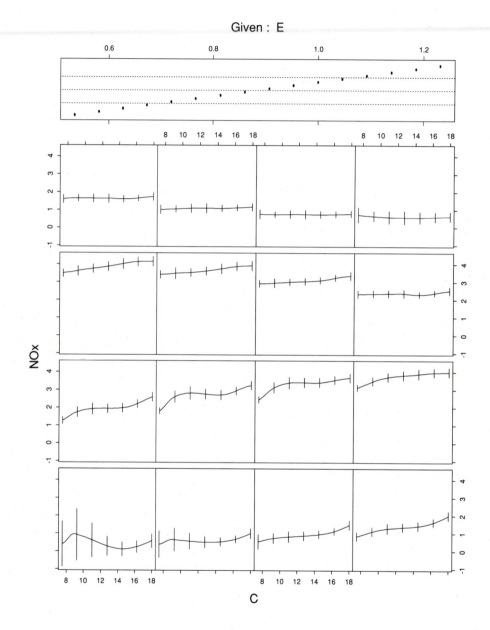

Figure 8.16: *Ethanol data—coplot of the local regression fit with pointwise 99% confidence intervals.*

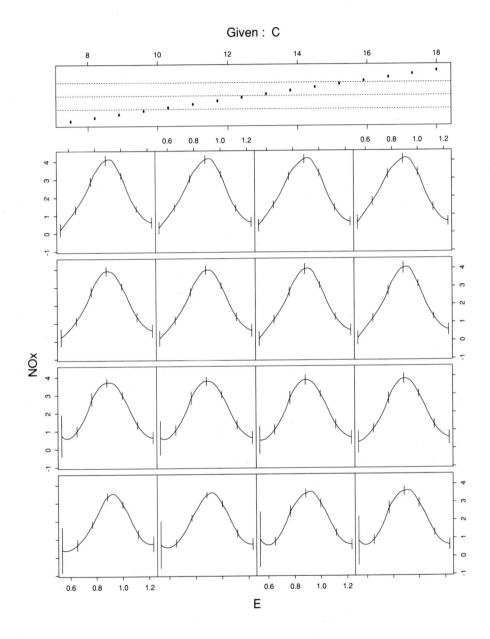

Figure 8.17: *Ethanol data—coplot of the local regression fit with pointwise 99% confidence intervals.*

```
Number of Observations:        88
Equivalent Number of Parameters: 21.6
Residual Standard Error:       0.1761
Multiple R-squared:            0.98
Residuals:
     min    1st Q  median   3rd Q    max
 -0.3975 -0.09133 0.00862 0.06417 0.3382
> ethanol.cp
Call:
loess(formula = NOx ~ C * E, span = 1/4, degree = 2,
parametric = "C", drop.square = "C")

Number of Observations:        88
Equivalent Number of Parameters: 18.2
Residual Standard Error:       0.1808
Multiple R-squared:            0.98
Residuals:
     min    1st Q   median   3rd Q    max
 -0.4388 -0.07436 -0.009093 0.06651 0.5485
```

The equivalent number of parameters has dropped by about 15%, the residual
standard error has increased insignificantly, and diagnostic plots, not shown here,
indicated no lack of fit. But the big gain is that we can now increase span to 1/2
without introducing lack of fit:

```
> ethanol.cp <- update(ethanol.cp, span = 1/2)
> ethanol.cp
Call:
loess(formula = NOx ~ C * E, span = 1/2, degree = 2,
parametric = "C", drop.square = "C")

Number of Observations:        88
Equivalent Number of Parameters: 9.2
Residual Standard Error:       0.1842
Multiple R-squared:            0.98
Residuals:
     min   1st Q  median   3rd Q    max
 -0.5236 -0.0973 0.01386 0.07345 0.5584
```

In so doing we have driven the equivalent number of parameters to less than half
of what it was originally and kept the residual standard error about the same. The
coplots in Figures 8.18 and 8.19 show the resulting fitted surface:

```
plot(ethanol.cp, given = 16, evaluation = 50, confidence = 7,
   coverage = 0.99)
```

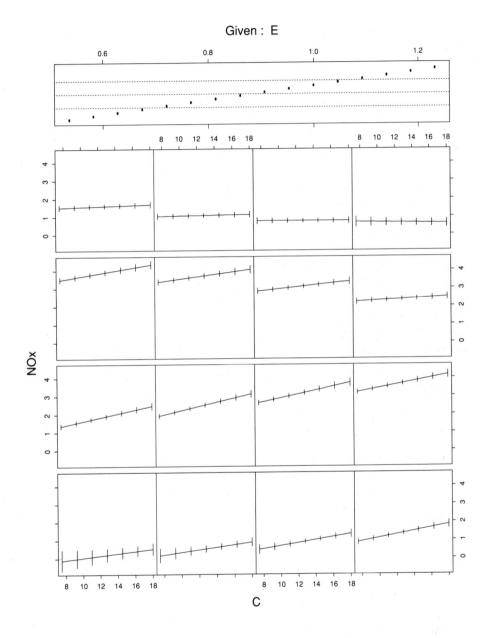

Figure 8.18: *Ethanol data—coplot of conditionally parametric local-regression fit with pointwise 99% confidence intervals.*

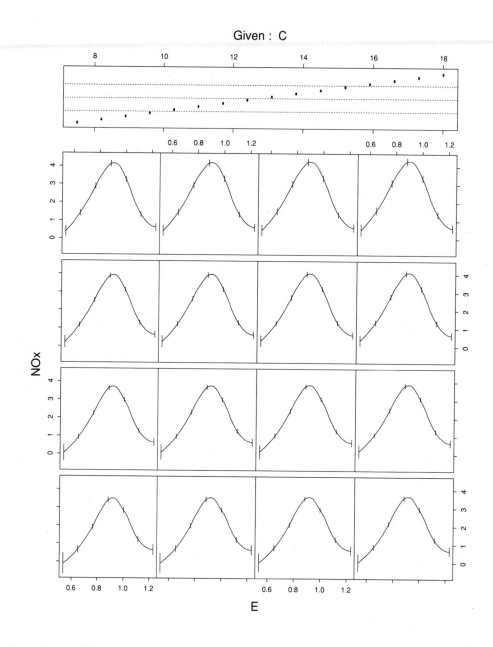

Figure 8.19: *Ethanol data—coplot of conditionally parametric local-regression fit with pointwise 99% confidence intervals.*

Computing the Fitted Surface and Confidence Intervals

We turn now to a further discussion of how `predict()` is used to evaluate a fitted surface and to compute information for confidence intervals. As we saw at the beginning of this section, for two or more predictors there are two data structures that can be given to the second argument, `newdata`, which specifies the points in the space of the predictors at which the evaluation takes place. The first data structure is a plain old data frame:

```
> newdata
      C   E
1   7.5 0.6
2   9.0 0.8
3  12.0 1.0
4  15.0 0.8
5  18.0 0.6
```

The following evaluates the ethanol surface at points in the space of the predictors given by the rows of `newdata`:

```
> predict(ethanol.m, newdata)
[1] 0.2815825 2.5971411 3.0667178 3.2555778 1.0637788
```

The result is a vector whose length is equal to the number of rows of the data structure. This is similar to our use of `predict()` for one predictor, but there is now one difference: the function must be able to match each column of `newdata` with a predictor specified by the `formula` of the `loess` object. The matching is done by looking at the column names of `newdata` and the names of the predictors in `formula`. In our example, the columns of `newdata` are C and E.

A second data structure can be used when the evaluation points form a grid. For example, to show the curves in Figure 8.18, the function `plot()` used `predict()` to evaluate the ethanol surface on a 50×16 grid of 800 points in the *(C, E)* plane. We refer to the 800 points as the *grid points* and the equally spaced values of C and E that define the grid as the *marginal grid points*. Let's see how this evaluation is carried out. First we compute the marginal grid points and put them in a list:

```
C.marginal <- seq(min(C), max(C), length = 50)
E.marginal <- seq(min(E), max(E), length = 16)
CE.marginal <- list(C = C.marginal, E = E.marginal)
```

Then we use the function `expand.grid()`:

```
CE.grid <- expand.grid(CE.marginal)
```

This creates a data frame whose rows are the coordinates of the grid; attributes of the data frame provide the information that the data structure describes a grid. Now we do the evaluation:

```
ethanol.fit <- predict(ethanol.cp, CE.grid)
```

Matching is done in the same way as for the first data structure. The evaluation is carried out at the grid values, and the result is a numeric array whose dimension is equal to the number of predictors:

```
> dim(ethanol.fit)
[1] 50 16
> names(dimnames(ethanol.fit))
[1] "C" "E"
```

The (i, j)th element of the array `ethanol.fit` is the evaluation at the ith marginal grid point of C and the jth marginal grid point of E.

As with one predictor, confidence-interval information can be computed at each point of `newdata` by setting `se.fit=TRUE`, but again we point out that this increases the computational intensity substantially. To get the intervals shown in Figure 8.19, we do the following:

```
C.marginal <- seq(min(C), max(C), length = 7)
E.marginal <- seq(min(E), max(E), length = 16)
CE.marginal <- list(C = C.marginal, E = E.marginal)
CE.grid <- expand.grid(CE.marginal)
ethanol.se <- predict(ethanol.cp, CE.grid)
ethanol.ci <- pointwise(ethanol.se, coverage = .99)
```

8.2.3 Air Data

We turn now to an application with three predictors. The data frame `air`, which is used as an example in Chapter 7, contains four variables:

```
> names(air)
[1] "ozone"       "radiation"   "temperature" "wind"
> dim(air)
[1] 111   4
```

These data are from an environmental study that analyzed how the air pollutant ozone depends on three meteorological variables: radiation, wind speed, and temperature. The data are daily measurements of the four variables for 111 days.

For three or more predictors, carrying out fitting and inference for local regression models in S is no more complicated than for two. What gets harder, of course, is graphing the data to explore and diagnose. The function `coplot()` can be used for three predictors since it allows plotting against one predictor, conditioning on two others. Thus, for three predictors, we can make three coplots, graphing against each predictor conditional on the other two. Figure 8.20 shows one of the three coplots for the `air` data:

```
attach(air)
w.given <- co.intervals(wind, 4, 0.5)
t.given <- co.intervals(temperature, 4, 0.5)
coplot(ozone ~ radiation | temperature * wind,
    given = list(temperature = t.given, wind = w.given),
    panel = function(x,y) panel.smooth(x, y, span = 1, degree = 1))
```

We have conditioned on wind and temperature. The dependence panels are the
4×4 matrix of panels. The given panels, one for each conditioning predictor, are to
the right and top. As we move up a column of dependence panels, the intervals of
wind speed increase, and as we move from left to right across a row of dependence
panels, the intervals of temperature increase. For example, the points on the (2,3)
panel of the coplot are observations for which the temperature measurements are
in the second interval and the wind speed measurements are in the third interval.
We omit the two remaining coplots, but in the analysis of these data they were
carefully studied.

Let's fit a local regression model to the data:

```
> air.m <- loess(ozone ~ radiation * temperature * wind, span = .8,
+     degree = 2)
> air.m
Call:
loess(formula = ozone ~ radiation * temperature * wind,
    span = 0.8, degree = 2)

Number of Observations:           111
Equivalent Number of Parameters: 15.7
Residual Standard Error:          0.4331
Multiple R-squared:               0.81
Residuals:
     min    1st Q   median  3rd Q     max
   -1.158  -0.2906  -0.05033 0.2229  0.8758
```

Diagnostic plots revealed that the specifications of air.m were reasonable assumptions. Let's now display the fitted surface:

```
> air.ranges <- list(radiation = quantile(radiation, c(1/4, 3/4)),
+     temperature = quantile(temperature, c(1/4, 3/4)), wind =
+     quantile(wind, c(1/4, 3/4)))
> plot(air.m, given = 5, confidence = 7, which.plots = "radiation",
+     ranges = air.ranges)
```

The argument which.plots has selected just one of the three possible coplots to be
graphed. The argument ranges has been used to specify the ranges of the evaluation points and given values to be the lower and upper quartiles of the predictor
observations rather than the default, which is the minima and the maxima. The
result is shown in Figure 8.21.

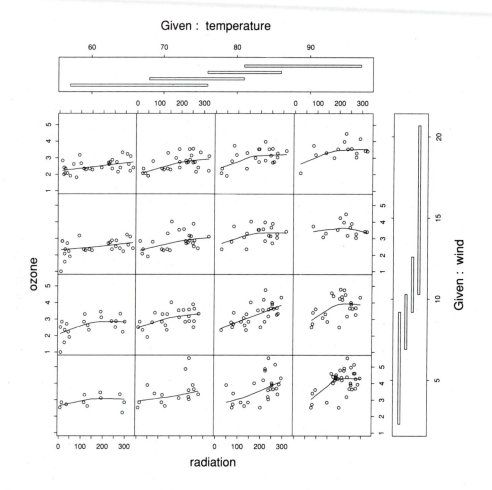

Figure 8.20: *Air data—coplot of ozone against solar radiation given wind speed and temperature with scatterplot smoothings.*

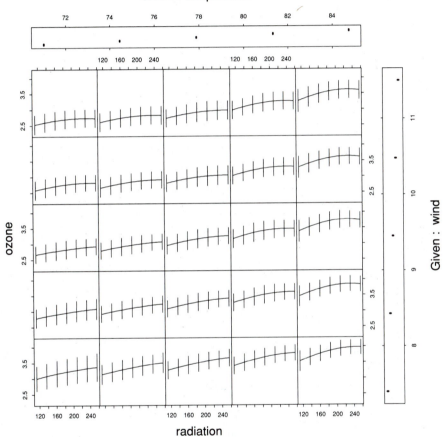

Figure 8.21: *Air data—local regression fit with pointwise 99% confidence intervals.*

8.2.4   Galaxy Velocities

NGC7531 is a spiral galaxy in the Southern Hemisphere with a very bright inner ring. When looked at from the earth, the galaxy takes up a small area on the celestial sphere. Figure 8.22 shows measurements of the radial velocity of the galaxy at 323 locations in this area (Buta, 1987). The positions have been jittered slightly to reduce overplotting. The horizontal scale of the graph is the east-west coordinate

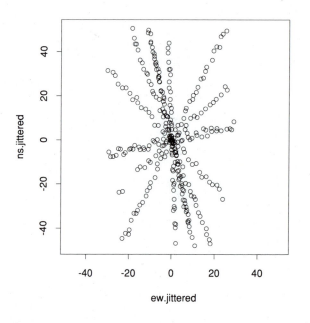

Figure 8.22: *Galaxy data—locations of velocity measurements.*

and the vertical scale is the north-south coordinate. Note that north is up and east is to the left because we are looking at the celestial sphere from the inside. Each measurement lies along one of seven slits that nearly intersect at a single point near the origin, (0,0).

The data are stored in a data frame `galaxy`:

```
> names(galaxy)
[1] "east.west"        "north.south"      "angle"
[3] "radial.position"  "velocity"
> dim(galaxy)
[1] 323    5
```

The first column contains the east-west positions of the measurements and the second contains the north-south positions. For each observation, the value in `angle` is the angle with the horizontal of the slit on which the observation lies; the units are degrees of counterclockwise rotation from horizontal:

```
> attach(galaxy)
> sort(unique(angle))
[1]   12.5  43.0  63.5  92.5 102.5 111.0 133.0
```

`radial.position` contains signed distances from the origin to the measurement locations; a distance is multiplied by -1 if the east-west coordinate is negative and by 1 if it is positive:

```
> range(radial.position)
[1] -52.4  55.7
```

Finally, `velocity`, whose units are km/sec, contains the velocity measurements:

```
> summary(velocity)
        Min   1Q Median   3Q  Max
  [1,] 1409 1522   1586 1669 1775
```

Data Exploration

Figure 8.22 was made by the following:

```
attach(galaxy)
ew.jittered <- jitter(east.west, factor = 1/2)
ns.jittered <- jitter(north.south, factor = 1/2)
lim <- range(ew.jittered, ns.jittered)
plot(ew.jittered, ns.jittered, xlim = lim, ylim = lim)
```

`xlim` and `ylim` were specified in `plot()` to keep the number of units per cm the same on the vertical and horizontal scales. Figure 8.23 uses `coplot()` to explore the velocities by graphing velocity against radial position for each slit:

```
coplot(velocity ~ radial.position | angle, given.values =
  sort(unique(angle)), panel = function(x, y) panel.smooth(x, y,
  span= 1/2, degree=2))
```

The figure shows that it is sensible to approach modeling velocity dependence by an overall smooth pattern with random variation superimposed.

Modeling

The goal in the analysis of these data is to understand how galaxy velocity varies over the measurement region. Thus, velocity is a dependent variable and there are

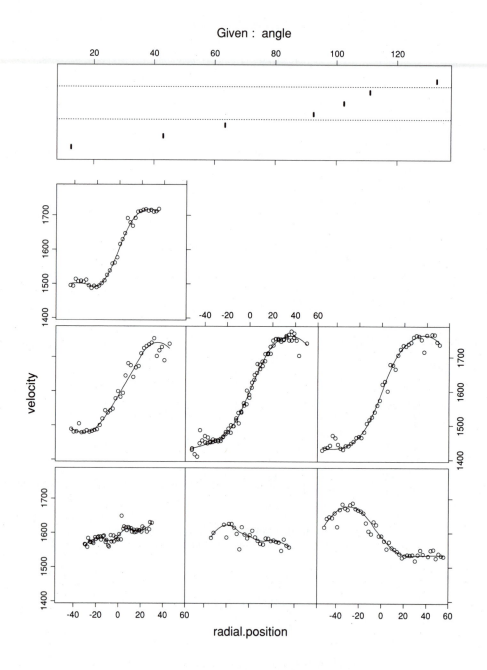

Figure 8.23: *Galaxy data—coplot of velocity against radial position given slit angle.*

two predictors: east-west position and south-north position. In Figure 8.23 the curvature of the underlying pattern is substantial; thus we will specify a locally-quadratic surface. Since many points appear to deviate substantially from the overall pattern compared to the deviations of the majority of points, it seems prudent to specify symmetric errors. Finally, it makes sense to preserve the spatial metric of the predictors and not normalize the variation in their measurements:

```
> galaxy.m <-loess(velocity ~ east.west * north.south, degree = 2,
+   span = 0.35, normalize = F, family = "symmetric")
> galaxy.m
Call:
loess(formula = velocity ~ east.west * north.south, span = 0.35,
    degree = 2, normalize = F, family = "symmetric")

Number of Observations:            323
Equivalent Number of Parameters:   19.6
Residual Scale Estimate:           12.3
Residuals:
    min  1st Q median 3rd Q    max
 -57.23 -5.898 0.2501 9.417 53.52
```

Let's evaluate the surface on a grid and then make a contour plot:

```
> galaxy.marginal <- list(east.west = seq(-29,29),
+   north.south = seq(-49,49))
> galaxy.fit <- predict(galaxy.m, expand.grid(galaxy.marginal))
> contour(galaxy.marginal$east.west, galaxy.marginal$north.south,
+   galaxy.fit, v = seq(1435, 1755, by = 40), xlim = c(-50, 50),
+   xlab = "EW", ylab = "NS")
> contour(galaxy.marginal$east.west, galaxy.marginal$north.south,
+   galaxy.fit, v = seq(1435, 1755, by = 20), labex=0, add=T)
```

The result is shown in Figure 8.24. Recall that we studied the fits to ethanol and air by coplots, but in this application it makes sense to use a contour plot since we want to see the surface as a whole entity—finding peaks, troughs, ridges, steep terrain, and so forth—and are not interested in conditional dependence.

Diagnostic Checking

Of course, we must carry out diagnostic checking to make sure we have not plotted nonsense in Figure 8.24. First, in Figure 8.25, we make a coplot of the residuals, displaying them as we did the original data:

```
coplot(residuals(galaxy.m) ~ radial.position | angle,
    given = sort(unique(angle)), panel = function(x, y)
    panel.smooth(x, y, span = 1, degree = 2, zero.line = TRUE))
```

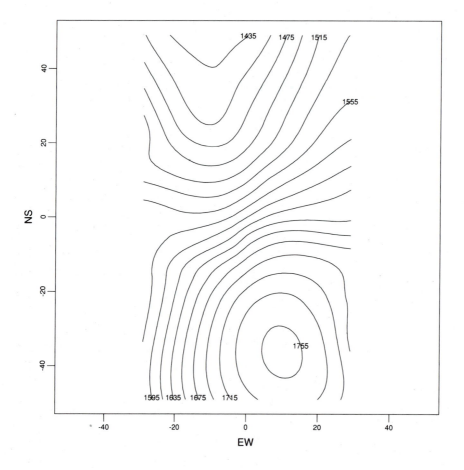

Figure 8.24: *Galaxy data—contour plot of local regression fit.*

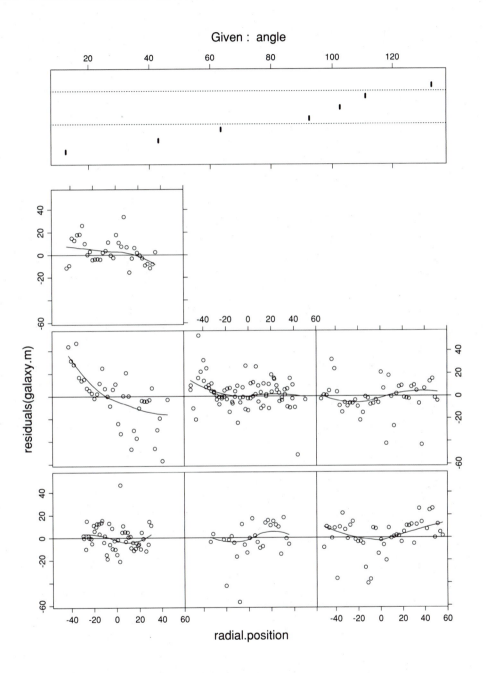

Figure 8.25: *Coplot of residuals with scatterplot smoothings.*

The (1,2) dependence panel shows some clear lack of fit. At the left extreme, the distortion is as large as 40 km/sec, which is more than we would like. But since the fraction of observations that are affected is small we push on, but noting that our results are somewhat tainted. Figure 8.26 is a normal probability plot of the residuals:

```
qqnorm(residuals(galaxy.m))
qqline(residuals(galaxy.m))
```

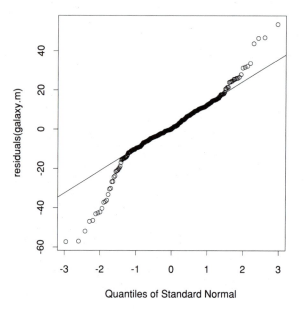

Figure 8.26: *Gaussian quantile plot of residuals.*

The distribution of the residuals is symmetric and strikingly leptokurtic. The robust estimation is clearly justified, and we should feel quite smug at having guessed correctly from the exploratory coplot.

Confidence Intervals

Figure 8.24 shows that the velocity surface has a backbone of sorts. Consider the line in the plane of the predictors that goes through the origin and through the position, (10, -37), where the maximum of the surface occurs. The surface is roughly symmetric in directions perpendicular to the line. Also, the line passes close to the minimum of the surface. Let's evaluate the surface along this line and compute confidence intervals at selected positions:

```
ns <- seq(-49, 49, length = 100)
ew <- ns/(-3.7)
fit.newdata <- data.frame(east.west = ew, north.south = ns)
spine.fit <- predict(galaxy.m, fit.newdata)
ns <- seq(-49, 49, length = 15)
ew <- ew/(-3.7)
limits.newdata <- data.frame(east.west = ew, north.south = ns)
spine.se <- predict(galaxy.m, limits.newdata, se.fit = TRUE)
spine.limits <- pointwise(spine.se, coverage = .99)
```

Figure 8.27 plots the fit against north-south position, and shows the 99% confidence intervals:

```
ylim <- range(spine.fit, spine.limits$upper, spine.limits$lower)
plot(fit.newdata$east.west, spine.fit, xlab = "North-South Coordinate",
  ylab = "Velocity", ylim = ylim, type = "l")
segments(limits.newdata$east.west, spine.limits$lower,
  limits.newdata$east.west, spine.limits$upper)
```

8.2.5 Fuel Comparison Data

For the gas data in Section 8.2.1, the fuel used in the engine experiment was gas, and for the ethanol data Section 8.2.2, the fuel was ethanol. In the first case, the compression ratio, C, was equal to 7.5, and in the second case, C took on five values, one of which was 7.5. To compare the two fuels we form a new data frame, fc, which consists of the 22 ethanol observations for which $C = 7.5$ and the 22 gas observations:

```
> summary(fc)
> summary(fc)
      NOx              E              Fuel
 Min.   :0.54    Min.   :0.63    ethanol:22
 1st Qu.:1.60    1st Qu.:0.78    gas    :22
 Median :2.70    Median :0.90
 Mean   :2.80    Mean   :0.91
 3rd Qu.:3.90    3rd Qu.:1.00
 Max.   :5.30    Max.   :1.20
```

Thus, the two predictors are E, which is numeric, and Fuel, which is a factor.

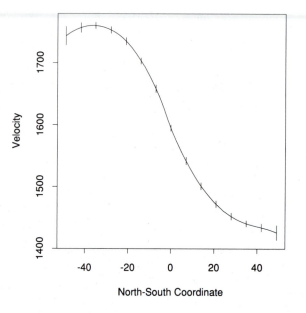

Figure 8.27: *Galaxy data—local regression fit along the backbone with pointwise 99% confidence intervals.*

Exploratory Data Display

The plotting function `coplot()` allows a given variable to be a factor, so we can use it to graph the `fc` data:

```
coplot(NOx ~ E | Fuel, given.values = unique(Fuel),
    columns = 2, rows = 1, data = fc)
```

The result is shown in Figure 8.28.

Modeling

As with the gas data, we fit a locally-quadratic model with **span** equal to 2/3:

```
> attach(fc)
> fc.m <- loess(NOx ~ E * Fuel, span = 2/3, degree = 2)
> fc.m
Call:
```

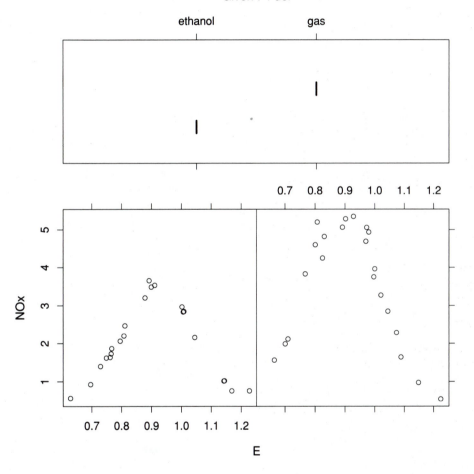

Figure 8.28: *Fuel comparison data—coplot of* NO_x *against E given fuel type.*

```
loess(formula = NOx ~ E * Fuel, span = 2/3, degree = 2)
```

```
Number of Observations:            44
Equivalent Number of Parameters: 11.1
Residual Standard Error:           0.2691
Multiple R-squared:                0.98
Residuals:
    min   1st Q   median   3rd Q     max
 -0.5604 -0.1283 -0.01978 0.06869 0.6234
```

Diagnostic Checking

Figure 8.29 is a coplot of the residuals:

```
coplot(residuals(fc.m) ~ E | Fuel, given.values = unique(Fuel),
   panel = function(x, y) panel.smooth(x, y, span = 1, degree = 1),
   zero.line = TRUE, columns = 2, rows = 1)
```

No lack of fit appears, but one serious problem does stand out glaringly in the figure; the gas residuals have a wider spread. That is, the specification of a constant variance, σ^2, appears incorrect. The estimate of σ, which is pooled over both levels of Fuel, is $s = 0.27$. The estimate from the observations, where Fuel is equal to "ethanol", is shown by the following:

```
> fc.ethanol.m
Call:
loess(formula = NOx ~ E * Fuel, subset = Fuel == "ethanol",
    span = 2/3, degree = 2)

Number of Observations:            22
Equivalent Number of Parameters: 5.6
Residual Standard Error:           0.1696
Multiple R-squared:                0.98
Residuals:
    min   1st Q  median   3rd Q     max
 -0.2487 -0.1022 -0.0126 0.05354 0.3094
```

The estimate with Fuel equal to "gas" is shown by the following:

```
> fc.gas.m
Call:
loess(formula = NOx ~ E * Fuel, subset = Fuel == "gas", span = 2/3,
    degree = 2)

Number of Observations:            22
Equivalent Number of Parameters: 5.5
```

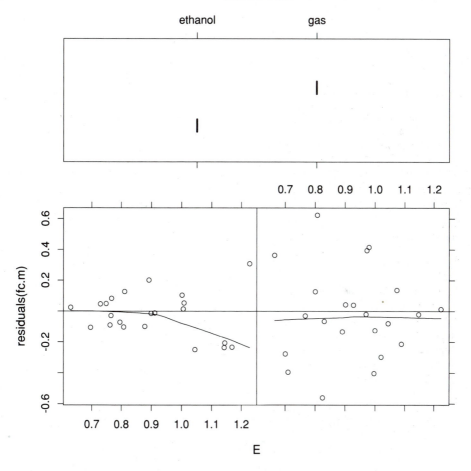

Figure 8.29: *Coplot of residuals with scatterplot smoothings.*

```
Residual Standard Error:        0.3404
Multiple R-squared:             0.96
Residuals:
     min  1st Q   median  3rd Q     max
  -0.5604 -0.213 -0.02511 0.1271 0.6234
```

Thus, the estimate of σ for "gas" is about double that for "ethanol". We clearly need to rethink the model.

A New Model

One action we can take is to analyze the gas data and the ethanol data separately. (We have already fitted a satisfactory model, gas.m, to the gas data in Section 8.2.1.) Instead, let's cheat a bit. Suppose that σ^2 for "gas" is four times that for "ethanol", and that we knew this a priori. We will fit a model using a priori weights that gives weight 4 to the ethanol observations and weight 1 to the gas observations:

```
> fc.weights <- c(1,4)[match(Fuel, c("gas","ethanol"))]
> fc.new.m <- update(fc.m, weights = fc.weights)
> fc.new.m
Call:
loess(formula = NOx ~ E * Fuel, weights = fc.weights, span = 2/3,
    degree = 2)

 Number of Observations:          44
 Equivalent Number of Parameters: 11.1
 Residual Standard Error:         0.3398
 Residuals:
      min   1st Q   median   3rd Q     max
  -0.5604 -0.1283 -0.01978 0.06869 0.6234
```

Notice that the estimate of σ is close to that obtained for the fit using just the gas data. The reason is that in our new weighted analysis we have defined the a priori weights to be 1 for "gas", so the error standard deviation for these runs is σ. Figure 8.30 shows the fitted curves and 99% confidence intervals:

```
plot(fc.new.m, confidence = 7, columns = 2, rows = 1)
```

Notice that the intervals for ethanol are smaller than those for gas; the reason, of course, is the larger error variance for gas.

Diagnostic Checking

Consider carrying out diagnostics for fc.new.m. First, the fit is the same as for fc.m because in the multiple-fit procedure for factors, curves are fitted separately and

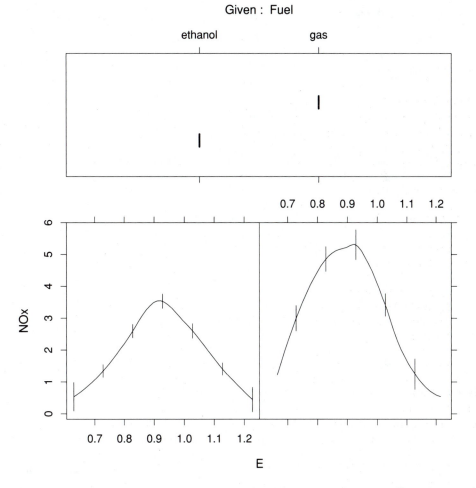

Figure 8.30: *Fuel comparison data—coplot of local regression fit with 99% confidence intervals.*

independently to the ethanol runs and to the gas runs. (*N.B.* This would not be true if we used a robust loess estimate, because m, the residual absolute deviation used in the robustness iterations, is based on all of the residuals.) Thus, Figure 8.29 is also the diagnostic residual plot for `fc.new.m`. As we have already concluded, the fit appears satisfactory. To investigate the distributional assumptions, we use the standardized residuals, $\hat{\varepsilon}_i^*$, which are computed by:

```
residuals(fc.new.m)*sqrt(fc.weights)
```

A normal probability plot of $\hat{\varepsilon}_i^*$ justified the assumption of normality, and a plot of $\sqrt{|\hat{\varepsilon}_i^*|}$ against $\hat{y}_i$ revealed no dependence of scale on the fitted values. Thus, all is well with `fc.new.m`, but in our elation we must not forget that the analysis is tainted by having used an estimated variance ratio as given a priori.

8.3   Specializing and Extending the Computations

8.3.1   Computation

In the examples of Section 8.2, the function `predict()` did not use the loess fitting method to compute surfaces directly at every evaluation point. Rather, to get very fast computation, a default algorithm was used that employs interpolation. In this algorithm, a set of points, typically small in number, is selected for direct computation using the loess fitting method, and a surface is evaluated using an interpolation method that is based on *blending functions*. The space of the predictors is divided into rectangular cells using an algorithm based on *k-d* trees. The loess fit is evaluated at the cell vertices, and then blending functions do the interpolation. The `loess` objects contain data structures that store the *k-d* trees and the fits at the vertices. This information is used by `predict()` to carry out the interpolation. Of course, the resulting interpolated surface is not exactly the same as that of a surface computed directly, but the agreement is typically excellent. Even when it is not, the interpolation method is a perfectly logical smoothing method that has a number of desirable properties. This approach is what allows us, for example, to rapidly compute the surface of the galaxy data at a grid of 5841 values. Doing a direct loess evaluation at all of these points would be expensive. The interpolation method, however, results in one restriction: the surface cannot be evaluated outside the range of the data; that is, the value of each numeric variable for an evaluation point must lie within the range of the observations of that variable in the data. This is not the case for the direct computation method, so evaluation can be done anywhere.

The local regression functions produce quantities that express in various ways information about degrees of freedom. `loess()` returns the equivalent number of parameters, `predict()` returns the degrees of freedom of *t*-intervals, and `anova` returns the numerator and denominator degrees of freedom of an F-test. In the examples of Section 8.2, these quantities, which are defined in Section 8.4, are computed by an approximation method that is described in Section 8.4. A supercomputer environment (or a user with a great deal of patience) would be needed to routinely compute these statistical quantities exactly.

Most users will not want to use direct computation of surfaces or exact computation of the statistical quantities. However, those who want to explore the computational and statistical methods of loess fitting can change the computational methods using the argument `control` of `loess()`. The argument is specified by the output of the function `loess.control()`:

```
my.control <- loess.control(surface = "direct", statistics = "exact")
gas.slower <- loess(NOx ~ E, data = gas, control = my.control)
```

In these expressions we have used the argument `surface` to switch the computation of the surface from `"interpolate"` to `"direct"`, and the computation of the statistical quantities from `"approximate"` to `"exact"`.

The function `loess.control()` can also be used to control two other computational matters. When interpolation is used, an argument `cell` controls the cell size of the k-d tree. The maximum fraction of points allowed inside a cell is `cell*span`; in the algorithm, a cell is divided if the maximum is exceeded. Also, the argument `iterations` specifies the number of iterations of the loess robust estimate.

8.3.2 Inference

We stressed in Section 8.2 that it is critical to carry out diagnostic methods to study, among other things, surplus and lack of fit. In some applications, however, a clearly identifiable lack of fit might be acceptable if the identified magnitude of the distortion is judged to be small for the purpose to which the fit is put. For example, we might want a distorted surface if it made communication simpler and the distortion did not interfere with the judgment of salient features. But one problem is that an estimate, s, of σ based on a distorted fit would be biased, and thus a confidence interval based on this estimate would not have the stated coverage. There is a remedy. Suppose we have two loess fits, `fit.biased` and `fit.unbiased`, the first distorted and the second not. We can use the value of s from the undistorted fit to form confidence intervals for the distorted fit. We do this by changing `fit.biased`:

```
fit.biased$inference <- fit.unbiased$inference
```

Now giving `fit.biased` to `predict()` gives correct confidence intervals. It should be appreciated that the intervals are not for the true surface, but rather for the expected value of the distorted estimate.

8.3.3   Graphics

In some cases, enough evaluation is done by plot() for loess objects that we want
to save the fit and confidence intervals for future renderings of the graph. This
can be done using the function preplot(), which saves the computations for future
plotting by plot():

```
ethanol.plot <- preplot(ethanol.cp, confidence = 7)
plot(ethanol.plot)
```

8.4   Statistical and Computational Methods

In this section we discuss computational and statistical methods in the fitting of
local regression models. In Section 8.4.1, we discuss the methods of inference that
arise from the loess fitting method. In Section 8.4.2, we discuss computational
methods that underlie loess fitting, and numerical problems that can arise. To
keep the discussion from being cumbersome, we suppose that the predictors are
all numeric. Extending the results to the case where factors are present is quite
obvious.

8.4.1   Statistical Inference

Initially, we will suppose that the errors have been specified to be Gaussian and the
variances have been specified to be constant.

One important property of a Gaussian-error loess estimate, $\hat{g}(x)$, is that it is
linear in y_i—that is,

$$\hat{g}(x) = \sum_{i=1}^{n} l_i(x) y_i$$

where the $l_i(x)$ do not depend on the y_i. This linearity results in distribution
properties of the estimate that are very similar to those for classical parametric
fitting.

Suppose that the diagnostic methods have been applied and have revealed no
lack of fit in $\hat{g}(x)$; we will take this to mean that $E\hat{g}(x) - g(x)$ is small. Suppose
further that diagnostic checking has verified the specifications of the error terms in
the model.

Estimation of σ

Since $\hat{g}(x)$ is linear in y_i, the fitted value at x_i can be written

$$\hat{y}_i = \sum_{j=1}^{n} l_j(x_i) y_j.$$

Let L be the matrix whose (i, j)th element is $l_j(x_i)$ and let

$$\bar{L} = I - L$$

where I is the $n \times n$ identity matrix. For $k = 1$ and 2, let

$$\delta_k = \text{tr}(\bar{L}'\bar{L})^k.$$

We estimate σ by the scale estimate

$$s = \sqrt{\frac{\sum_{i=1}^{n} \hat{\varepsilon}_i^2}{\delta_1}}.$$

Confidence Intervals for $g(x)$

Since

$$\hat{g}(x) = \sum_{i=1}^{n} l_i(x)y_i,$$

the standard deviation of $\hat{g}(x)$ is

$$\sigma(x) = \sigma \sqrt{\sum_{i=1}^{n} l_i^2(x)}.$$

We estimate $\sigma(x)$ by

$$s(x) = s \sqrt{\sum_{i=1}^{n} l_i^2(x)}.$$

Let

$$\rho = \frac{\delta_1^2}{\delta_2}.$$

The distribution of

$$\frac{\hat{g}(x) - g(x)}{s(x)}$$

is well approximated by a t distribution with ρ degrees of freedom; we can use this result to form confidence intervals for $g(x)$ based on $\hat{g}(x)$. Notice that the value δ_1 by which we divide the sum-of-squares of residuals is not the same as the value ρ used for the degrees of freedom of the t distribution. For classical parametric fitting, these two values are equal. For loess, they are typically close but not close enough to ignore the difference. We will refer to ρ as the *look-up degrees of freedom* since it is the degrees of freedom of the distribution that we look up to get the confidence interval.

Analysis of Variance for Nested Models

We can use the analysis of variance to test a null local regression model against an alternative one. Let the parameters of the null model be $\alpha^{(n)}$, $\lambda^{(n)}$, $\delta_1^{(n)}$, and $\delta_2^{(n)}$. Let the parameters of the alternative model be α, λ, δ_1, and δ_2. For the test to make sense, the null model should be *nested* in the alternative; we will define this concept shortly. Let rss be the residual sum-of-squares of the alternative model, and let $rss^{(n)}$ be the residual sum-of-squares of the null model.

The test statistic, which is analogous to that for the analysis of variance in the parametric case, is

$$F = \frac{(rss^{(n)} - rss)/(\delta_1^{(n)} - \delta_1)}{rss/\delta_1}.$$

F has a distribution that is well approximated by an F distribution with denominator look-up degrees of freedom ρ, defined earlier, and numerator look-up degrees of freedom

$$\nu = \frac{(\delta_1^{(n)} - \delta_1)^2}{\delta_2^{(n)} - \delta_2}.$$

The concept of a null model being nested in the alternative expresses the idea that the alternative is capable of capturing any effect that the null can capture, but the definition is more precisely a specification of when it makes sense to use the analysis of variance to compare two models. The null is nested in the alternative if the following conditions hold:

(1) $\alpha^{(n)} \geq \alpha$.

(2) $\lambda^{(n)} \leq \lambda$.

(3) If the square of a numeric predictor is dropped from the alternative model, then it must not be present in the null model; the converse need not be true.

(4) The models must have the same numeric predictors with the following exception: a conditionally parametric predictor in the alternative need not be present in the null; if present, though, it must also be conditionally parametric.

Conditions (2) to (4) can be expressed in a different way. To explain, we need to differentiate *neighborhood variables*—the predictors used to determine the neighborhoods in the loess fitting—and *fitting variables*—the predictors that are fitted locally by weighted least squares. Let's take a specific example. Suppose there are three numeric predictors: u, v, and w. Suppose $\lambda = 2$, u is taken to be conditionally parametric, and the square of w is dropped. The neighborhood variables are v and w. The fitting variables are a constant, u, u^2, v, v^2, w, uv, and vw. Now we can reexpress (2) to (4) by the following:

(2)$'$ The null and alternative models have the same neighborhood variables.

(3)$'$ The fitting variables of the null model are a subset of the fitting variables of the alternative model.

The Equivalent Number of Parameters

Let

$$\mu = \text{tr}(L' L).$$

If the $\hat{y}_i$ are the fitted values, then

$$\mu = \frac{\sum_{i=1}^n \text{Variance}(\hat{y}_i)}{\sigma^2}.$$

We will call μ the *equivalent number of parameters* since if the $\hat{y}_i$ were the fitted values for a linear model, the right side of the last equation would be the number of estimated parameters. μ is greater than or equal to τ, the number of fitting variables, and approaches τ as α tends to infinity. The equivalent number of parameters is one measure of the amount of smoothing. (Chapter 7 has another.) Strictly speaking, μ depends on α, on the values of the predictors, and on the choices of the neighborhood and fitting variables. However, having selected all of these factors except α, we can get, approximately, a desired value μ by taking α to be $1.2\tau/\mu$, where τ is the number of fitting variables.

Symmetric Errors

When the error distribution is specified to be symmetric, inferences are based on *pseudo-values*. Let the robustness weights and the median absolute residual used in the final update of the fit, $\hat{g}(x)$, be r_i and m, respectively, and let $\psi(u; b) = uB(u; b)$. The pseudo-values are

$$\ddot{y}_i = \hat{y}_i + cr_i\hat{\varepsilon}_i$$

where $\hat{y}_i$ are the fitted values, $\hat{\varepsilon}_i$ are the residuals, and

$$c = \frac{n}{\sum_{i=1}^n \psi'(\hat{\varepsilon}_i; 6m)}.$$

Inferences are carried out by applying the inference procedures of the Gaussian case but replacing the observations of the response y_i by the pseudo-values $\ddot{y}_i$. For example, suppose we want to compute a confidence interval for $g(x)$ about the robust estimate, $\hat{g}(x)$. Using the pseudo-values as the response, we compute a Gaussian-error estimate, ρ, and $s(x)$ as described above. The confidence interval for $g(x)$ is the $\hat{g}(x)$ plus and minus $s(x)$ times a t value with ρ degrees of freedom. The true coverage using this procedure is well approximated by the nominal coverage. For the analysis of variance, we proceed in a similar fashion using the pseudo-values from the alternative model and carrying out the Gaussian-error procedures. For small samples, the approximation is not as good as for confidence intervals and produces optimistic results, but work is under way to find methods for adjusting degrees of freedom that will improve the approximations.

Errors with Unequal Scales

Suppose we have specified that the random errors ε_i in the model have the property that $a_i\varepsilon_i$ are identically distributed where the *a priori weights*, a_i, are positive and known. Then various modifications are made to the methods of inference.

For the Gaussian-error estimate, the operator matrix L is, of course, different from that in the equal-variance case, but δ_1 and δ_2 are defined in terms of L as before. The estimate of σ becomes

$$s = \sqrt{\frac{\sum_{i=1}^n a_i\hat{\varepsilon}_i^2}{\delta_1}}$$

and the estimate of the standard deviation of $\hat{g}(x)$ becomes

$$s(x) = s\sqrt{\sum_{i=1}^n l_i^2(x)/a_i}.$$

For the analysis of variance, all residual sum-of-squares are modified by adding the terms a_i, as done above for s.

For the robust estimate, the median absolute residual is defined using the *standardized residuals*

$$\hat{\varepsilon}_i^* = \sqrt{a_i}\hat{\varepsilon}_i$$

That is,

$$m = \text{median}(|\hat{\varepsilon}_i^*|).$$

Similarly, the robustness weights are

$$r_i = B_{6m}(\hat{\varepsilon}_i^*).$$

The pseudo-values are

$$\ddot{y}_i = \hat{y}_i + cr_i\hat{\varepsilon}_i$$

where c is now

$$c = \frac{n}{\sum_{i=1}^n \psi'(\hat{\varepsilon}_i^*; 6m)}.$$

8.4.2 Computational Methods

Interpolation by k-d Trees and Blending

The k-d tree is a particular data structure for partitioning space by recursively cutting cells in half by a hyperplane orthogonal to one of the coordinate axes (Bentley, 1975). For our application, the k in the name refers to the number of neighborhood variables, those predictors that are used to define the neighborhoods.

Here is how the k-d tree is formed. Start with a rectangular cell just containing the values of the neighborhood variables. Pick the predictor whose spread is the greatest and divide the cell in half at the median along the axis of that predictor. Recursively apply the same division procedure to each subcell. If a cell contains fewer than βn points, where β is a small fraction, do not refine it. Figure 8.31 shows a k-d tree for two predictors, $n = 500$, and $\beta = 0.05$.

Once the k-d tree is built, $\hat{g}(x)$ is directly computed at the vertices. By "vertex," we just mean a corner of a cell; "vertex" seems a better term than "corner" because a vertex of one cell typically lies in the middle of a side of an adjacent cell. In addition to computing $\hat{g}(x)$ at a vertex, a derivative of $\hat{g}$ at the vertex is approximated by the derivative of the locally-fitted surface. This derivative is a natural by-product of the least-squares computation and costs nothing extra to obtain.

Typically, the number of vertices, v, will be much smaller than n. This is at least true asymptotically, because the number of cells needed to achieve a certain accuracy of approximation depends on the smoothness of $\hat{g}(x)$, not n. In Figure 8.31 there are 66 vertices, so we solve 66 least-squares problems instead of one problem per evaluation of $\hat{g}(x)$. (Recall that for the galaxy surface we carried out 5841 evaluations to make a contour plot.) The amount of work in general to construct the k-d tree, including vertex coefficients, is $O(v((1.5 + \alpha\tau)n + \tau^3))$. After building the tree, each interpolation costs $O(\log v)$. Since τ is fixed and v is asymptotically bounded, the total running time is linear in the size of the input and output.

Let's turn now to the scheme used to build a piecewise polynomial approximation to $\hat{g}$. To simplify the discussion, we will suppose that there are two neighborhood variables. For our k-d tree, the boundary of each rectangular cell is cut into segments by vertices. (There are four sides, some of which will likely contain internal vertices, breaking them into more segments.) On each segment, the surface is interpolated using the unique cubic polynomial determined by the fits and derivatives at the vertices. To interpolate in the interior of the cell, we apply blending functions, also known as transfinite interpolants (Cavendish, 1975). This technique, well known in computer-aided design, takes a certain combination of univariate interpolants in each variable separately to build a surface. In effect, each cell is subdivided and on each piece a cubic polynomial in two variables is constructed although the computation is not actually done this way.

For one and two neighborhood variables, the interpolation function is C^1, but

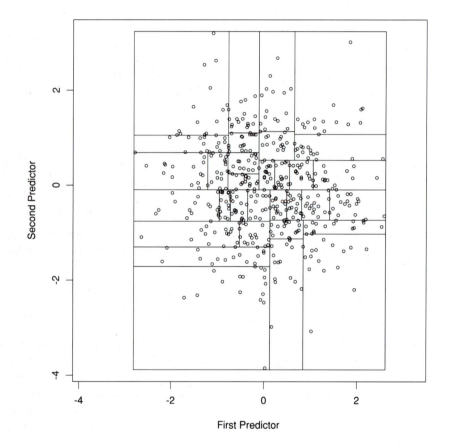

Figure 8.31: *A k-d tree.*

for more variables, the present code does not use enough vertices to guarantee a consistent approximation across cell facets. Hence the overall approximation may not be C^1 or even C^0. This defect will be removed in a future implementation.

Computing δ_i

Three statistical quantities are described in Section 8.4.1 that provide information about degrees of freedom—μ, ρ, and ν. These three quantities are functions of δ_1, δ_2, and n. Straightforward computation of the δ_i is horrendously expensive, so we have developed methods of approximation. First, we generated a large number of datasets, each with a response and one or more predictors, and computed the δ_i for each. We discovered, through substantial graphical analysis, that the δ_i could be predicted to within a few percent by the following predictors: λ, n, τ, and

$$\zeta = \frac{\sqrt{\tau/\operatorname{tr}(L)} - \sqrt{\tau/n}}{1 - \sqrt{\tau/n}}.$$

The model that was fitted is semiparametric, involving both parametric functions and a local regression model.

Error Messages from the Bowels of Loess

Although loess fitting is based on sound numerical methods, some delicate situations can arise that require the judgment of the user. When problems are detected by the loess FORTRAN routines, messages are transmitted up to the S user.

One class of messages involves the smoothing parameter α. In order for the least-squares problem in a direct computation of $\hat{g}(x)$ to be well posed, α must be large enough that there are as many data points in the neighborhood as fitting variables, τ, in the local regression. Moreover, since neighborhood weights drop to 0 at the boundary, at least τ of these points must be strictly inside the neighborhood. If α is too small, the fix is to increase it or reduce τ by lowering λ or dropping squares.

The sample points must be sufficiently well distributed as well as sufficiently numerous. For example, consider locally quadratic fitting in one predictor. If, because of multiplicities, there are only two distinct sample locations inside a neighborhood, then a quadratic polynomial is not uniquely determined.

When numerical problems arise because of poor conditioning of the design matrix of the local regression, small eigenvalues are set to zero and a pseudo-inverse message is sent. None of this means the fit has a problem, but a pseudo-inverse message is a caution that extra alertness must be used in examining the diagnostic displays.

Mathematically, $\operatorname{tr}(L)$ is greater than or equal to τ, the number of fitting variables. Numerically, however, if eigenvalues are set to zero, $\operatorname{tr}(L)$ can drop below τ, which causes the method of computing δ_i approximately to abort. If this indicator of an eigenvalue meltdown occurs, the coded message "Chernobyl" is sent up to the S user.

Finally, when the interpolation method is used, the FORTRAN code must allocate space based on a prediction from the number of observations, the number of numeric predictors, and the specification of the surface and errors. If this allocated space is too small, the k-d tree division is truncated and a warning message sent up. In some cases the problem is extreme enough that the fit is not carried out; this necessitates increasing the value of α.

Bibliographic Notes

Local regression models are treated in detail in a new book by Cleveland and Grosse (forthcoming). But methods of local fitting date back at least to the 1920s. Initial applications were to smooth a time series (Macauley, 1931). An early use of local fitting for the general regression problem was investigated by Watson (1964). The method amounted to fitting a constant locally—in other words, taking the polynomial degree λ to be zero. This came to be known as kernel smoothing. It leads to very interesting theoretical work but is not of use in practice since it is hard to coax the method into following the patterns in most datasets. More serious attempts at local fitting were suggested by McLain (1974), who fitted quadratic polynomials, and Stone (1977), who fitted linear polynomials. The method of fitting used here was described by Cleveland (1979) for one predictor, and is the basis of the S function lowess, which has now been upgraded to the function scatter.smooth(). Cleveland and Devlin (1988) extended the method to two or more predictors and investigated the sampling properties in the Gaussian case. (Sampling properties in the symmetric case are still under development.) The computational methods described in Sections 8.3 and 8.4, which are crucial to local regression being useful in practice, are due to Cleveland and Grosse (1991).

Chapter 9

Tree-Based Models

Linda A. Clark
Daryl Pregibon

This chapter describes S functions for tree-based modeling. Tree-based models provide an alternative to linear and additive models for regression problems and to linear logistic and additive logistic models for classification problems. The models are fitted by binary recursive partitioning whereby a dataset is successively split into increasingly homogeneous subsets until it is infeasible to continue. The implementation described in this chapter consists of a number of functions for growing, displaying, and interacting with tree-based models. This approach to tree-based models is consistent with the data-analytic approach to other models, and consists primarily of fits, residual analyses, and interactive graphical inspection.

9.1 Tree-Based Models in Statistics

Tree-based modeling is an exploratory technique for uncovering structure in data. Specifically, the technique is useful for classification and regression problems where one has a set of classification or predictor variables ($\boldsymbol{x}$) and a single-response variable (y). When y is a factor, decision or classification rules are determined from the data—for example,

      if   ($x_1 \leq 2.3$) and ($x_3 \in \{A, B\}$)
   then y is most likely to be in level 5.

When y is numeric, regression rules for description or prediction are of the form

      if   ($x_2 \leq 413$) and ($x_9 \in \{C, D, F\}$) and ($x_5 \leq 3.5$)
   then   the predicted value of y is 4.75.

A classification or regression tree is the collection of many such rules determined by a procedure known as *recursive partitioning*, which is discussed in detail in Section 9.4. This form of classification or prediction rule is very different from that given by more classical models, such as logistic and linear regression analyses, where *linear combinations* are the primary mode of expressing relationships between variables. Indeed, this difference is both the strength of the method and also its weakness.

Statistical inference for tree-based models is in its infancy and far behind that for logistic and linear regression analyses. This is partly because a particular type of *variable selection* underlies tree-based models (e.g., each rule contains only a subset of the available classification or predictor variables, and some may not be used at all). Despite the lack of formal procedures for inference, the method is gaining widespread popularity as a means of devising prediction rules for rapid and repeated evaluation, as a screening method for variables, as a diagnostic technique to assess the adequacy of linear models, and simply for summarizing large multivariate datasets. Some possible reasons for its recent popularity are that:

- in certain applications, especially where the set of predictors contains a mix of numeric variables and factors, tree-based models are sometimes easier to interpret and discuss than linear models;

- tree-based models are invariant to monotone reexpressions of predictor variables so that the precise form in which these appear in a model formula is irrelevant;

- the treatment of missing values (NAs) is more satisfactory for tree-based models than for linear models; and

- tree-based models are more adept at capturing nonadditive behavior; the standard linear model does not allow *interactions* between variables unless they are prespecified and of a particular multiplicative form.

Among the other models covered in this book, tree-based models provide the only means of analysis for factor response variables at more than two levels.

Tree-based models are so-called because the primary method of displaying the fit is in the form of a binary tree. We now provide several examples to motivate the range of application of the methods. The examples are organized according to the type (numeric or factor) of the response variable (y) and the classification or predictor variables (x) involved.

9.1.1   Numeric Response and a Single Numeric Predictor

Figure 9.1 displays two views of a tree-based model relating mileage to weight of automobiles in the `car.test.frame` data frame. The left panel of the figure is the

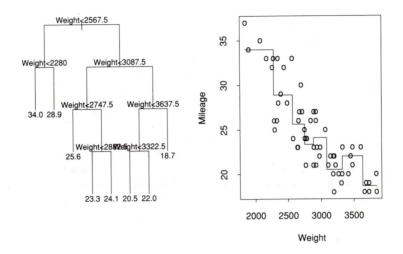

Figure 9.1: *Displays of a tree-based model relating mileage to automobile weight. The plot in the left panel shows how a tree is typically displayed, whereby successive partitions of the data into homogeneous subsets are shown with the rule labeling each split. The overplotting of labels is a common occurrence with this type of display. The plot in the right panel shows a function plot of the same tree together with the actual data values. This representation is only practical for at most two predictor variables.*

standard method of displaying a tree-based model. The idea is that in order to predict mileage from weight, one follows the path from the top node of the tree, called the *root*, to a terminal node, called a *leaf*, according to the rules, called *splits*, at the interior nodes. Automobiles are first split depending on whether they weigh less than 2567.5 pounds. If so, they are again split according to weight being less than 2280 pounds, with the lighter cars having predicted mileage of 34 miles/gallon and the heavier cars having slightly lower mileage of 28.9 miles/gallon. For those automobiles weighing more than 2567.5 pounds, six weight classes are ultimately formed, with predicted mileage varying from 25.6 miles/gallon to a gas-guzzling low of 18.7 miles/gallon. The relationship between mileage and weight seems to behave according to intuition, with heavier cars having poorer mileage than the lighter cars. It appears that doubling the weight of an automobile roughly halves its mileage.

The right panel displays the tree-based model in a more specialized form and one that is more conventional for data of this sort. Here the data themselves and the fitted model are displayed together. As a function of automobile weight, the fitted model is a step function. The height of each step corresponds to the average mileage for automobiles in the weight range under that step. There are a total of eight steps, one for each of the terminal nodes in the tree in the left panel.

9.1.2   Factor Response and Numeric Predictors

The data in this section are from the `kyphosis` data frame introduced in Chapter 6 and analyzed further in Chapter 7. Recall that in those chapters linear and additive logistic models predict the probability of developing `Kyphosis` from the variables `Age`, `Start`, and `Number`. The resulting prediction equations are smooth functions of the first two predictors. By contrast, we now demonstrate tree-based prediction equations that are not smooth but share the essential features of these more traditional analyses.

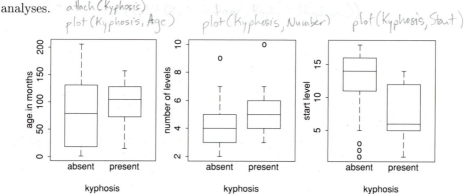

Figure 9.2: *Boxplots of the three numeric predictor variables in the* `kyphosis` *data frame. For each variable, the distribution of individuals with and without* `Kyphosis` *are displayed side by side. The predictor* `Start` *exhibits the greatest difference in these distributions since the lower quartile of those without* `Kyphosis` *is just below the upper quartile of those with* `Kyphosis`.

The distributions of the predictor variables are plotted as a function of `Kyphosis` in Figure 9.2. Of the three predictors, `Start` appears to be the best single predictor since there is a much greater propensity of `Kyphosis` for individuals having `Start`≤ 12 than those with `Start`> 12. The algorithm underlying tree-based prediction determines this cutoff more objectively (by optimization) as 12.5. Moreover, the method then applies the same principle separately to individuals with `Start`≤ 12.5 and those with `Start`> 12.5—namely, comparing the distributions of the predictors as functions of `Kyphosis`. The result of repeated application of this idea leads to the tree displayed in Table 9.1. This semigraphical representation is different from those used in Figure 9.1. It is most useful when the details of the fitting procedure are of interest.

node), split, n, deviance, yval, (yprob)
    * denotes terminal node

(handwritten annotations: = prob (Kyphosis is absent), = prob (Kyphosis is present), .21 = 17/81, # deviance)

```
 1) root 81 83.234001 absent (0.790 0.2100)
   2) Start<12.5 35 47.804001 absent (0.571 0.4290)
     4) Age<34.5 10 6.5019999 absent (0.900 0.1000) *
     5) Age>34.5 25 34.296001 present (0.440 0.5600)
      10) Number<4.5 12 16.301001 absent (0.583 0.4170)
        20) Age<127.5 7 8.3760004 absent (0.714 0.2860) *
        21) Age>127.5 5 6.73 present (0.400 0.6000) *
      11) Number>4.5 13 16.048 present (0.308 0.6920)
        22) Start<8.5 8 6.0279999 present (0.125 0.8750) *
        23) Start>8.5 5 6.73 absent (0.600 0.4000) *
   3) Start>12.5 46 16.454 absent (0.957 0.0435)
     6) Start<14.5 17 12.315 absent (0.882 0.1180)
      12) Age<59 5 0 absent (1.000 0.0000) *
      13) Age>59 12 10.813 absent (0.833 0.1670)
        26) Age<157.5 7 8.3760004 absent (0.714 0.2860) *
        27) Age>157.5 5 0 absent (1.000 0.0000) *
     7) Start>14.5 29 0 absent (1.000 0.0000) *
```

Table 9.1: *A tree-based model for predicting* Kyphosis. *The first number after the split is the number of observations. The second number is the* deviance, *which is the measure of node heterogeneity used in the tree-growing algorithm. A deviance of zero corresponds to a perfectly homogeneous node. This term is defined more precisely in Section 9.4.*

The split on Start partitions the 81 observations into groups of 35 and 46 individuals (nodes 2 and 3), with probability of Kyphosis of 0.429 and 0.0435, respectively. This first group is then partitioned into groups of 10 and 25 individuals (nodes 4 and 5), depending on whether Age is less than 34.5 years or not. The former group, with probability of Kyphosis of 0.10, is not subdivided further. The latter group is subdivided into groups of 12 and 13 individuals (nodes 10 and 11), depending on whether or not Number is less than 4.5. The respective probabilities for these groups are 0.417 and 0.692. This procedure continues, yielding nine distinct probabilities of Kyphosis ranging from 0.0 to 0.875. Clearly, as the partitioning continues, our trust in the individual estimated probabilities decreases as they are based on less and less data. Many of the tools discussed in Section 9.2 are aimed at assessing the degree of over- or underfitting of a tree-based model.

9.1.3   Factor Response and Mixed Predictor Variables

The data are from the `market.survey` data frame introduced in Chapter 3 and subsequently analyzed in Chapters 6 and 7. Here we briefly review the available data, which were obtained from a survey of 1000 people; for now, we concentrate on the 759 individuals for whom complete data were obtained. The aim of the survey was to identify segments of the residential long-distance market, where AT&T should concentrate its marketing efforts. The variables collected include household income (`income`), number of household moves in the past five years (`moves`), age of respondent (`age`), education level (`education`), employment category (`employment`), average monthly usage (`usage`), whether the respondent has a nonpublished phone number (`nonpub`), whether the respondent participates in the Reach Out America Plan (`reach.out`), whether the respondent holds a calling card (`card`), and the respondent's chosen long-distance carrier (`pick`).

The tree in Figure 9.3 provides a particularly simple prediction rule for long-distance carrier. For average usage of more than \$12.50 per month, the preferred choice is AT&T. For average usage of less than \$12.50 per month, the choice depends on whether the respondent has a nonpublished directory listing. If so, then AT&T is again the preferred choice, but if the directory listing is published, then an "other common carrier" (OCC) is preferred. (Evidently the OCC folks did some telemarketing themselves!)

9.2   S Functions and Objects

Our approach is not to have a single function for tree-based modeling, but rather a collection of functions, which, together with existing S functions, form a basis for building and assessing this new class of models. Our implementation centers around the idea of a tree object. This object provides commonality among functions to grow, manipulate, and display trees.

9.2.1   Growing a Tree

There is a single function to grow a tree, named `tree()`. The expression

```
> z.auto <- tree(Mileage ~  Weight, car.test.frame)
```

grows a regression tree using the variables `Mileage` and `Weight` from the data frame `car.test.frame` and gives the name `z.auto` to the resulting tree object. Similarly, the expression

```
> z.kyph <- tree(Kyphosis ~ Age + Number + Start, kyphosis)
```

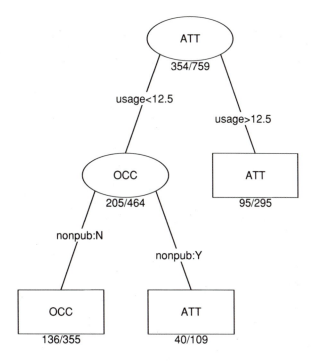

Figure 9.3: *A display of a tree fitted to the long-distance marketing data. This form of tree display is primarily for presentation purposes as it conceals the details of the tree-growing process. The edges connecting the nodes are labeled by the left and right splits. Interior nodes are denoted by ellipses and terminal nodes by rectangles, with the predicted value of the response variable centered in the node. The number under each terminal node is the misclassification error rate; for example, in the rightmost node, which is labeled ATT, 95 out of the 295 respondents in the node actually picked OCC.*

grows a classification tree using the variables from the data frame `kyphosis` and gives the name `z.kyph` to the resulting tree object. The function `tree()` automatically distinguishes between regression and classification trees according to whether the response variable is numeric or a factor. It implements a binary recursive partitioning algorithm described in Section 9.4. The only detail relevant to the present discussion is that the algorithm adds nodes until they are homogeneous or contain too few observations (≤ 5, by default).

The function `tree()` takes two arguments, a `formula` object and a `data.frame`, either of which can be missing. As with all modeling functions, a missing `data.frame` argument simply means that the functions expect the variables named in `formula` to be in the search list. If `formula` is missing, then it is constructed automatically from the `data.frame` using the first variable as the response. For example, an equivalent expression defining `z.kyph` is `tree(kyphosis)`. Valid formulas for trees allow all standard manipulations of variables such as `cut()`, `log()`, `I()`, etc. These are seldom used on the right side of a formula since trees are invariant to monotone reexpressions of individual predictor variables. The only meaningful operator in a formula for trees is " + ," indicating which variables are to be included as predictors. This is so because trees capture interactions without explicit specification. Given these points, it may seem that formulas for trees are a gross overkill as a means of specifying the terms used in the model. Nonetheless, they provide a convenient means to specify reexpressions of the response variable and, more importantly, to facilitate applying quite different models to the same data.

A tree object contains information regarding the partitioning of the predictor variables into homogeneous regions that is required by subsequent functions for manipulating and displaying trees. Predictably, a tree object has class `"tree"`. Generic functions such as `summary()`, `print()`, `plot()`, `residuals()`, and `predict()` work as expected for objects of class `"tree"`. A summary of a fitted tree-based model is available by the `summary()` function:

```
> summary(z.auto)

Regression tree:
tree(formula = Mileage ~  Weight, car.test.frame)
Number of terminal nodes:  8
Residual mean deviance:   4.208 = 218.819 / 52
Distribution of residuals:
   Min. 1st Qu. Median   Mean 3rd Qu.   Max.
 -3.889 -1.111   0.000  0.000  1.167   4.375

> summary(z.kyph)

Classification tree:
tree(formula = Kyphosis ~ Age + Number + Start, kyphosis)
Number of terminal nodes:  9
Residual mean deviance:  0.594 = 42.742 / 72
Misclassification error rate: 0.123 = 10 / 81
```

Notice that there is some difference in the summary depending on whether the tree is a classification or a regression tree.

A tree prints using indentation as a key to the underlying structure. Since `print()` is invoked upon typing the name of an object, a tree can be printed simply by typing its name. The example given in Table 9.1 was constructed with the expression `z.kyph`. The amount of information displayed by `print()` relative to `summary()` might seem disproportionate for objects of class `"tree"`, but the philosophy that `print()` should provide a quick look at the object is maintained, as it does little more than format the contents of a tree object. The `summary()` function on the other hand does involve computation that can result in less than instantaneous response.

Subtrees

A subtree of a tree object can be selected or deleted in a natural way through subscripting; for example, a positive subscript corresponds to selecting a subtree and a negative subscript corresponds to deleting a subtree. This implies that there is an ordering or index to tree objects that permits identification by number. Indeed, nodes of a tree object are numbered to succinctly capture the tree topology and to provide quick reference. An example of the numbering scheme is that given in Table 9.1 for the tree grown to the `kyphosis` data. Descendants of node number 3 can be removed, or a new subtree can be rooted at node 3, as follows:

```
> z.kyph[-3]
node), split, n, deviance, yval, (yprob)
      * denotes terminal node

  1) root 81 83.234 absent (0.790 0.2100)
    2) Start<12.5 35 47.804 absent (0.571 0.4290)
      4) Age<34.5 10 6.502 absent (0.900 0.1000) *
      5) Age>34.5 25 34.296 present (0.440 0.5600)
       10) Number<4.5 12 16.301 absent (0.583 0.4170)
         20) Age<127.5 7 8.376 absent (0.714 0.2860) *
         21) Age>127.5 5 6.73 present (0.400 0.6000) *
       11) Number>4.5 13 16.048 present (0.308 0.6920)
         22) Start<8.5 8 6.028 present (0.125 0.8750) *
         23) Start>8.5 5 6.73 absent (0.600 0.4000) *
    3) Start>12.5 46 16.454 absent (0.957 0.0435) *


> z.kyph[3]
node), split, n, deviance, yval, (yprob)
      * denotes terminal node

  3) Start>12.5 46 16.454 absent (0.957 0.0435)
    6) Start<14.5 17 12.315 absent (0.882 0.1180)
     12) Age<59 5 0 absent (1.000 0.0000) *
     13) Age>59 12 10.813 absent (0.833 0.1670)
       26) Age<157.5 7 8.376 absent (0.714 0.2860) *
       27) Age>157.5 5 0 absent (1.000 0.0000) *
    7) Start>14.5 29 0 absent (1.000 0.0000) *
```

Implicit in our discussion above is that a subtree of a tree object is itself a tree object. This allows a subtree to be printed with the same ease as the original tree.

The importance of tree subscripting becomes apparent as tree size gets larger. For example, consider growing a tree to the long-distance marketing data:

```
> z.survey <- tree(market.survey, na.action = na.omit)
```

The tree displayed earlier in Figure 9.3 is a particularly terse summary of this tree obtained with the expression `z.survey[-c(4,5,3)]`. The complete tree, `z.survey`, is displayed using the `plot()` function in Figure 9.4. The function displays a tree as an unlabeled dendrogram, rooted at the top of the figure. The `plot.tree()` method takes an optional argument, `type=`, which controls node placement. The default is nonuniform spacing whereby the vertical position of a node pair is a function of the importance of the parent split. It is particularly appropriate during analysis where the primary consideration is often one of tree simplification. The alternate (`type="u"`) behavior uses node depth to guide vertical placement of nodes. This results in a uniform layout that is useful for subsequent labeling. The tree displayed in the left panel of Figure 9.4 was obtained with the default node spacing, e.g., `plot(z.survey)`, while that in the right panel was obtained by `plot(z.survey, type = "u")`. In the former plot, the importance of the first few splits is readily apparent. This insight is at the expense of reduced resolution at the leaves of the tree, where detail is arguably of lesser importance.

Labeling a tree is distinct from plotting a tree. The size of the tree displayed in Figure 9.4 demonstrates why two separate functions are required; once the tree is plotted, labeling may or may not follow depending on its topology. The `text()` method for trees provides a means to label the dendrogram displayed by `plot()`. The user has control over what components of the tree object are used as labels at interior or leaf nodes. The tree displayed in the left panel of Figure 9.1 was labeled with `text(z.auto)`.

Tree-based modeling is similar in many ways to that discussed in previous chapters. An important similarity is the degree to which tools to diagnose model adequacy are applied. Figure 9.5 displays two commonly used plots for regression models as applied to the automobile mileage example—namely, a scatterplot of residuals versus fitted values and a normal probability plot of residuals. The fitted values are obtained with the expression `predict(z.auto)`. The residuals, *observed−fitted*, are obtained by subtracting the fitted values from the response variable, or directly with the expression `residuals(z.auto)`. The normal probability plot does not suggest any unusual patterns, but the plot of residuals versus fitted values demonstrates heteroscedasticity. This pattern, together with the moderate curvature demonstrated in Figure 9.1, suggests that a reexpression of the response variable, say from miles per gallon to gallons per mile, might be more appropriate.

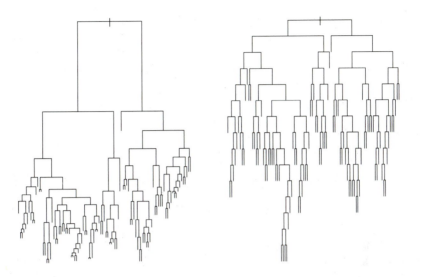

Figure 9.4: *Dendrograms of the tree* z.survey *grown to the long-distance marketing data. The dendrogram on the left uses the change in deviance to guide the vertical positioning of each pair of nodes. Resolution at the leaves of the tree is sacrificed to provide a visual cue of split importance. The dendrogram on the right uses node depth to guide the placement of each node. (The root has depth 0.)*

Pruning and Shrinking

Another aspect of assessing a fitted tree-based model is the extent to which it can be simplified without sacrificing goodness-of-fit. This is also an important consideration for prediction. Since tree size is intentionally not limited in the growing process, a certain degree of overfitting has occurred. There are two ways to address this problem; the one to choose depends upon whether the primary concern is parsimonious description or accurate prediction.

Figure 9.6 displays three variations of z.kyph, the classification tree grown to the kyphosis data. The first panel is the dendrogram for the full tree with nine terminal nodes. The second panel is a *pruned* version with three terminal nodes. The third panel is a *shrunken* version with nine *actual* terminal nodes and about three *effective* terminal nodes. Note that the pruned tree shares the same estimated probabilities as the full tree but that apart from the root node, those of the shrunken tree are completely different. Summaries of the pruned and shrunken trees are:

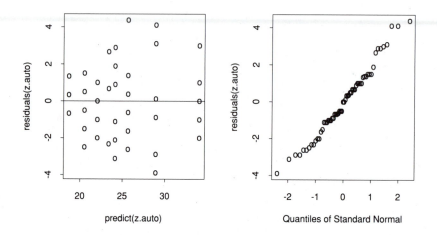

Figure 9.5: *Two standard diagnostic plots for regression data as applied to the fit described by* z.auto. *The plot in the left panel is that of residuals versus fitted values. The plot in the right panel is a normal probability plot of residuals. These plots suggest that there are no apparent outliers but that the variance seems to increase with level.*

```
> summary(zp.kyph)

Classification tree:
prune.tree(tree = z.kyph, k = 5)
Variables actually used in tree construction:
[1] "Age"    "Start"
Number of terminal nodes:  3
Residual mean deviance:  0.734 = 57.252 / 78
Misclassification error rate: 0.173 = 14 / 81

> summary(zs.kyph)

Classification tree:
shrink.tree(tree = z.kyph, k = 0.25)
Number of terminal nodes:  9
Effective number of terminal nodes:  2.8
Residual mean deviance:  0.739 = 57.754 / 78.2
Misclassification error rate: 0.136 = 11 / 81
```

Which tree is better? In one sense, the pruned tree, since it provides a much more

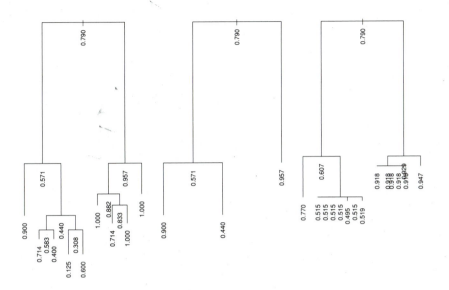

Figure 9.6: *Three variations of the tree grown to the* `kyphosis` *data. All plots are on a common scale, and nonuniform (vertical) spacing of nodes is used. The nodes are labeled with the estimated probability that* `Kyphosis==absent`. *The node labels have been rotated to improve readability. The first panel is the full tree* `z.kyph`; *it is a graphical representation of the tabular version presented in Table 9.1. The second panel is a pruned version of* `z.kyph`, *whereby the least important splits have been* pruned *off. Note that the estimated probabilities and node heights match those of the full tree. The third panel is a shrunken version of* `z.kyph`, *whereby the estimated probabilities have been* pulled back *or shrunken toward the root. Apart from the root, neither the estimated probabilities nor the node heights match those of the full tree. The squashing of the dendrogram at the bottom indicates that these nodes have been shrunk completely to their parents.*

succinct description of the data (note that only two out of the three predictors remain). In another sense, the shrunken tree, since its misclassification error rate is lower than that of the pruned tree. Thus, there is no hard and fast rule on which is better; the choice depends on where your priorities lie (simplicity versus accuracy). We now proceed to describe these methods in more detail.

The function `prune.tree()` takes a tree object as a required argument. If no additional arguments are supplied, it determines a nested sequence of subtrees of the supplied tree by recursively snipping off the least important splits. Importance is captured by the cost-complexity measure:

$$D_\alpha(T') = D(T') + \alpha size(T')$$

where $D(T')$ is the deviance of the subtree T', $size(T')$ is the number of terminal nodes of T', and α is the cost-complexity parameter. For any specified α, cost-complexity pruning determines the subtree T' that minimizes $D_\alpha(T')$ over all sub-trees of T. The optimal subtree for a given α is obtained by supplying `prune.tree()` with the argument `k=`α. For example, the tree displayed in the second panel of Figure 9.6 was obtained by `prune.tree(z.kyph, 5)`. If `k=`α is a vector, the sequence of subtrees that minimize the cost-complexity measure is returned rather than a tree object.

The function `shrink.tree()` takes a tree object as a required argument. If no additional arguments are supplied, it determines a sequence of trees of the supplied tree that differ in their fitted values. A particular tree in the sequence is indexed by α, which defines *shrunken* fitted values according to the recursion:

$$\hat{y}(node) = \alpha\bar{y}(node) + (1 - \alpha)\hat{y}(parent)$$

where $\bar{y}(node)$ is the usual fitted value for a node, and $\hat{y}(parent)$ is the *shrunken* fitted value for the node's parent—that is, it was obtained by applying the same recursion. The function `shrink.tree()` uses a particular parametrization of α that *optimally* shrinks children nodes to their parent based on the magnitude of the difference between $\bar{y}(node)$ and $\bar{y}(parent)$. The sequence is anchored between the full tree ($\alpha = 1$) and the root node tree ($\alpha = 0$). A heuristic argument allows one to map α into the number of *effective* terminal nodes, thereby facilitating comparison with pruning. The tree for a given α is obtained by supplying `shrink.tree()` with the argument `k=`α. For example, the tree displayed in the third panel of Figure 9.6 was obtained by `shrink.tree(z.kyph, .25)`. If `k=`α is a vector, the sequence of trees that are determined by these shrinkage parameters is returned rather than a tree object.

Figure 9.7 displays the sequences for pruning and shrinking `z.survey`. These are obtained by omitting the `k=` argument and plotting the resulting object. These objects have class `"tree.sequence"` for which a `plot()` method exists. Each panel displays the deviance versus size (the number of terminal nodes or the number of *effective* terminal nodes) for each tree in the sequence. An additional (upper) axis shows the mapping between size and `k` for each method. By construction, the deviance decreases as tree size increases, a common phenomenon in model-fitting (i.e., the fit improves as parameters are added to the model). This limits the usefulness of the plot except in those situations where a dramatic change in deviance occurs at a particular value of `k`.

It should not be surprising that the sequences produced by these methods provide little guidance on what size tree is adequate. The same data that were used to grow the tree are being asked to provide this additional information. But since the tree was optimized for the supplied data, the tree sequences have no possible alternative but to behave as observed. There are two ways out of this dilemma:

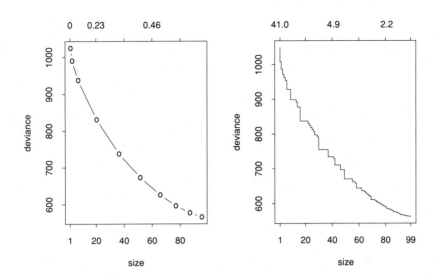

Figure 9.7: *Plots of deviance versus size (number of terminal nodes) for sequences of subtrees of* z.survey. *The left panel is based on optimal shrinking while the right panel is based on cost-complexity pruning. The former is plotted as a continuous function to reinforce its continuous behavior. The latter is plotted as a step function because optimal subtrees remain constant between adjacent values of* k. *Each panel has an additional axis along the top indicating the values of* k *that correspond to the different sized subtrees in the sequence.*

one is to use new (independent) data to guide the selection of the right size tree, and the other is to reuse the existing data by the method of *cross-validation*. In either case, the issue of tree-based prediction of new data arises. Let's pursue this diversion before returning and concluding our discussion of choosing the right size tree.

Prediction

An important use of tree-based models is predicting the value of a response variable for a known set of predictor variables. By prediction we mean to evaluate the splits describing a tree-based model for a set of predictor variables and defining the yval at the deepest node reached as the prediction. Normally this corresponds to a leaf node of the tree, but we adopt the convention that a prediction may reside in a nonterminal node if, in following along the path defined by the set of predictor variables for a new observation, a value of a predictor is encountered that has

never been seen at that node in the tree-growing process. The classic case of this
is encountering a missing value (NA) when only complete observations were used
to grow the tree. More generally and more subtly, this condition occurs for factor
predictors whenever a split is encountered where the value goes neither left nor right
(e.g., if $x = B$ and the left and right splits at a node are, respectively, $x \in \{A, C\}$
and $x \in \{D, E\}$).

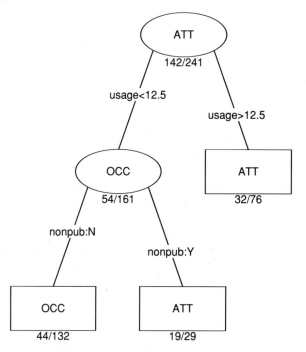

Figure 9.8: *Tree representation of prediction from the classification tree* zs.survey. *The
node labels are the predicted values of* pick. *The numbers displayed under each node represent the misclassification error rate for the new data* na.market.survey. *The overall
misclassification error rate is quite high (0.41). Four of the respondents remain at the root
node due to missing values in the predictor* usage.

We return to the long-distance marketing example where we illustrate prediction
using an additional 241 survey respondents. These respondents were part of the
initial survey but were omitted from the preliminary analysis because of missing
values in the variables. The data are collected in the data frame na.market.survey.
Predictions from the tree in Figure 9.3 are displayed in Figure 9.8. The figure shows
the disposition of the 241 observations along the prediction paths of the tree. Of
the 241 observations, 161 are directed to the left (OCC), 76 to the right (ATT), and

4 remain at the root node (due to missing values for `usage`). Of the 161, 132 are directed to the left (OCC) and 29 to the right (ATT). The misclassification error rate associated with these predictions is quite high (41%). This error rate varies from leaf node to leaf node, from 33% (leftmost leaf), to 66% (middle leaf), to 42% (rightmost leaf).

The tree displayed in Figure 9.8 was obtained with the expression:

```
zd.survey <- predict(zs.survey, na.market.survey, type = "tree")
```

The `predict()` method takes a tree object and a data frame. The tree object is likely to be a simplified version of that provided by `tree()`. The names of the variables in the data frame must include the predictors in the formula used to construct the tree. The function returns the values predicted by the tree for the data in the data frame, either as a vector (the default) or as a tree object, `type="tree"`. If a data frame is not supplied, `predict()` returns the fitted values for the data used to construct the tree; we used this feature in our earlier discussion of residual plots.

Cross-validation

We now return to the topic of choosing the right size tree based on data not used to grow the tree. Test data can be supplied to the functions `prune.tree()` and `shrink.tree()` with the `newdata=` argument. The functions return an object of class `"tree.sequence"` containing the sequence evaluated on the test data. Figure 9.9 illustrates this functionality for the market survey data, where the new data consist of those held back due to missing values. These plots span a wide range of tree sizes, but the most promising are those with fewer than a dozen nodes. The range can be restricted by suitable specification of the argument `k`. Panel 1 of Figure 9.10 demonstrates such a restriction for `k` in the range 0.05 to 0.20 for the optimal shrinking sequence. Evidently, either a very small tree is called for or the data with NAs are not drawn from the same population as those without.

The function `cv.tree()` can be used to address this ambiguity by applying a procedure described in Section 9.3 called *cross-validation*. The basic idea is to divide the original data into mutually exclusive sets. For each set, a tree is grown to the remaining sets and a subtree sequence obtained; the set *held out* is then used to evaluate the sequence. Deviances from each set are accumulated (as a function of `k`) and returned as an object of class `"tree.sequence"`. A plot of the cross-validated deviance versus tree size is seldom monotone decreasing since data used to evaluate the sequences were not used to construct them. A common feature of the plot is a fairly flat minimum, and trees in this region are candidates for further consideration. The result of tenfold cross-validation of the tree `z.survey` is displayed in the right panel of Figure 9.10. The plot was obtained by the expressions

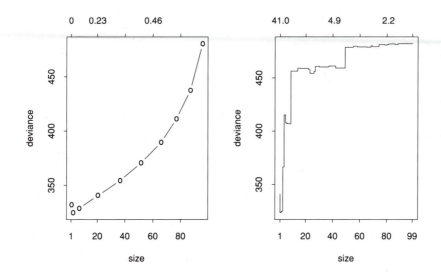

Figure 9.9: *Plots of deviance versus size for sequences of subtrees of* z.survey *evaluated on new data. The new data are from the data frame* market.survey *but were omitted from the fit due to missing values in some of the predictors. The* na.tree.replace() *function was used to replace NAs with an additional factor level. Since the original tree was constructed from data without missing values, this in effect means that when the new level* "NA" *is encountered, the deviance at that node is used. The left panel is based on optimal shrinking while the right panel is based on cost-complexity pruning. Comparison with Figure 9.7 highlights the differences in these sequences when based on training and independent test data. This figure suggests that either a very simple tree (at most three nodes) be used to summarize these data, or that the two datasets, those with and those without NAs, are qualitatively different.*

```
> k <- seq(.05, .20, length = 10)
> cv.survey <- cv.tree(z.survey, r.survey, k = k)
> plot(cv.survey, type = "b")
```

The dataset r.survey contains a random permutation of the integers 1 to 10, of length length(pick), denoting the assignment of the observations into 10 mutually exclusive sets. The function cv.tree() will determine a permutation by default, but it is often useful to specify one, especially if comparison with another sequencing method is desired. The final argument, FUN=, specifies which sequencing function is to be used; the default is shrink.tree.

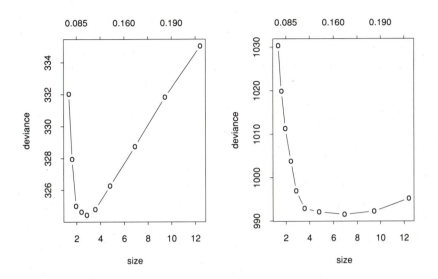

Figure 9.10: *Plots of deviance versus size for sequences of shrunken trees of* z.survey. *The range of trees considered was restricted to values of* k *between 0.05 and 0.20, corresponding to trees with effective size from 1 to 12. The left panel is based on evaluating the sequence on new data while the right panel is based on cross-validation. The left panel provides sharp discrimination in tree size, strongly suggesting a three-node tree. The right panel is not so sharp and is typical of sequences computed by cross-validation. Even so, a modest seven-node tree is suggested.*

9.2.2 Functions for Diagnosis

Residual analysis is important and not peculiar to a single class of models. In the case of trees, it is natural to exploit the very representation that is used to capture and describe the fitted model—namely, the dendrogram—as the primary means of diagnosis. We now introduce functions that utilize the tree metaphor to facilitate and guide diagnosis. The functions divide themselves along the natural components of a tree-based model—namely, *subtrees, nodes, splits,* and *leaves.* Most of the methods involve *interacting* with trees, and by this we usually mean *graphical interaction.* We note parenthetically, and sometimes explicitly below, that all the functions can be used noninteractively (by including a list of node numbers as an argument), but their usefulness seems to be significantly enhanced when used interactively.

In certain of the figures in this section, a (new) general mechanism to obtain multiple figures within the S graphics model is used. The *split-screen mode* is an alternative to `par(mfrow)` that allows arbitrary rectangular regions (called *screens*) to be specified for graphics input and output. We use this mechanism rather than the standard multifigure format not only to attain a more flexible layout style, but also because the order in which screens are accessed is under user control. It is able, for example, to arbitrarily receive graphics input from one screen and send graphics output to another. We have attempted to restrict our use of the split-screen mode to minimize the introduction of too much ancillary material. A single function `tree.screens()`, called without arguments, will set up a generic partition of the figure region used by the tree-specific functions that we provide. See the detailed documentation of `split.screen()` for further information.

9.2.3   Examining Subtrees

The function `snip.tree()` allows the analyst to *snip off* branches of a tree either through a specified list of nodes, or interactively by graphic input. For the former, the subset method for tree objects described earlier, `"[.tree"()`, is a convenient shorthand. For example, the expression `z.auto[-2]` is equivalent to the expression `snip.tree(z.auto, 2)`. This usage requires knowing the number of the node or nodes in question; the interactive approach obviates this need. It is most convenient when working at a high-resolution graphics terminal and provides a type of *what-if* analysis on the displayed tree. The graphical interface is such that a single click of the graphics input device (e.g., a mouse) informs the user of the change in tree deviance that would result if the subtree rooted at the selected node is snipped off; a second click on the same node actually does the snipping. By snipping, we mean that the tree object is modified to reflect the deleted subtree and also that the portion of the plotted dendrogram corresponding to the subtree rooted at the selected node is "erased." The process can be continued, and, on exit, what remains of the original tree is returned as a tree object. An example of the textual information displayed during this process is as follows:

```
> zsnip.survey <- snip.tree(z.survey)
node number:   4
   tree deviance =   562.518
   subtree deviance =   741.663
node number:   10
   tree deviance =   741.663
   subtree deviance =   786.214
node number:   7
   tree deviance =   786.214
   subtree deviance =   962.767
```

Here we first selected and then reselected nodes 4, 10, and 7 of the tree `z.survey`. Note how the subtree deviance at one stage becomes the tree deviance at the next stage. The graphical result of this process is displayed in Figure 9.11. The second

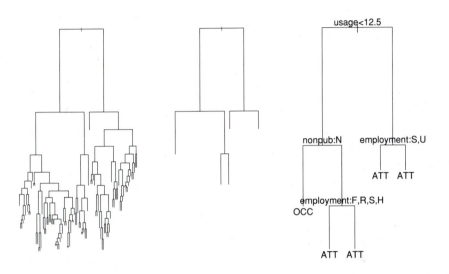

Figure 9.11: *An illustration of interactive snipping of subtrees. The full tree* z.survey *is plotted in the first panel. Upon selection of a node, the change in deviance that would result by snipping off the subtree rooted at that node is displayed. If it is reselected, the subtree is snipped off, which has the side effect of erasing the subtree from the dendrogram. The second panel shows what remains of the tree after the subtrees rooted at nodes 4, 10, and 7 are snipped off. The final panel replots and labels the snipped tree.*

panel shows the result of snipping off the subtrees rooted at nodes 4, 10, and 7. The final panel replots the snipped tree zsnip.survey and labels it. This points out one reason for snipping—gaining resolution at the top of the tree so that it can be usefully labeled. The node numbers of the branches that were snipped off are collected together and pasted into the call component of the tree object to inform the user that the result was obtained by snipping nodes so-and-so from tree such-and-such. For example, the call component of zsnip.survey is

```
> zsnip.survey$call
snip.tree(tree = z.survey, nodes = c(4, 10, 7))
```

The function select.tree() is the dual of snip.tree(). It allows individual subtrees of a specified tree to be selected and assigned. For each node number supplied, the function returns a tree object rooted at that node. If no nodes are supplied, the function expects them to be selected by graphical interaction. When more than one node is specified or selected, the subtrees are organized as a list, with the node number naming the individual elements. One might reasonably call

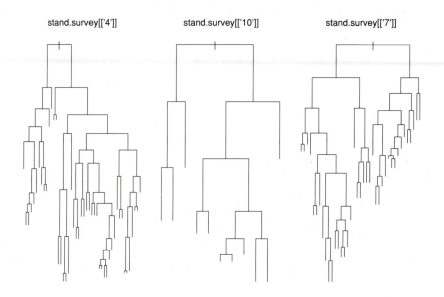

Figure 9.12: *An illustration of a stand of trees. The three panels contain the subtrees of*
z.survey *that were* snipped off *in Figure 9.11. Each tree in the stand is amenable to all
methods for tree objects, including plot methods. The panels in the figure were obtained by
applying the* plot() *method to the stand* stand.survey.

such a list a *stand* (of trees). An interesting feature of stands results from the fact
that the trees it contains are bona fide tree objects. Thus, they are amenable to
any and all display and analysis functions for trees. A useful way to peruse a stand
is by applying a function to it using apply(). For example, Figure 9.12 is obtained
by the expression

```
> stand.survey <- select.tree(z.survey, nodes = c(4, 10, 7))
> sapply(stand.survey, plot)
```

Like snip.tree(), the subset method for tree objects, "[.tree"(), is a convenient
shorthand for select.tree(). For example, z.survey[c(4, 10, 7)] is equivalent
to the expression given above for stand.survey. Also like snip.tree(), the call
component of a selected subtree is constructed to inform the user that the result
was obtained by selecting subtree so-and-so from tree such-and-such.

9.2.4   Examining Nodes

Much information concerning a fitted tree resides in the nodes. It is important
that this information be readily available, and yet, there is too much information to

usefully label a dendrogram with. We now introduce some tree-specific functions to encourage users to browse the nodes of a fitted tree-based model. Let's introduce a new example based on the data frame `cu.summary` described in Section 3.1.1. The data are summarized as follows:

```
summary(cu.summary)
        Price               Country         Reliability
Min.    : 5866      USA        :49     Much worse :18
1st Qu.:10090      Japan      :31     worse      :12
Median :13150      Germany    :11     average    :26
Mean   :15740      Japan/USA: 9       better     : 8
3rd Qu.:19160      Sweden    : 5      Much better:21
Max.   :41990      Korea     : 5      NAs        :32
                   (Other)   : 7

        Mileage              Type
Min.    :18.00     Compact:22
1st Qu.:21.00     Large   : 7
Median :23.00     Medium :30
Mean   :24.58     Small  :22
3rd Qu.:27.00     Sporty :26
Max.   :37.00     Van    :10
NAs    :57
```

The model we entertain addresses the relationship of automobile characteristics to automobile reliability. The fitted tree-based model is obtained by the expression

```
> f.cu <- formula(Reliability ~ Price + Country + Mileage + Type)
> z.cu <- tree(f.cu, cu.summary, na.action = na.tree.replace)
```

and is plotted in Figure 9.13. Since this is a classification tree with a five-level response variable, much information has been suppressed in the labeled dendrogram. Node contents may be inspected with the `browser()` method for trees, which takes a tree object as a required argument and an optional list of nodes. If the latter is omitted, the function waits for the user to select nodes with the graphics input device. For example, clicking on the left-child of the root node of the tree `z.cu` yields:

```
> browser(z.cu)
node number: 2
 split: Country:Japan,Japan/USA
 n: 27
 dev: 36.9219
 yval: Much better
Much worse worse    average      better Much better
        0        0 0.1111111 0.1111111   0.7777778
```

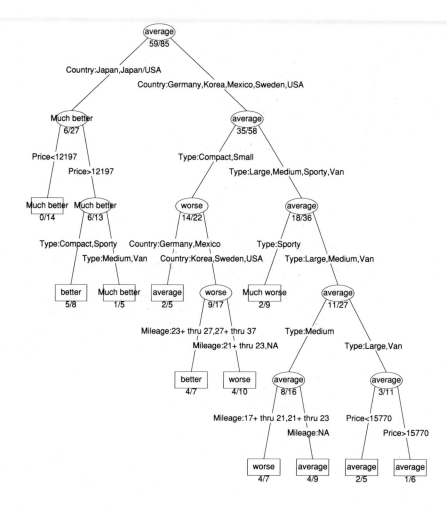

Figure 9.13: *A display of a tree fitted to the automobile reliability data. The response variable has levels* Much Worse, worse, average, better, Much Better. *The predicted value of the response variable is centered in the node. The number under each terminal node is the misclassification error rate. The split at the root node suggests that Japanese cars, whether manufactured here or abroad, have much better perceived reliability than cars of other nationalities.*

The `identify()` method also takes a tree object as a required argument and an optional list of nodes. If the latter is omitted, the function waits for the user to select nodes from the dendrogram. The function returns a list, with one component for each node selected, containing the names of the observations falling in the node. For example, clicking on the leftmost node of the tree `z.cu` yields:

```
> identify(z.cu)
node number: 4
    Acura Integra 4
    GEO Prizm  4
    Honda Civic 4
    Mazda Protege 4
    Nissan Sentra 4
    Subaru Loyale 4
    Toyota Corolla 4
    Toyota Tercel 4
    Honda Civic CRX Si 4
    Honda Accord 4
    Nissan Stanza 4
    Subaru Legacy 4
    Toyota Camry 4
```

The "4" following each automobile name is actually part of the name (these are all four-cylinder cars) and has nothing to do with the fact that node 4 was selected. If the result of `identify()` is assigned, these names can then be used as subscripts to examine data specific to individual nodes. The following expressions demonstrate how the predictor `Price` varies for observations in nodes 2 and 3:

```
> node2.3 <- identify(z.cu, 2:3)
> quantile(Price[node2.3[["2"]]])
[1]  6488.00  9730.50 12145.00 17145.25 24760.00
> quantile(Price[node2.3[["3"]]])
[1]  5899.0  9995.0 13072.5 20225.0 39950.0
```

quantile gives 5 number summary — min, lower quartile, median, upper quartile, max

Nodes 2 and 3 are the left and right children, respectively, of the root node. Given that the more reliable cars follow the left path rather than the right, apart from the least expensive automobiles, it appears that you pay more for more troublesome cars!

The function `path.tree()` allows the user to obtain the *path* (sequence of splits) from the root to any node of a tree. It takes a tree object as a required argument and an optional list of nodes. If the latter is omitted, the function waits for the user to select nodes from the dendrogram. The function returns a list, with one component for each node specified or selected. The component contains the sequence of splits leading to that node. In interactive mode, the individual paths are (optionally) printed out as nodes are selected. The function is useful in those cases where tree size or label lengths are such that severe overplotting results if the tree is labeled indiscriminately. For example, selecting one of the deep nodes of the tree z.cu yields:

```
> path.tree(z.cu)
node number: 26
    root
    Country:Germany,Korea,Mexico,Sweden,USA
    Type:Compact,Small
    Country:Korea,Sweden,USA
    Mileage:23+ thru 27,27+ thru 37
```

By examining the path, we can see that the automobiles in this node consist of those manufactured in Korea, Sweden, and USA, which are compact or small, and for which the reported mileage is between 23 and 37 mpg.

9.2.5 Examining Splits

The tree grown to the automobile reliability data suggests that Japanese cars, whether manufactured here or abroad, are more reliable than cars of other nationalities. Should we believe this? The answer in general is no; the recursive partitioning algorithm underlying the tree() function is just that: an algorithm. There may well be other variables, or even other partitions of the variable Country, that discriminate reliable from unreliable cars, but these just miss out being the "best" split among all possible. The function burl.tree() allows the user to select nodes and observe the competition for the best split at that node. For numeric predictors, a high density plot is used to show the *goodness-of-split* at each possible cut-point split. For factor predictors, a scatterplot plot displays goodness-of-split versus a decimal equivalent of the binary representation of each possible subset split; the plotting character is a string labeling the left split. Figure 9.14 provides an example for the tree z.cu. The plots under the dendrogram show a clear preference for splits involving the variable Country. Figure 9.15 is an enlargement of the scatterplot for Country. We see that the candidate splits divide into two groups, one of which (top) discriminates better than the other (bottom). Among those in the top portion, that labeled ef=Japan, Japan/USA is the best; moreover, it is the common intersection of all the candidate splits in the top portion. Given this information, we are more likely to believe that this split is meaningful.

The function hist.tree() also focuses on splits at specified or interactively selected nodes by displaying side-by-side histograms of supplied variables. Specifically, the histogram on the left displays the distribution of the observations on that variable following the left split, while the histogram on the right displays the distribution of the observations following the right split. It is similar to burl.tree() in that it displays a variable's discriminating ability, but is different in that it allows

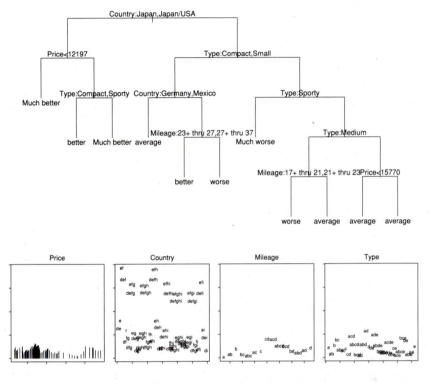

Figure 9.14: *An illustration of burling a tree-based model. The top panel displays the labeled dendrogram of* z.cu; *initially, the lower portion is empty. Upon selection of the root node, the plots in the lower four panels are displayed. These show, for each predictor in the model formula, the goodness-of-split criterion for each possible split. The goodness-of-split criterion is the difference in deviance between the parent (in this case the root node) and its children (defined by the tentative split); large deviance differences correspond to important splits. For numeric predictors, a high-density plot conveys the importance of each possible cut-point split. For factor predictors, an arbitrary ordering is used along the abscissa (x-axis) to separate different subset splits; the left split is used as a plotting character. The ordinate (y-axis) of all plots is identical. These plots show that, at the root node,* Country *is the best discriminator of automobile reliability. It also shows that there are many good subset splits on* Country, *the "best" being the one labeled* ef *in the upper left. Upon selection of another node in the dendrogram, the lower portion of the screen is erased and refreshed with four new panels displaying the splits relevant at that node.*

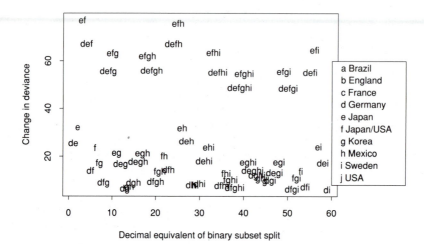

Figure 9.15: *A scatterplot of the competing subset splits on* Country *at the root node of the tree* z.cu. *The plotted character strings are the left splits; none contain* j *since it is the last level of* Country *and, by construction, resides in right splits only. No subsets contain* abc *since automobiles from these countries were omitted due to missing values of the response variable* Reliability; *this occurred silently by* na.tree.replace() *when* z.cu *was grown. There are no singleton splits for* d, g, h, *or* i *since these countries have fewer than five automobiles in the model frame and the algorithm has a minimum subset size of five. The splits seem to divide into two groups: those having good discriminating power (upper portion), and those having mediocre to poor power (lower portion). The former all contain* ef, *supporting its selection as the best discriminating subset.*

variables other than predictors to be displayed. Figure 9.16 provides an example for the tree z.cu fitted to the automobile reliability data. This example resulted from the expression:

```
> hist.tree(z.cu, Reliability, Price, Mileage, nodes = 1)
```

At a glance we see the complete distribution of the response variable Reliability for nodes 2 and 3 (the children nodes of the root). It is interesting that not a single Much Better car follows the right split. The second panel (Price) graphically conveys what our earlier analysis using identify() suggested: that the most reliable cars are not the most expensive ones. It appears that status and reliability are incompatible in these data.

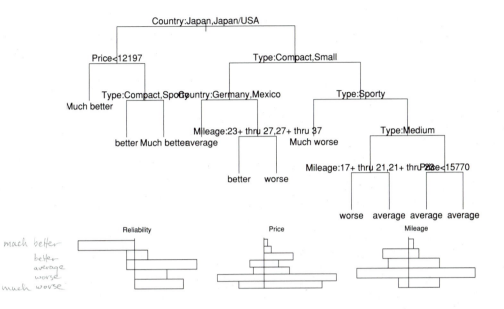

Figure 9.16: *A illustration of the function* `hist.tree()` *at the root node of the automobile reliability tree* `z.cu`*. The upper portion of the plot contains the labeled dendrogram. The lower portion displays a side-by-side histogram for each of the variables* `Reliability`*,* `Price`*, and* `Mileage`*. The left-side histogram summarizes the observations following the left split, and similarly for the right. The figure shows that Japanese cars manufactured here or abroad tend to be more reliable, less expensive, and more fuel efficient than others.*

9.2.6 Examining Leaves

Often it is useful to observe the distribution of a variable over the leaves of a tree. Two related (noninteractive) functions encourage this functionality. They are noninteractive since they do not depend on user selection of a particular node; their intended effect is across all terminal nodes. The function `tile.tree()` augments the bottom of a dendrogram with a plot that shows the distribution of a specified factor for observations in each leaf. These distributions are encoded into the widths of tiles that are lined up with each leaf. If numeric variables are supplied, they are automatically quantized. One use of this function is for displaying class probabilities across the leaves of a tree. An example is displayed in Figure 9.17. A related function `rug.tree()` augments the bottom of a dendrogram with a (high-density) plot that shows the average value of the specified variable for observations in each leaf. These averages are encoded into lengths of line segments that are lined up with each leaf. The function takes an optional argument, `FUN=`, so that summaries other than simple averages (e.g., trimmed means) can be obtained. Figure 9.18 displays

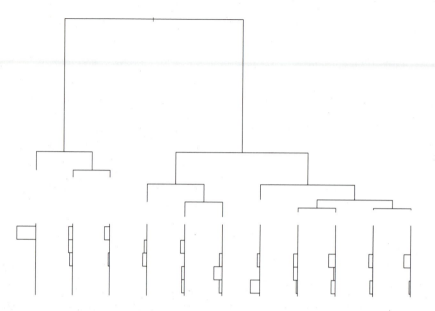

Figure 9.17: *The dendrogram of the automobile reliability tree* `z.cu` *enhanced with a tiling of the variable* `Reliability`. *The distribution of* `Reliability` *over the leaves of the tree is readily discerned. Successive calls to* `tile.tree()` *with other variables is encouraged by not replotting the dendrogram—only the new tiling is plotted after the bottom screen is "erased".*

the distribution of the variable `usage` for the tree grown to the market survey data. Recalling that the split at the root node was `usage` $\leq$ 12.5, the general shape of the rug is as expected: lower on the left and higher on the right. Somewhat unexpected is the fact that the heavier users are, by and large, much heavier users.

9.3   Specializing the Computations

As described in the preceding section, the tree object is a repository for a number of by-products of the tree-growing algorithm. The named components of a tree object are

```
> names(z.survey)
 [1] "frame"    "where"     "terms"  "call"
```

The `frame` component is a data frame, one row for each node in the tree. The row labels, `row.names(frame)`, are node numbers defining the topology of the tree. Nodes of a (full) binary tree are laid out in a regular pattern:

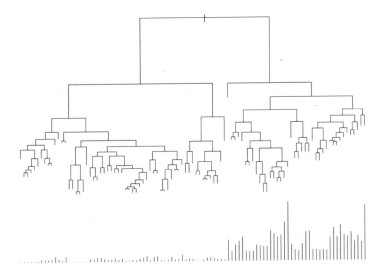

Figure 9.18: *The dendrogram of the long-distance marketing tree* z.survey *enhanced with a rug of the variable* usage. *The distribution of this variable over the leaves of the tree is readily discerned. Successive calls to* rug.tree() *with other variables is encouraged by not replotting the dendrogram—only the new rug is plotted after the bottom screen is erased.*

More generally, nodes at depth d are integers n, $2^d \le n < 2^{d+1}$. Of course, any specific tree is not full and consists of a subset of all possible nodes. The ordering of the nodes in the frame corresponds to a depth-first traversal of the tree according to this numbering scheme.

The elements (columns) of frame contain the following node-specific information:

- the variable used in the split at that node (var)

- the number of observations in the node (n)

- a measure of node heterogeneity (dev)

- the fitted value of the node (yval)

- the matrix of left and right split values (splits).

Routine application of the functions in this chapter does not require users to manipulate this object directly, but for concreteness we display the 21 row z.cu$frame here:

	var	n	dev	yval	splits.left	splits.right
1	Country	85	260.997544	average	:ef	:dghij
2	Price	27	36.921901	Much better	<12197	>12197
4	<leaf>	14	0.000000	Much better		
5	Type	13	26.262594	Much better	:ae	:cf
10	<leaf>	8	17.315128	better		
11	<leaf>	5	5.004024	Much better		
3	Type	58	146.993133	average	:ad	:bcef
6	Country	22	59.455690	worse	:dh	:gij
12	<leaf>	5	6.730117	average		
13	Mileage	17	42.603572	worse	:cd	:be
26	<leaf>	7	15.105891	better		
27	<leaf>	10	19.005411	worse		
7	Type	36	68.976020	average	:e	:bcf
14	<leaf>	9	9.534712	Much worse		
15	Type	27	50.919255	average	:c	:bf
30	Mileage	16	33.271065	average	:ab	:e
60	<leaf>	7	14.059395	worse		
61	<leaf>	9	16.863990	average		
31	Price	11	12.890958	average	<15770	>15770
62	<leaf>	5	6.730117	average		
63	<leaf>	6	5.406735	average		

This example illustrates a labeling convention specific to trees whereby levels of factor predictors are assigned successive lower-case letters. Thus, the first right split, :dghij (on Country), is shorthand for :Germany,Korea,Mexico,Sweden,USA. Such a convention is necessary in order to provide meaningful information about splits in a limited amount of space. The problem is particularly acute for labeling plotted dendrograms but is also important in tabular displays such as that resulting from print(). The labels() method for trees allows full control over which style of labels is desired; it is usually invoked by printing and plotting functions rather than called directly by the user.

In the case of classification trees, an additional component of the frame object is the matrix (yprob) containing the class probability vectors of the nodes labeled by the levels of the response variable. We omitted this in the above display of z.cu$frame in order to conserve space.

The where component of a tree object is a vector containing the row number (in frame) of the terminal node that each observation falls into. It has a names attribute that corresponds to the row.names of the model frame used to grow or otherwise define the tree. Like the frame component, it is heavily used in many of the functions that manipulate trees. For example, the vector of fitted values is obtained as z$frame[z$where, "yval"]. The remaining components, "terms" and "call", are identical to those described in previous chapters.

We emphasize that for the most part you will not have to look directly at the values of these components. However, in order to modify the behavior of any of the supplied functions, or to construct new ones, you should first feel comfortable manipulating these components. For example, consider the following function (provided in the library):

```
meanvar.tree() <- function(tree, xlab = "ave(y)",
    ylab = "ave(deviance)", ...) {
      if(!inherits(tree, "tree"))
            stop("Not legitimate tree")
      if(!is.null(attr(tree, "ylevels")))
            stop("Plot not useful for probability trees")
      frame <- tree$frame
      frame <- frame[frame$var == "<leaf>",  ]
      x <- frame$yval
      y <- frame$dev/frame$n
      label <- row.names(frame)
      plot(x, y, xlab = xlab, ylab = ylab, type = "n", ...)
      text(x, y, label)
      invisible(list(x = x, y = y, label = label))
}
```

This function uses only the `frame` component to produce a plot of the within-node variance (`dev/n`) versus the within-node average (`yval`) for numeric responses. The node number is used as the plotting character. This plot is useful for assessing the assumption of constant variability throughout predictor space. If trend is apparent in the plot, a reexpression of the response variable y is recommended for proper trees to be grown.

The functions we provide are intended to make the task of modeling data with binary trees more pleasant and at the same time more powerful. The examples in the previous sections showed how the user might directly use these functions during an analysis. Of course, the functions can also be called by other functions and thus form the building blocks for more specialized functions or even more complicated manipulations of tree-based models.

The single best example illustrating the power of using the functions as primitives in a more complicated function is given by the technique known as *cross-validation*. Specifically, consider the problem of selecting the optimal tree in a pruning or shrinking sequence. The general idea is that the deviances, used as a measure of predictive ability, for any of the trees in the sequence are far too optimistic—that is, too small—as they are based on the same data used to construct the tree. It would be better—that is, less biased—to use an independent sample with which to assess the predictive ability of any specific tree. Cross-validation is an attempt to do just this where the original dataset is carved into K mutually exclusive subsets, each of which will serve as an independent test set for trees grown on learning sets composed of the union of the $K - 1$ remaining subsets. For each of the learning sets, a tree must be grown and a pruning or shrinking sequence determined. The corresponding test set must then be dropped down the trees in the sequence and some measure of goodness computed (e.g., misclassification error rate or deviance—we use the latter). These are then summed over the induced replications and displayed. An implementation is as follows:

```
cv.tree <- function(tree, rand, FUN = shrink.tree, ...)
{
    if(!inherits(object, "tree"))
        stop("Not legitimate tree")
    m <- model.frame(object)
    p <- FUN(object, ...)
    if(missing(rand))
        rand <- sample(10, length(m[[1]]), replace = T)
    which <- unique(rand)
    cvdev <- 0
    for(i in which) {
        tlearn <- tree(model = m[rand != i, ])
        plearn <- FUN(tlearn, newdata = m[rand == i, ], p$k, ...)
        cvdev <- cvdev + plearn$dev
    }
    p$dev <- cvdev
    p
}
```

Apart from some initialization steps, the function first sequences the original tree and assigns the result to `p`. In the `for` loop, we use two different high-level tree manipulation functions. We first use `tree()` to grow a tree to the learning model, `m[rand != i, ]`. This is followed by a call to the sequencing function, `shrink.tree()` by default, to produce the sequence for the learning tree and to evaluate the sequence for the model containing the test data, `m[rand == i, ]`. Finally, the deviances are summed across samples and returned for subsequent plotting.

Other functions for tree-based modeling are included in the library that have not been explicitly mentioned in the text. Some are low-level utility functions that are called by the high-level functions accessed directly by the user. Others are high-level functions that are specialized for certain numerical or graphical purposes. The function `basis.tree()` is an example of the former whereby an orthogonal basis for a fitted tree is computed. There is one basis vector for each split and one for the root (the unit vector). A linear model fitted to this basis yields fitted values identical to those from the tree. This linear model representation of a fitted tree-based model is sometimes useful for suggesting new methods for understanding trees (e.g., shrinkage estimation.) The functions `post.tree()` and `partition.tree()` are examples of special purpose graphics functions. The function `post.tree()` does not require

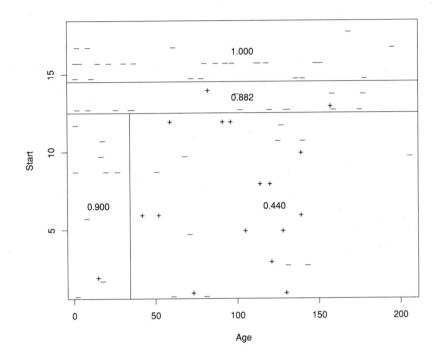

Figure 9.19: *A display of* z.kyph[-c(5, 6)], *a subtree of* z.kyph *depending on the variables* Age *and* Start. *The plot was obtained with the expression* partition.tree(z.kyph[-c(5, 6)], label = "absent"). *The data values appear on the plot as the plotting characters "-" and "+." These were added with the expression* text(Age, Start, ifelse(Kyphosis == "absent", "-", "+")). *Three of the four regions are quite homogeneous; no apparent structure is discernible in the remaining one.*

activation of a graphics device, but rather that the user has access to a printer compatible with the PostScript page-description language. The trees displayed in Figures 9.3, 9.8, and 9.13 were produced by post.tree(). This "pretty printed" display of a tree uses uniform vertical spacing of nodes and is more appropriate for presentation than for diagnosis.

The function partition.tree() is peculiar to trees that depend on at most two predictor variables. For a single predictor, partition.tree() displays the tree as a step function, each step corresponding to a terminal node of the tree. This display sacrifices the information in the tree object concerning the sequence of splits leading to the leaf nodes, but gains familiarity of expression when one regards y

as a function of x. The example in the right panel of Figure 9.1 was created with `partition.tree()`. For two predictors, `partition.tree()` displays the partition of the plane into homogeneous regions, each rectangular region corresponding to a terminal node of the tree. In certain cases it is possible to reconstruct the sequence of splits giving rise to the partition from the display, although this is not the primary intended purpose. An optional argument, `label`, allows the user to specify the labels associated with the partition, the default being the fitted value `yval`. For classification trees, a specific level of the response factor can be specified. Figure 9.19 demonstrates a two-variable example based on the subtree `z.kyph[-c(5, 6)]`.

Certain enhancements to the display functions are desirable so that more information can be displayed subject to the constraint of minimal overplotting. For example, the `text()` method for trees introduced in Section 9.2 allows an argument, `FUN=`, to encourage users to explore interactive labeling. Suppose a user had a function, say `brush()`, which allowed one to paint on labels (say with button 1) as well as erase them (say with button 2). By paint we mean that buttons are depressed and held rather than simply clicked. Then one could selectively label a plotted dendrogram in those cases where unrestricted labeling would conceal the dendrogram itself.

A somewhat different specific proposal that we considered was displaying a histogram or a boxplot of the distribution of y at each node of the tree. This would allow comparison of scale and shape changes as nodes are split in addition to location differences, as is currently done. A function `zoom.tree()` might then be written so that selecting a node might *zoom in* or otherwise provide an enlargement of the histogram. This would necessitate some device-specific graphics functions, which we have attempted to avoid.

9.4 Numerical and Statistical Methods

Tree-based models are defined most precisely by the algorithm used to fit them. The algorithm attempts to partition the space of predictor variables (X) into homogeneous regions, such that within each region the conditional distribution of y given x, $f(y|x)$, does not depend on x. We first present the algorithm and then discuss the three essential components as regards our implementation.

Initialize: current node = root = $\{y_i , i = 1, \ldots, n\}$
    stack = NULL

Recurse: for current node $\neq$ NULL

    Loop: for each x_j partition x into two sets X_{LEFT} and X_{RIGHT} such that $f(y|X_{LEFT})$ and $f(y|X_{RIGHT})$ are most different

Split node: split current node into Y_{LEFT} and Y_{RIGHT} according to the x_j and the associated split that is best among all x's

Test: if *ok* to split Y_{RIGHT}

push Y_{RIGHT} onto stack

if *ok* to split Y_{LEFT}

current node $= Y_{LEFT}$

else *pop* stack

Partitioning the Predictors

Predictor variables appropriate for tree-based models can be of several types: factors, ordered factors, and numeric. Partitions are governed solely by variable type and therefore do not require explicit specification by the user.

If x is a factor, with say k levels, then the class of splits consists of all possible ways to assign the k levels into two subsets. In general, there are $2^{k-1}-1$ possibilities (order is unimportant and the empty set is not allowed). So, for example, if x has three levels (a, b, c), the possible splits consist of $a|bc$, $ab|c$, and $b|ac$.

If x is an ordered factor with k ordered levels, or if x is numeric with k distinct values, then the class of splits consists of the $k - 1$ ways to divide the levels/values into two contiguous, nonoverlapping sets. These splits can be indexed by the midpoints of adjacent levels/values, which we call *cutpoints*. By convention, we implicitly extend the range beyond the observed data, so that at the left-most cutpoint, c_L defines the split $-\infty < x \le c_L$, and similarly for the right-most cutpoint. Note that the *values* of a numeric predictor are not used in defining splits, only their *ranks*. Indeed, it is this aspect of tree-based models for numeric predictors that render them invariant under monotone transformations of x.

Comparing Distributions at a Node

We depart slightly from most previous authors on recursive partitioning methods in that our view is more closely akin to classical models and methods for regression and classification data. Our view is that we are estimating a step function $\tau(\boldsymbol{x})$ that is simply related to a primary parameter in the conditional distribution of $y|\boldsymbol{x}$. The likelihood function provides the basis for choosing partitions. Specifically, we use the *deviance* (likelihood ratio statistic) to determine which partition of a node is "most likely" given the data. The implementation is such that the type of the response variable is the sole determinant of whether a classification tree (factor response y) or a regression tree (numeric y) is grown. The current implementation ignores any possible ordering of an ordered factor response variable; arguably, this should be exploited in the fitting.

The model we use for classification is based on the multinomial distribution where we use the notation, for example,

$$y = (0, 0, 1, 0)$$

to denote the response y falling into the third level out of four possible. The vector $\mu = (p_1, p_2, p_3, p_4)$, such that $\sum p_k = 1$, denotes the probability that y falls into each of the possible levels. In the terminology of Chapter 6, the model consists of the stochastic component,

$$y_i \sim \mathcal{M}(\mu_i), \ i = 1, \ldots, N$$

and the structural component

$$\mu_i = \tau(\boldsymbol{x}_i).$$

The deviance function for an observation is defined as minus twice the log-likelihood,

$$D(\mu_i; y_i) = -2 \sum_{k=1}^{K} y_{ik} \log(p_{ik}).$$

The model we use for regression is based on the normal (Gaussian) distribution, consisting of the stochastic component,

$$y_i \sim \mathcal{N}(\mu_i, \sigma^2), \ i = 1, \ldots, N$$

and the structural component

$$\mu_i = \tau(\boldsymbol{x}_i).$$

The deviance function for an observation is defined as

$$D(\mu_i; y_i) = (y_i - \mu_i)^2,$$

which is minus twice the log-likelihood scaled by σ^2, which is assumed constant for all i.

At a given node, the mean parameter μ is constant for all observations. The maximum-likelihood estimate of μ, or equivalently the minimum-deviance estimate, is given by the node proportions (classification) or the node average (regression).

The deviance of a node is defined as the sum of the deviances of all observations in the node $D(\hat{\mu}; y) = \sum D(\hat{\mu}; y_i)$. The deviance is identically zero if all the y's are the same (i.e., the node is pure), and increases as the y's deviate from this ideal. Splitting proceeds by comparing this deviance to that of candidate children nodes that allow for separate means in the left and right splits,

$$D(\hat{\mu}_L, \hat{\mu}_R; y) = \sum_L D(\hat{\mu}_L; y_i) + \sum_R D(\hat{\mu}_R; y_i)$$

The split that maximizes the change in deviance (goodness-of-split)

$$\Delta D = D(\hat{\mu}; y) - D(\hat{\mu}_L, \hat{\mu}_R; y)$$

is the split chosen at a given node.

Limiting Node Expansion

The above discussion implies that nodes become more and more pure as splitting progresses. In the limit a tree can have as many terminal nodes as there are observations. In practice this is far too many, and some reasonable constraints should be applied to reduce the number. We use two different criteria for deciding if a node is suitable for splitting. *Do not split:*

- if the node deviance is less than some small fraction of the root node deviance (say 1%); and

- if the node is smaller than some absolute minimum size (say 10).

These limits are implemented through the arguments `mindev` and `minsize`, respectively, in the function `tree.control()`. The current defaults are given above in parentheses.

The default is quite liberal and will still result in an overly large tree with roughly $N/10$ terminal nodes. This is intentional and mimics "best current practice" in recursive partitioning methods. Indeed, the major problem of early tree-building algorithms was deciding when to stop expanding nodes. It was indeed critical as the tree was built in a forward stepwise manner, and once the final node was expanded, modeling was complete. The approach we adopt is not to limit node expansion in the tree-growing process. Instead, an overly large tree is grown, and one must decide which branches to prune off or find some other way account for overfitting (e.g., recursive shrinking). The difference in the approaches is similar to that between forward and backward stepwise selection of variables in linear models. Forward methods can be fooled when the best early split does not meet the criterion of splitting and tree growth is halted—when in fact this split is necessary to clear the field for very important succeeding splits. The example of looking for interactions in linear model residuals provides an illustration.

The design of our functions had this concept in mind from its inception, providing a simple interface to growing a large tree, while providing a collection of interactive functions to inspect nodes, identify observations, snip branches, select subtrees, etc. Our recommended approach to tree building is far less automatic than that provided by other software for the same purpose, as the unbundling of procedures for growing, displaying, and challenging trees requires user initiation in all phases. We now turn to another issue that also requires the user to get involved in the modeling process.

9.4.1 Handling Missing Values

Tree-based models are well suited to handling missing values and several possibilities exist for building trees and predicting from them in the face of NAs. For tree building itself, the current implementation of tree() only permits NAs in predictors, and only if requested by the special na.action() for trees, na.tree.replace(). The effect of this function is to add a new level named "NA" to any predictor with missing values; numeric predictors are first quantized. The net effect of using na.tree.replace() is that the new variable is treated like any other factor as regards determination of the optimal split. If x has three levels (a, b, c), the candidate splits accommodating missing values are $NA|abc$, $NAa|bc$, $NAab|c$, $NAb|ac$, $NAc|ab$, $NAbc|a$, and $NAac|b$. Other possible ways to adapt tree() to allow missing values in ordinal and numeric variables would likely require changes in the underlying algorithm.

As described earlier on page 392, the approach we adopt for prediction is that once an NA is detected while dropping a (new) observation down a fitted tree, the observation "stops" at that point where the missing value is required to continue the path down the tree. This is equivalent to sending the observation down both sides of any split requiring the missing value and taking the weighted average of the vector of predictions in the resulting set of terminal nodes. We chose this method over that based on so-called *surrogate splits* because we believe it to be less affected by nonresponse bias. A surrogate split at a given node is a split on a variable other than the optimal one that best predicts the optimal split. If a new observation is being predicted that has a missing value on the split-defining variable, then prediction continues down the tree so long as there is data on the variable given by the surrogate split.

We note in passing another function concerned with missing values. The function na.pattern() enumerates the distinct pattern of missing values in a data frame, together with the number of occurrences. For example,

```
> na.pattern(market.survey)
0000000000 0000000011 0000000100 0000100000 0001000000 0010000000
       759         16          4          3          2          1

0100000000 0100000011 0100010000 0100100000 0100110000 0101000000
       168          1          2          5          4         10

0101100000 0101110000 0110000000 0110100000 0111110000
         1          8          2          2         12
```

indicates that all but 241 observations were complete, and of these 168 had information missing on the second variable (income!) alone. The remaining 73 observations have a variety of patterns of missing values; of these, all but 26 have income among the missing fields.

9.4.2 Some Computational Issues

It should be clear that a fair amount of computation is required to select the best split at a given node. The algorithm underlying tree-based models is computationally intensive. Although it is possible to implement it entirely in the S language, we chose instead to write several of the underlying routines in C. Most have to do with the actual tree-growing process (`grow.c`, `splitvar.c`, and `vsplit.c`), others are for sequencing (`prune.c` and `shrink.c`), another for efficient prediction (`pred.c`), and, finally, others for character manipulation (`btoa.c`) and printing labels (`prlab.c`). Most users will not have to deal with this underlying code, but there are cases where it is unavoidable and even desirable to modify code at this level for some desired effect. Ultimately, such changes need to be compiled and loaded into S.

Our implementation is efficient in the sense that excessive computation is avoided by *updating*, whereby the assessment of split optimality (ΔD) is done incrementally after it has been done once for a particular split. Further computational improvement is possible for splits of factor predictors (where it is needed most!) provided that y is numeric or has at most two levels. If this is the case, then the average value of y in each level of the factor can be used to order the levels so that the best split is among the $k - 1$ contiguous splits after reordering. This fails for factor responses with more than two levels since it is unclear how a reordering is to be effected.

9.4.3 Extending the Computations

Tree-based models can be extended to response variables from the exponential family of distributions $f(y; \mu)$ described in Chapter 6. This results in the class of generalized tree-based models (GTMs), whereby the stochastic component of a response is assumed to be an exponential family member and the structural component is described by a tree structure. Thus, for exponential family distributions, there is a logical progression of models of the structural component afforded by linear predictors (GLMs, Chapter 6), additive predictors (GAMs, Chapter 7), and tree-based predictors (GTMs). In principle, the extension is quite straightforward as the only change to the existing software is in the form of the deviance function. Note in particular that specification of a link function is not necessary since the estimate of μ in each node is the within-node average for all exponential family distributions. However, link specification would be necessary in the event that an *offset* is used. More importantly, an offset induces iteration in the calculation of the within-node fitted value. For computational efficiency, one would determine splitting rules using an approximation to the deviance, say the *score function*, and only iterate to convergence once a candidate variable and splitting rule have been determined. This would increase the amount of computation by only a trivial amount relative to the current implementation for classification and regression.

Another possible generalization is the enlargement of the class of splitting rules allowed by our tree-growing algorithm. Specific possibilities include linear combination splits for selected sets of numeric predictors, as well as *boolean combinations* whereby splits on individual factor predictors are *AND*ed and *OR*ed to form a single split at a node. A convenient user interface is obtained by allowing a `matrix` data type in the formula expression supplied to `tree()`, such that columns of the matrix represent the individual variables to be combined: a matrix of numeric variables for linear combination splits, and a logical matrix for boolean combination splits. Thus, splits for these variable types are defined implicitly just as they are for numeric predictors and factors. The computational complexity of such splitting is unwieldy, and only suboptimal selections using *heuristics* are likely to be feasible.

Another interesting possibility is to consider hierarchical or conditional variables that are typical of surveys. For example, depending on whether or not a person is head of household, certain sections of a survey are not completed by the respondent. For others, the values for the entries in these sections are missing, not at random, but because of the structure of the instrument. Tree-based models are particularly adept at capturing these types of data since by decomposing the sample into homogeneous subgroups, the responses to the conditional part of these questions are appropriate once the primary variable has been used in a split. It would seem that a useful way to implement such variables is through an activation bit, which is on for all primary variables, but gets turned on for the secondary ones only when their primary variable is used in a split.

Bibliographic Notes

The introduction of tree-based models in statistics, particularly statistics for the social sciences, is due to Sonquist and Morgan (1964). An implementation of their ideas was realized in the computer program AID (Automatic Interaction Detection), which served to stimulate much subsequent research, such as THAID (Morgan and Messenger, 1973) and CHAID (Kass, 1980). These methods differed primarily in the stopping rules used to halt tree growth.

The inclusion of a chapter on tree-based modeling in this book is due to the influence of the work on classification and regression trees by Breiman et al. (1984). Besides masterfully presenting the material to the mainstream statistical audience, they are responsible for several important pioneering ideas that have redefined the state-of-the-art of tree-based methods. The primary innovation was not to limit node expansion in the tree-growing process. They recommended growing an overly large tree and spending one's effort deciding which branches to prune off. Their method of determining a pruning sequence, based on the concept of *minimal cost complexity*, forms the basis for the function `prune.tree()`. Subsequent work by Chou et al. (1989) generalizes this concept to other tree functionals besides tree size. Their other important innovation was the introduction of surrogate splits to provide a mechanism to grow trees and make predictions in the presence of NAs and also to provide a measure of variable importance.

Our methodology parallels that of Ciampi et al. (1987) in the use of the likelihood function as the basis for choosing partitions. This is a departure from that of Breiman et al. who use a variety of measures for tree growing and subsequent pruning. The precise definition of the shrinkage scheme discussed in Section 9.2 is also based on the likelihood (deviance) function. Recursive shrinking of tree-based models is a relatively new application of shrinkage estimators due to Hastie and Pregibon (1990). It has not been used as extensively as cost-complexity pruning nor have extensive comparisons been performed with it.

The computational shortcut for enumerating subset splits for factors and numeric responses dates back to Fisher (1958). This shortcut extends to binary responses but not to factor responses with more than two levels. Chou (1988) suggests a heuristic that restricts search to a (possibly) nonoptimal set of partitions. The split produced by the heuristic gets closer to the optimal split as the number of the levels of the factor increase—exactly the case where exhaustive search is infeasible. The current implementation of `tree()` does not incorporate this heuristic.

Chapter 10

Nonlinear Models

Douglas M. Bates
John M. Chambers

This chapter discusses the analysis of data using nonlinear models such as nonlinear regression, general likelihood models, or Bayesian estimation.

Throughout this book, statistical models have been defined by a three-part paradigm:

- a *formula* that specified the structural form of the models;

- *data* that corresponded to the variables in the formula;

- further specifications, such as probabilistic assumptions, that completed the definition of the model sufficiently to allow fitting.

We first introduced the paradigm in Chapter 2, in a specialized form. For linear models, the formula could use a shorthand that omitted explicit mention of the parameters to be estimated and used special interpretations of some S operators to allow compact specification of commonly occurring models. The formula represented an additive prediction from one or more terms. Further, the expressions for the terms, when evaluated using the data supplied, always produced vectors or matrices with elements or rows corresponding to the same set of observations. Use of data frame objects went along with this specialization of models. Subsequent chapters dealt with a variety of models more general than ordinary linear models, but which could still use the specialized version of the paradigm, along with some additional specifications.

We now must use a more general interpretation of formulas, to deal with the more general models considered in this chapter. These model formulas contain the parameters of the model explicitly, no restrictions are put on the data, and the criteria for fitting the models are essentially unlimited. However, model formulas are often quite similar to the special cases, and data frames can still be used, often in an extended form.

This generality does mean that the user needs to supply more information. In many examples, the computations will also be more difficult, and successful numerical solution to the estimation problem will not be guaranteed. So nonlinear models come with a cautionary warning that getting answers out may not be as easy as before. What one buys with the extra difficulty is a completely unrestricted range of models. While models of specialized types might be more convenient or numerically easier, if they conflict either with the data or with the subject-matter understanding of the problem, you should try to fit a model you believe to be more appropriate. Here are the techniques that may make that possible.

The estimation techniques for nonlinear models differ from those in many other chapters in that the techniques to determine parameter estimates are explicitly iterative. The desired parameter estimates are required to optimize some objective function, such as the sum of squared residuals or the likelihood function. The advantage of using S for nonlinear model applications is that expressions and functions can be described easily in S. The basic paradigm remains: model formulas are S expressions, data are organized into data frames, and the functions of this chapter organize the information to set up and carry out the iterative fitting required.

The primary S functions described in this chapter provide an interface to nonlinear optimization routines and to nonlinear regression. We describe the use of these functions for some common types of nonlinear models, summaries of these models, and methods for studying the variability in the estimates. Some special cases, such as partially linear models, will be discussed. As in earlier chapters, the S functions and underlying software can be used for more advanced or specialized applications. The statistical summaries and diagnostics in this chapter are more rudimentary than in most earlier chapters, partly because the range of nonlinear models is so large that little statistical theory can be assumed. Applications with a more limited range of models may be able to design specialized summaries based on more specific assumptions.

10.1   Statistical Methods

The statistical models to be considered use various general fitting criteria. In practice, two kinds of criteria occur most frequently: minimizing sums of contributions from observations, and the specialization of this to the case of nonlinear regression by least squares. There are plenty of other criteria, and the numerical techniques of the chapter can be adapted to them, but these two organize the statistical information from the model in a form that facilitates summaries and diagnostics. They also retain many of the concepts developed in earlier chapters, in extended or approximate form.

Typical minimum-sum fitting criteria arise from probability models, in which parameters are estimated by maximizing the likelihood or by some other computationally similar criterion. A model in which n independent observations are distributed with probability densities $p_i(\theta)$ for some vector of parameters θ leads to maximum-likelihood estimation framed in terms of minimizing the negative log-likelihood:

$$\ell(\theta) = \sum_{i=1}^{n}(-\log(p_i(\theta)))$$

The individual probabilities generally depend, of course, on the data.

As an example, consider some data on the results of table tennis matches. The United States Table Tennis Association assigns each of its members a numerical rating, based on the member's performance in tournaments. Winning a match boosts the winner's rating and lowers the loser's rating by some number of points, depending on their current ratings. The intuitive notion is that players with a higher rating should tend to win over players with a lower rating, and the greater the difference in rating, the more likely the higher-rated player is to win. Colin Mallows fitted a probability model to the results of 3017 matches to study the relation between rating and chance of winning. The model assumes a logistic distribution in which $\log(p/(1-p))$ was proportional to the difference in rating between the winner and loser:

$$p_i = \frac{e^{D_i\alpha}}{1 + e^{D_i\alpha}} \tag{10.1}$$

where $D_i = W_i - L_i$, the difference between the ratings of the winner and loser of the ith match. This is about the simplest nontrivial model. It has one parameter, α, representing the effect of a unit difference in the ratings. The point of main interest was whether in fact this effect was the same for all levels of play, whatever the average rating of the two players. We can add this into the model with a second parameter. Letting $R_i = .5(W_i + L_i)$,

$$p_i = \frac{e^{D_i\alpha + R_i\beta}}{1 + e^{D_i\alpha + R_i\beta}}$$

To fit the model, we minimize the negative log-likelihood,

$$\sum(-log(p_i)) = \sum -D_i\alpha + log(1 + e^{D_i\alpha}) \tag{10.2}$$

in the case of one parameter. This model is, in fact, treatable as a generalized linear model, as in Chapter 6. However, it is a very simple model presented as is, and will help to illustrate a number of techniques.

As a second example, consider the data presented in Chapter 1 on visible skips in an industrial experiment on wave soldering. We have already analyzed these data by a variety of models, but in fact only in this chapter can we tackle directly models that fully reflect the physical intuition and the observed behavior of the data, as we hinted at the end of Chapter 1.

Physical theory and intuition suggest a model in which the process is in either a "perfect" or "imperfect" state. In the perfect state, no defects will occur. In the imperfect state, there may or may not be defects, manifesting themselves as skips in the soldering. Both the probability of being in the imperfect state and the distribution of skips in that state depend on the factors in the experiment. One form of the model can be described by postulating that some "stress" depending on the factor levels induces the process to be in the imperfect state and also increases the tendency to generate skips when in the imperfect state.

For the ith experimental run, the corresponding factor levels determine a stress, say $\mathcal{S}_i$. The stress is itself a parametric function of the levels of the factors chosen for inclusion in the model, exactly as in Chapter 5. The stress is a linear function,

$$\mathcal{S}_i = \sum_{j=1}^{p} x_{ij}\beta_j$$

where β is the vector of parameters resulting from some suitable coding of qualitative factors (and possibly their interactions). The probability of being in the imperfect state is monotonically related to the stress by a logistic distribution:

$$\frac{1}{1 + e^{-\tau \mathcal{S}_i}}$$

As the stress increases, this probability approaches 1. Given that the process is in the imperfect state, the probability of k_i skips is modeled by the Poisson distribution with mean, say λ_i:

$$e^{-\lambda_i}\frac{\lambda_i^{k_i}}{k_i!}$$

For $y_i = 0$, the probability that $y = y_i$ is the probability of the perfect state *plus* the probability of being in the imperfect state and having 0 skips. For $y_i > 0$, it is the probability of being in the imperfect state and having y_i skips:

$$\text{Prob}(y = y_i) = \begin{cases} \frac{e^{-\tau \mathcal{S}_i}}{1+e^{-\tau \mathcal{S}_i}} + \frac{e^{-\lambda_i}}{1+e^{-\tau \mathcal{S}_i}} & \text{if } y_i = 0 \\ \frac{1}{1+e^{-\tau \mathcal{S}_i}}e^{-\lambda_i}\frac{\lambda_i^{k_i}}{k_i!} & \text{if } y_i > 0 \end{cases}$$

The mean skips in the imperfect state is always positive and modeled in terms of the stress by

$$\lambda_i = e^{\mathcal{S}_i}$$

Since the stress is an arbitrarily-scaled linear function, we only need one scale parameter, τ, which we can apply to either the logistic or the Poisson part of the model. We can now proceed to estimate β and τ by maximizing the likelihood, or equivalently by minimizing the negative log-likelihood. As in the previous example, we can write a formula for this in terms of the data and the parameters, from the probability specified above. The ith element of the negative log-likelihood can be written

$$\ell_i(\beta, \tau) = log(1 + e^{-\tau S_i}) - \begin{cases} log(e^{-\tau S_i} + e^{-e^{S_i}}) & \text{if } y_i = 0 \\ e^{S_i} - y_i S_i & \text{if } y_i > 0 \end{cases} \tag{10.3}$$

omitting expressions that do not involve the parameters of the model. This model does *not* reduce to any of the techniques of earlier chapters. Fitting it with sizable quantities of data is a challenging task. We will study it in the following sections; for a full discussion, see Lambert (1991). see p. 449

As a third example, consider a nonlinear regression model. Data from a biochemical experiment where the initial velocity of a reaction was measured for different concentrations of the substrate are given in the data frame Puromycin. The data came from two runs, one on cells treated with the drug Puromycin and one on cells without the drug. The three variables in the data frame are the concentration of the substrate, the initial velocity of the reaction, and an indicator of treated or untreated. The experimenters expected a Michaelis–Menten relationship between the reaction velocity and the concentration, modeled by

$$V = \frac{V_{\max} c}{K + c} + \varepsilon \tag{10.4}$$

where V is the velocity, c is the enzyme concentration, $V_{\max}$ is a parameter representing the asymptotic velocity as $c \to \infty$, K is the Michaelis parameter and ε is experimental error. Furthermore, they expected that the treatment with the drug would change $V_{\max}$ but would not change K appreciably.

By plotting velocity against concentration separately for the two levels of treatment, we can see the general pattern directly (Figure 10.1). The plot can be made as follows:

```
> attach(Puromycin)
> plot(conc, vel, type="n")
> text(conc, vel, ifelse(state == "treated", "T", "U"))
```

There does indeed seem to be a change in the asymptotic velocity for the two different runs. It is a little more difficult to tell about the Michaelis parameter, K, since it determines the shape of the curve. This parameter is the concentration at which the velocity becomes half the asymptotic velocity.

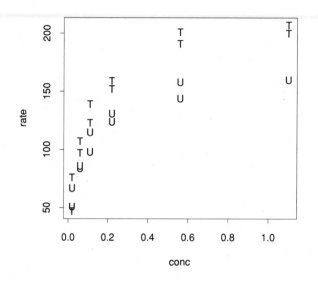

Figure 10.1: *Initial velocity of the enzymatic reaction versus the concentration of the sub-strate. Treated and untreated runs are plotted as* "T" *and* "U".

By analogy with the linear model (Chapter 4), the parameter estimates $\hat{V}_{\max}$ and $\hat{K}$ can be chosen to minimize the sum-of-squares of the residuals:

$$S(V_{\max}, K) = \sum \left(V_i - \frac{V_{\max} c_i}{K + c_i} \right)^2$$

The parameter K enters the expression for the fit nonlinearly. The sum-of-squares criterion can be derived as a special case of a sum of contributions to the negative log-likelihood if we assume ε in (10.4) is Gaussian with constant variance. Since the plot indicates that the variability for replicate observations is reasonably constant across the range of the data, the use of nonlinear least squares appears warranted. As with a linear model, nonlinear least squares estimates are often useful even when the error term is not assumed to be Gaussian with constant variance.

Unlike the linear model, nonlinear regression needs starting estimates for the parameters. These can be obtained from the plot which suggests that $V_{\max}$ is near 200 for the treated cells and near 160 for the untreated cells. Since the value of K is the concentration at which V reaches $V_{\max}/2$, we expect this to be near 0.1 for both runs.

Examples such as these lead us to formulate models to be estimated either by minimizing a function (typically the negative log-likelihood) or by formulating a nonlinear least-squares criterion. As with linear models, maximum likelihood and nonlinear regression produce estimates for the parameter. In addition, the computations lead to a description of the objective function or the model surface in the neighborhood of the parameter estimates. Such descriptions can sometimes be used to give approximations to quantities such as the standard errors or correlations of the parameter estimates, as in earlier chapters. The theory supporting these approximations is weaker than for ordinary linear models, however. The approximation used in nonlinear regression is to replace the nonlinear model by its linear Taylor series approximation at the parameter estimates and to use methods for linear statistical models on the approximation. These results are called the *linear approximation results*. For likelihood models, the distributional results are asymptotic; namely, maximum-likelihood estimates tend, for large samples, toward a normal distribution with mean equal to the true parameter and variance matrix given by the inverse of the *information matrix*, the negative of the matrix of second derivatives of the log-likelihood, given suitable regularity assumptions about the model. It is not possible to make any precise distributional statement in general about finite-sample distributions in either case. Statistical assumptions underlying any model need always to be questioned and tested; for nonlinear models, extra caution is called for.

Nonlinear regression models are obviously special cases of the general minimum-sum fitting criteria, so one might think of them as redundant. The specialization is worthwhile, however, both because the numerical fitting in this case is often more efficient and because the direct use of the linearized approximation helps in summarizing the model.

10.2 S Functions

Now we proceed to describe and illustrate the software for fitting and summarizing nonlinear models. Section 10.2.1 presents the functions that fit the models; this section needs some careful reading, even if you are familiar with earlier chapters, because of the more general form of nonlinear models. Section 10.2.2 covers the summary and diagnostic functions. Section 10.2.3 discusses a topic specific to this chapter: the computation of derivatives for nonlinear models. Section 10.2.4 extends the summary techniques to *profiling*, refitting the model with some parameters held fixed, to show the variability of the parameter estimates more directly. Finally, Section 10.2.5 covers an important special case of nonlinear regression, in which some of the parameters enter linearly.

10.2.1   Fitting the Models

This chapter has two fitting functions, one for general minimization models and one for nonlinear regression. In typical use, they both have three arguments, specifying the model, the data, and starting estimates for the parameters.

```
ms(formula, data, start)
nls(formula, data, start)
```

As usual, these functions return objects describing the fitted model. The `formula` argument gives the structural form of the model. If provided, `data` is a frame containing the data referenced in the model. The optional argument `start` specifies starting values for the parameters to be estimated. The object returned has class `"ms"` or `"nls"`. Each of the three arguments involves some new ideas, so let's consider them in turn.

Formula

The `formula` in nonlinear models is an expression in S, involving data, parameters in the model, and any other relevant quantities. As in earlier chapters, the operator $\sim$ marks off the prediction and, for nonlinear regression, the response. For instance, the nonlinear regression model (10.4) on page 425 can be written:

```
vel ~ Vm*conc/(K + conc)
```

As usual, we read this as: "Model `vel` as ...". Also as usual, the left and right sides of the $\sim$ represent response and predictor. The key difference, however, is that the expression on the right includes all the parameters as well as the data. The formula contains both variables like `vel` and `conc`, and parameters like `Vm` and `K`. This must be so since we no longer assume a linear or additive model that would define the coefficients implicitly. When operators like + or / appear in nonlinear formulas, they mean just what they mean in ordinary S expressions; they do *not* imply the special shorthand used in formulas in earlier chapters.

Minimization models have no explicit response. Instead, the formula is written with the $\sim$ symbol at the left. In the table tennis example, equation (10.2) on page 423 corresponds to a model formula

```
~ - D * alpha + log(  1 + exp( D * alpha) )
```

where `D` is a variable in the data and `alpha` is the parameter to fit. This can be read the same way as the previous formula: "Model as `-D * alpha` ...".

What specifically do these formulas compute? The nonlinear least-squares formula defines the *response* as the left operand of $\sim$ and the *prediction* as the right operand. These must evaluate to numeric objects of the same length. The `nls()` function tries to estimate parameters to minimize the sum of squared differences between response and prediction. The function `nls()` can also handle formulas where only the right side of the ∼ operator is supplied. In this case the formula is interpreted as the residual vector. See Section 10.3.3 for an example.

The minimization formula computes some numeric vector, and `ms()` estimates parameters to minimize the sum of this vector. The concept here is linked to maximum-likelihood models, with the formula defined to compute the vector of elements of the negative log-likelihood, as the examples will illustrate. The computational form, however, does not depend on this concept. The elements can be anything and there need not be more than one of them, so that any optimization problem can be presented to `ms()`. The advantage of having individual elements is only that they may be convenient for purposes of summaries and diagnostics, if they help to point out the contribution of individual observations.

The evaluated model formulas can include *derivatives* with respect to the parameters. The derivatives are supplied as attributes to the vector that results when the right operand of ∼ is evaluated. The derivative values are used by the fitting algorithms. When explicit derivatives are not supplied, the algorithms will use numeric approximations. These approximations usually increase the amount of computation needed and sometimes may introduce numeric problems. Accurate derivatives are sometimes crucial to success in numerical estimation for nonlinear models. However, expressions for derivatives are often difficult to get right. If you can try out some initial examples without computing derivatives, you will get a feeling for nonlinear models more easily. Section 10.2.3 explains how to supply derivatives and provides some tools to help construct the necessary expressions.

Data; Parametrized Data Frames

In most nonlinear modeling, the relevant data include not only variables similar to those encountered earlier, but also other quantities such as initial estimates for parameters or fixed values occurring in the model formula. These should usually go along with the data but they are not columns of the data frame. For this purpose *parametrized* data frames are convenient. These were introduced in Section 3.3.4. The function `parameters()` can be used to extract or to set the parameter attributes of a data frame to any list of named values. Setting parameters automatically promotes the data frame to be a `pframe` object. Attaching such an object automatically makes the parameters, as well as the variables, available for computation by name.

Parameters of a data frame can appear anywhere in a model formula. Unlike variables, however, parameters can have any length or mode, and no checking or coercion is done. For example, the table tennis model needed to estimate `alpha` and `p`. If some computations have produced a0 and p0 as initial values that we would like to carry along with the `pingpong` data frame, they will be set as follows:

```
> parameters(pingpong) <- list(alpha = a0, p = p0)
```

This assignment turns `pingpong` into a `pframe` object. Another, perhaps more typical, approach to introducing starting estimates as parameters is shown on page 431. Throughout this chapter, when we refer to a data frame we expect that in fact this will be a parametrized data frame.

Starting Values; Identifying Parameters

The fitting functions `nls()` and `ms()` have to know which names in the formula correspond to parameters to be estimated. Starting values must be supplied for these parameters. The functions apply the following two rules:

1. If the argument `start` is supplied, its names are the names of the parameters to be estimated, and the values are the corresponding starting estimates. The object can be either numeric or a list; in the latter case, more than one value can be associated with each name.

2. If `start` is missing, the parameters attribute of the `data` argument defines the parameter names and values.

We recommend using an explicit `start` argument to name and initialize parameters, for most applications. You can easily see what starting values were supplied and, as we will show, you can arrange to keep particular parameters constant when that makes sense. Keeping the starting values with the data frame is sometimes convenient and acceptable as an alternative.

Examples

A nonlinear regression corresponding to (10.4) can be fitted to the treated data in the `Puromycin` frame as follows:

```
> Treated <- Puromycin[Puromycin$state == "treated", ]
> Purfit1 <- nls(vel ~ Vm*conc/(K + conc), Treated,
+   list(Vm = 200, K = 0.1))
```

`Treated` is a new data frame with only the treated observations from `Puromycin`, and the `start` argument is a list with two elements for the two parameters to be estimated.

Let's look at a second example, and illustrate typical calculations to come up with starting estimates. To fit the model for the table tennis data, we need an initial estimate for `alpha`. A very crude estimate would come from replacing all the differences in ratings by $\pm\bar{d}$, where $\bar{d}$ is the mean difference, say. Then for each match, the probability from the model that the winner had a higher rating always satisfies

$$\bar{d} * \alpha = \log(p/(1-p))$$

We can solve this for an initial estimate of α if we replace p by the observed frequency with which the player with the higher rating wins.

The difference in the ratings, say D, will be required every time we evaluate the likelihood, so it pays to precompute it and save it in the data frame. Initially, the frame contains only the ratings of the winner and loser, along with a category identifying the matches. The following calculations turn it into a parametrized data frame with an additional variable and two parameters.

```
> param(pingpong, "p") <- 0
> attach(pingpong, 1)
> D <- winner - loser
> p <- sum(winner>loser)/length(winner)
> p
[1] 0.8223401
> alpha <- log(p/(1-p))/mean(D)
> alpha
[1] 0.007660995
> detach(1, save = "pingpong")
```

The first assignment just converts the data frame into a pframe, if it wasn't before. Now we attach it as the working data and do some calculations to create D, alpha, and p to provide the new variable and the initial values for the fitting. Detaching and saving will convert the latter two to parameters, since they are of the wrong length to be variables. Saving back on top of the original data frame is rather bold, and we don't recommend it as a general practice. We can now proceed to fitting with ms(), omitting the start argument, since we arranged for the parameters to be in the data frame.

Where a nonlinear model is at all complicated, you should organize it as a simple expression involving one or more S functions that do all the hard work. Even in this simple example, a little preliminary work will be worthwhile. Notice that the expression D * alpha appears twice in the formula on page 428. We can write a general function for the log-likelihood in any similar model in terms of this quantity.

```
                        argument
lprob <- function(lp)log(1 + exp(lp)) - lp
```
                    expression

If you have read the chapter on generalized linear models, you may recognize D * alpha as the linear predictor. If we added more terms and more parameters to our model, the argument to lprob() would be expanded accordingly, but lprob() would not change.

Isolating the nontrivial computations in a separate function is slightly more efficient in most cases, but more importantly it lets us concentrate on those computations and see where some care needs to be taken. Even in this model, some care is indeed needed, as we will note later. For the moment, however, we can plunge in:

```
> fit.alpha <- ms( ~ lprob(D * alpha), pingpong)
```

```
> fit.alpha
value: 1127.635
parameters:
      alpha
 0.0111425
formula: ~ lprob(D * alpha)
3017 observations
call: ms(formula = ~ lprob(D * alpha), data = pingpong)
```

We will come back to look in more detail at this fit in Section 10.2.3.

10.2.2   Summaries

As with fitted model objects in previous chapters, there is a special printing method
for ms and nls objects that prints out the information in the object suitable for
looking at directly. The output from printing the fit was illustrated above. There
are also summary() methods for both classes of objects.

Fitting the treated data from the Puromycin data frame produced the model
Purfit1. Suppose Purfit2 is the result of fitting the same model to the untreated
data from the same source. The two summaries for these models are:

```
> summary(Purfit1)
Formula: vel ~ (Vm * conc)/(K + conc)
Parameters:
          Value Std. Error  t value
Vm 212.6830000 6.94709000 30.61460
 K   0.0641194 0.00828075  7.74319
Residual standard error: 10.9337 on 10 degrees of freedom
Correlation of Parameter Estimates:
     Vm
K 0.765
> summary(Purfit2)
Formula: vel ~ (Vm * conc)/(K + conc)
Parameters:
          Value Std. Error  t value
Vm 160.2770000 6.48000000 24.73400
 K   0.0477027 0.00778116  6.13054
Residual standard error: 9.773 on 9 degrees of freedom
Correlation of Parameter Estimates:
     Vm
K 0.777
> (0.0641194 - 0.0477027)/sqrt(0.00828075^2 + 0.00778116^2)
[1] 1.444753
```

The last calculation shows that the difference in the fitted values of K is about 1.5
times its standard error, so the experimenters' feeling that treatment with the drug

should not change K may be warranted. We should check this by actually fitting data from both runs using a common K, as follows:

```
> Purboth <- nls(vel ~ (Vm + delV*(state=="treated"))*conc/
+ (K + conc), Puromycin, list(Vm=160, delV=40, K=0.05))
> summary(Purboth)
Formula: vel ~ ((Vm + delV*(state=="treated"))*conc)/(K + conc)
Parameters:
           Value Std. Error   t value
  Vm 166.6030000 5.80737000 28.68820
delV  42.0254000 6.27209000  6.70038
   K   0.0579696 0.00590999  9.80875
Residual standard error: 10.5851 on 20 degrees of freedom
Correlation of Parameter Estimates:
           Vm      delV
delV -0.5410
   K  0.6110    0.0644
> combinedSS <- sum(Purfit1$res^2) + sum(Purfit2$res^2)
> Fval <- (sum(Purboth$res^2) - combinedSS)/(combinedSS/19)
> 1-pf(Fval, 1, 19)
[1] 0.2055524
```

A detailed explanation of the statistical evaluation of whether the Ks could be equal is beyond the scope of this book. Briefly, we can say that the last three calculations are to determine the p value for an F-test of identical Ks versus different Ks. Since the p value is 20%, the identical Ks appear reasonable.

Further study of nonlinear models often involves calculations specific to the particular model. General statistical techniques such as Monte-Carlo sampling, resampling, and cross-validation are particularly valuable for nonlinear models. The technique of *profiling*—refitting holding all but one of the parameters constant—is another important mechanism. Its application to nonlinear regression is described in Section 10.2.4. These techniques differ from the asymptotic summaries in the important sense that they can provide some direct, although approximate, information about the behavior of the model in finite samples. The price is that these techniques nearly always require much more computation.

10.2.3 Derivatives

Numerical methods for fitting nonlinear models typically can make use of the derivatives of the objective function (in optimization) or of the predictor (in nonlinear least-squares) with respect to the parameters to be estimated. While the algorithms can proceed by using numerical estimates of these derivatives instead, these numerical estimates typically require more computation. Particularly in the case of models with many parameters, figuring out the derivatives analytically often speeds

up the computation. Even if efficiency is not a concern, numerical accuracy may
still suffer when the derivatives are estimated numerically. Some examples fail to
converge numerically if the derivatives are not computed analytically. So providing
analytical derivatives may be necessary: it is also a frequent source of aggravation
and human error. Fortunately, some computing tools are available that make the
work somewhat easier and the errors somewhat less likely.

Let's go back to the negative log-likelihood for the table tennis example, in its
simplest form:

$$\sum log(1 + e^{D_i \alpha}) - D_i \alpha$$

The corresponding S formula was

```
~ log(  1 + exp( D * alpha) ) - D * alpha
```

Differentiating with respect to α and simplifying a little gives

$$\sum -D_i/(1 + e^{D_i \alpha})$$

and the corresponding S expression would be:

```
-D / (  1 + exp( D * alpha) )
```

This model has only one parameter; usually, there would be derivatives for each of
the parameters in the model. Keep in mind that evaluating the formula in a data
frame with n observations produces the vector of n values, whose sum is the negative
log-likelihood. Similarly, evaluating the derivative expression gives n values for each
parameter. The gradient expression for a model with p parameters should evaluate
to an n by p matrix. This is the same shape as a matrix corresponding to p numeric
x variables in a linear model. In fact, the gradient matrix in a nonlinear model
plays the role of a "local" linear model in many respects, as we will see.

The gradient is supplied to the nonlinear fitting functions as an attribute. The
attribute can be attached directly to the formula; for example,

```
> fg.alpha <- ~  log(  1 + exp( D * alpha) ) - D * alpha
> attr(fg.alpha, "gradient")  <- ~ -D / (  1 + exp( D * alpha) )
> fg.alpha
  ~  log(1 + exp(D * alpha)) - D * alpha
Gradient:  ~   -D/(1 + exp(D * alpha))
```

The object fg.alpha has class "formula" and can be supplied to ms() as the formula
in fitting our table tennis model. The presence of the gradient attribute tells ms()
to use derivatives.

Most models are too complicated to write out the expression for the values and
the gradient explicitly. In this case, as we illustrated before, one writes a function
that captures the computations for the model, often in a more general form so the

same computations can be used in other, similar models. Gradients follow along in this case in just the same way: the function should return a value for the model as before, but attach to it an attribute containing the gradient matrix. Consider our function lprob() on page 431. We can arrange for it to compute the derivative with respect to alpha as well. By observing the expression for the gradient, we note that it can be computed from the quantities used in the previous version, plus the same quantity that appeared as the multiplier of alpha. This leads to the following function:

```
lprob2 <- function(lp, X){
    elp <- exp(lp)
    z <- 1 + elp
    value <- log(z)-lp
    attr(value, "gradient") <- -X/z
    value
}
```

Here lp is again the linear predictor and X is, in general, the data in that linear predictor. In our one-parameter example, it reduces to D. Notice that z was used for both the value and the gradient. Such gains in efficiency are common, and are one reason to prefer computing derivatives where possible. With the gradient computations carried out inside lprob2(), we will not be giving a separate gradient expression in the explicit formula. Instead, the model-fitting functions will look at the *evaluated* model for a gradient attribute and behave appropriately. Given p parameters and n observations, the gradient needs to be an n by p matrix—not a problem in this case, since the corresponding X will be a matrix of that dimension. Generally, gradient computations need to take some care to get dimensionality correct.

The fitting algorithm for minimization can use second derivatives of the model as well. The procedure is entirely analogous: in this case, an attribute "hessian" is provided as well as the "gradient" argument. The hessian expression should produce an n by p by p array of computed second derivatives. Here's how it works:

```
lprob3 <- function(lp, X){
    elp <- exp(lp)
    z <- 1 + elp
    value <- log(z) - lp
    attr(value, "gradient") <- -X/z
    if(length(dx <- dim(X))==2) {
        n <- dx[1]; p <-dx[2]
    }
    else { n <- length(X); p <- 1 }
    xx <- array(x, c(n, p, p))
    attr(value, "hessian") <- xx * aperm(xx, c(1, 3, 2)) * elp/z^2
    value
```

```
    }
```

Need we add that all the cautions about checking the computations apply to computing second derivatives as well, only more so?

The mathematics required to compute the gradient in most models encountered in practice will be a good deal harder than it was in these examples. Computing tools should be used to assist in this chore, even though they cannot take over completely. Two kinds of assistance are available: symbolic differentiation of S expressions and routines to approximate derivatives numerically. Symbolic differentiation does not work on all expressions and the results nearly always need to be examined to see potentially faster computations, but it can be a useful starting point that saves human error. Comparing numerical approximations to the evaluated versions of the exact gradient expression will catch most remaining errors.

A symbolic differentiation function, D(), was defined in an example in [S] (page 298). It returns an expression representing the derivative of its first argument with respect to the name or names specified in its second argument. Let's see what it can do with our formula:

```
> D( substitute(log(1+exp(D*alpha)) - D*alpha), "alpha")
(exp(D * alpha) * D)/(1 + exp(D * alpha)) - D
```

Not too bad, but as is typical, this is the expression *without* "simplifying a little".

For nonlinear models, we provide a somewhat more sophisticated version of symbolic differentiation via the function deriv(). This does some of the elimination of common expressions in the formula and its gradient. It also produces an expression in the form expected for nonlinear models.

```
> formula1 <- ~ log(1+exp(D*alpha))-D*alpha
> deriv(formula1, "alpha")
expression({
    .expr1 <- D * alpha
    .expr2 <- exp(.expr1)
    .expr3 <- 1 + .expr2
    .value <- (log(.expr3)) - .expr1
    .grad <- array(0, c(length(.value), 1), list(NULL, "alpha"))
    .grad[, "alpha"] <- ((.expr2 * D)/.expr3) - D
    attr(.value, "gradient") <- .grad
    .value
}
)
```

The value of deriv() is an S expression object. Evaluating this expression will produce both the formula value and its gradient, with most of the common subexpressions evaluated only once. Alternatively, deriv() will create a function object that, when called, produces the appropriate values and gradients. To do this, give deriv() a third argument that identifies the names of the arguments to the function you want to create. In our example, let's produce a function lprobg(), with arguments D and alpha, to evaluate formula1 and its derivative:

```
> lprobg <- deriv( formula1, "alpha", c("D", "alpha"))
> lprobg
function(D, alpha)
{
    .expr1 <- D * alpha
    .expr2 <- exp(.expr1)
    .expr3 <- 1 + .expr2
    .value <- (log(.expr3)) - .expr1
    .grad <- array(0, c(length(.value), 1), list(NULL, "alpha"))
    .grad[, "alpha"] <- ((.expr2 * D)/.expr3) - D
    attr(.value, "gradient") <- .grad
    .value
}
```

The third argument could also have been a function definition. This acts as a dummy version of the function to be returned, and is useful if we want to give default values to the arguments. For example,

```
deriv(formula1, "alpha", function(D=1, alpha=0)NULL)
```

will produce a function with the default values as shown. The body of the function in the argument is irrelevant. Comparing lprobg() to lprob2() above shows that, in this example, the mechanical derivatives do nearly as well as the hand-coded ones (and without the errors humans tend to produce). The expressions generated by deriv() use names beginning with "." to avoid conflicting with names chosen by the user. For more complicated problems, it is sometimes possible to do a better job of coding the derivatives "by hand" but the expressions from deriv() provide a good starting point for determining the derivatives yourself.

The second argument to deriv() can be a vector of parameter names:

```
> deriv(vel ~ Vm * (conc/(K + conc)), c("Vm","K"))
expression({
    .expr1 <- K + conc
    .expr2 <- conc/.expr1
    .value <- Vm * .expr2
    .grad <- array(0, c(length(.value), 2), list(NULL, c("Vm", "K")))
    .grad[, "Vm"] <- .expr2
    .grad[, "K"] <- - (Vm * (conc/(.expr1^2)))
    attr(.value, "gradient") <- .grad
    .value
})
```

The symbolic differentiation interprets each parameter name as a scalar. Functions such as `lprob2()` on page 435 cannot be produced directly from `deriv()`; the generalization from a scalar to a vector parameter must be done by hand.

Use of parentheses can help `deriv()` to isolate relevant subexpressions. This is desirable if you know that the subexpression will appear as part of the gradient. In our example, we put a redundant set of parentheses around `conc/(K + conc)`, forcing this to be a single expression, since this expression is the derivative with respect to `Vm`.

10.2.4   Profiling the Objective Function

The methods presented in Section 10.2.2 for assessing the uncertainty in the parameter estimates were based on a local quadratic approximation to the log-likelihood, or a local linear approximation to the nonlinear least-squares predictor. In both these cases, the approximation results in a local quadratic approximation to the *objective function*, which is either the negative log-likelihood or the residual sum-of-squares.

A more accurate picture of the uncertainty in the parameter estimates can be obtained by examining the objective function directly. When there are only two parameters, contours of the objective function can be plotted by generating a grid of values and using the `contour` function in S. When there are more than two parameters, direct examination of the objective function becomes much more difficult. Fixing all but two of the parameters at their estimated value and creating a grid of the objective function by varying the remaining two gives a "slice" through the higher-dimensional contour. However, this is not the appropriate computation. When assessing the uncertainty in the parameter estimates we usually want to see the *projections* of the higher-dimensional contour instead of such slices.

Although getting two-dimensional or three-dimensional projections of the objective function contours would often be too time consuming, in most cases it is feasible to look at one-dimensional projections or *profiles* of the objective function. These can be used to reconstruct some of the important features of the two- or three-dimensional projections; in particular, the extent of the contours can be determined.

To profile the objective function, we choose a parameter, say `delV` in the last example of Section 10.2.2, and fix it at a value different from the estimate, say 40 instead of 42.025. We then optimize the objective with respect to the remaining parameters:

```
> Pur.pro <- Puromycin
> parameters(Pur.pro) <- list(delV = 40)
> Pur.40 <- nls(vel ~ (Vm +delV*(state=="treated"))*conc/(K + conc),
+   Pur.pro, list(Vm = 160, K = 0.05))
> sum(Pur.40$res^2)
[1] 2252.551
```

The minimum residual sum-of-squares with `delV = 40` is greater than at the least-squares estimate, as it should be:

```
> sum(Purboth$res^2)
[1] 2240.891
```

By repeating the procedure of fixing `delV` at a value and minimizing the objective function with respect to the other parameters, we could build up the table

```
        delV profile.SS
[1,] 36.00000   2344.253
[2,] 38.00000   2286.983
[3,] 40.00000   2252.551
[4,] 42.02542   2240.891
[5,] 44.00000   2251.934
[6,] 46.00000   2285.585
```

Plotting these values shows that the profile sum-of-squares is very close to quadratic over this range.

Since it is easier to evaluate deviations from straight line behavior than to evaluate deviations from quadratic behavior, the profiled objective function is converted to the "signed square-root" scale. If we write the profiled objective as S and the parameter as θ, the converted profile is

$$\text{sign}(\theta - \hat{\theta})\sqrt{S - \hat{S}} \tag{10.5}$$

When the objective is quadratic in the parameters, this quantity will be linear in θ.

For the special case of the linear regression model, scaling (10.5) by $1/s$ produces the t statistic for the parameter θ. By analogy, when S is the residual sum-of-squares, we use

$$\tau = \left(\text{sign}(\theta - \hat{\theta})\sqrt{S - \hat{S}}\right)/s$$

as the nonlinear t statistic. If S is the negative log-likelihood, then

$$\zeta = \text{sign}(\theta - \hat{\theta})\sqrt{S - \hat{S}}$$

would be the analogue of the z statistic for θ. Even if neither of these statistical interpretations holds, (10.5) provides a way of visually assessing the validity of the quadratic approximation. In addition, it provides an empirical transformation of θ, say by fitting a spline to the (θ, ζ) pairs, under which the objective function is much closer to being quadratic. This can be useful for Monte-Carlo techniques such as importance sampling.

The function `profile` is used to automate the process of creating the profiles. The arguments are:

- the object returned by ms() or nls();

- the index of the parameter or parameters to be profiled;

- a vector of initial step sizes;

- a cut-off value on the scale of ζ for an ms object or τ for an nls object to indicate the magnitude at which the profiling should be terminated.

This function returns a list of data frames corresponding to the parameters being profiled. Each data frame contains a variable for the ζ (or τ) values and a matrix variable containing the parameter values. The values of the parameters being optimized are called the *profile traces*. These provide additional valuable information about the behavior of the objective function and can be used to reconstruct two-dimensional projections of contours of the objective (Bates and Watts, 1988, Appendix 6).

When the object from the nonlinear fit contains enough information to construct a quadratic approximation to the objective function, defaults are available for the step sizes and the cutoff. The indices always default to selecting all the parameters. The default profile for the Purboth fit is as follows:

```
> Pur.pro <- profile(Purboth)
> names(Pur.pro)
[1] "Vm"    "delV" "K"
> plot(Pur.pro)
```

The plot produced by the plotting method for profiles is shown in Figure 10.2. For each parameter being profiled, the method plots the τ or ζ value for each fit against the corresponding parameter value, using a smooth curve through the fitted points.

10.2.5   Partially Linear Models

We have said that all the parameters in a model must be given initial estimates, either through the start argument or within the data frame. This is not always the case in nonlinear least squares. There is an alternate form of the nls() function for *partially linear* regression models where some of the parameters appear linearly in the predictor. When values for the other parameters are chosen, the conditional least squares estimates for these linear parameters can be easily calculated using linear least squares. This is done automatically for the starting values and at every other stage in the iterations.

The Michaelis–Menten model is an example of a partially linear model since the parameter $V_{\max}$ behaves as a linear parameter. We can refit the model to the treated enzyme data as shown on page 430:

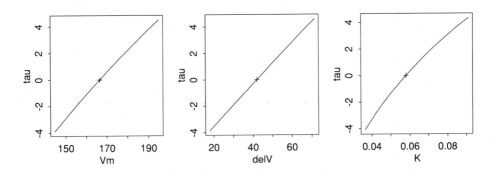

Figure 10.2: *The profile of the three-parameter fit* Purboth *shown on page 433. Each of the three panels shows the fit, on the square-root scale* (τ)*, when the parameter on the x-axis ranges over the values shown, and the other two parameters are optimized. Over this range, the model appears linear in* delV*, nearly so in* Vm*, and slightly less so in* K*. The cross (+) shows the global fit.*

```
> Untreated <- Puromycin[Puromycin$state == "untreated", ]
> Purfit.pl <- nls(vel ~ conc/(K + conc), Treated,
+    list(K = 0.1), alg = "plinear")
> Purfit.pl
Residual sum-of-squares : 1195.449
parameters:
          K      .lin1
 0.06411777 212.6815
formula: vel ~ conc/(K + conc)
12 observations
```

full model: vel ~ Vm*conc /(K+conc)

Although it appears that K is the only parameter being varied, the current optimal value of $V_{\max}$ is being used at each iteration and is available after convergence. Since this parameter is an implicit parameter, it is given the name .lin1. More details on the actual algorithm are given in Section 10.4.

The formula is interpreted slightly differently for partially linear models. The expression on the right of the $\sim$ must evaluate to a vector or a matrix. If it is a vector, it is implicitly converted to a matrix with as many rows as the length of the response on the left.

Another example may make this clearer. The data frame Lubricant contains the logarithm of the viscosity of a lubricant at various temperatures and pressures (Bates and Watts, 1988, Appendix A1.8), as shown in Figure 10.3:

```
> attach(Lubricant)
> unique(tempC)
```

```
[1]  0.00000 25.00000 37.77778 98.88889
> plot(pressure, viscos, type = "n")
> text(pressure, viscos, match(tempC, unique(tempC)))
```

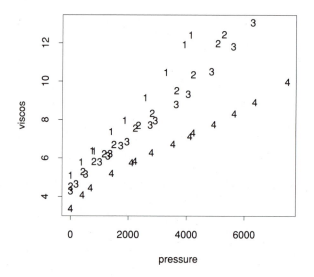

Figure 10.3: *Logarithm of the viscosity of a lubricant versus pressure for different temper-atures.*

The experimenters proposed a model of

$$
v = \frac{\theta_1}{\theta_2 + T} + \theta_3 p + \theta_4 p^2 + \theta_5 p^3 + (\theta_6 + \theta_7 p^2)p \exp\left(\frac{-T}{\theta_8 + \theta_9 p^2}\right)
$$

where v is the log viscosity, p is the pressure, and T is the temperature (Celsius). Six of the parameters $(\theta_1, \theta_3, \theta_4, \dots, \theta_7)$ behave linearly. Since the expression for the model is more complicated than we would want to type in a call to nls(), we define a function that returns a vector of the multipliers of these parameters:

```
Lub.mod <- function(temp, press, t2, t8, t9)
{
    efac <- press * exp(( - temp)/(t8 + t9 * press^2))
    c(1/(t2 + temp), press, press^2, press^3, efac, efac * press^2)
}
```

There are 53 observations of the viscosity. The value returned by this function has
length $318 = 53 \times 6$, giving a 53 by 6 model matrix for the 6 linear parameters.

To fit this model we only need initial estimates for the nonlinear parameters θ_2,
θ_8, and θ_9. Even guessing just these three could be difficult, however, and we have
to make some coarse approximations. Roughly we could say that θ_2 must be of the
same order of magnitude as the temperature to have any effect. Very small values
of θ_2 will be dominated by T in $\theta_2 + T$. Very large values of θ_2 will dominate T
and change the model to a different form. Using this reasoning, we could start θ_2
at 100. Similar reasoning would give a value around 100 for θ_8. Turning to θ_9, it
is not even clear if this parameter should be positive or negative. We may start it
out at 0 and see what happens. Adding these as parameters to the data frame, we
can try to fit the model:

```
> parameters(Lubricant) <- list(t2 = 100, t8 = 100, t9 = 0)
> Lubfit <- nls(viscos ~ Lub.mod(tempC,pressure,t2,t8,t9),
+    Lubricant, alg = "plinear")
Error in call to "nls": singular gradient matrix
Dumped
```

This failed to converge. If we repeat the call to nls with tracing enabled (see
Section 10.3.1), we find that even though the sum-of-squares is being reduced, the
parameter t8 is taking on unreasonable values. Eventually, the gradient matrix
becomes singular.

Our coarse starting estimates are inadequate and must be refined. We can use
nls() to help up do this by keeping some of the parameters fixed while refining the
values of the others. For example, we could see if 100 is actually a good estimate
for θ_2 in the lubricant data when $\theta_8 = 100$ and $\theta_9 = 0$. Since the parameters listed
in the start argument override the parameters in the data frame, we do not need
to modify the data frame; we simply list the parameters that we want to be varied
in the start argument. The others will retain their values from the data frame. For
example,

```
> Lubfit <- nls(viscos ~ Lub.mod(tempC,pressure,t2,t8,t9),
+    Lubricant, list(t2 = 100), alg = "plinear")
> Lubfit$parameters
        t2 ...
   218.979 ...
```

It appears that the value of θ_2 should be closer to 200 than to 100. Next we optimize
over both θ_2 and θ_8, then finally over all three nonlinear parameters:

```
> Lubfit <- nls(viscos ~ Lub.mod(tempC,pressure,t2,t8,t9),
+    Lubricant, list(t2=219, t8=100), alg = "plinear")
> Lubfit$parameters
        t2      t8 ...
```

```
  209.514 47.7296 ...
> Lubfit <- nls(viscos ~ Lub.mod(tempC,pressure,t2,t8,t9),
+    Lubricant, list(t2=210, t8=48, t9=0), alg = "plinear")
> Lubfit
Residual sum-of-squares : 0.08702405
parameters:
       t2        t8             t9     .lin1         .lin2
  206.5461 57.40411 -4.766996e-07 1054.542 0.001460311

        .lin3          .lin4          .lin5          .lin6
  -2.596517e-07 2.257323e-11 0.0004013854 3.528393e-11
formula: viscos ~ Lub.mod(tempC, pressure, t2, t8, t9)
53 observations
```

Now we have estimates for all nine parameters in the model.

The general approach of building from simpler models to more complex models, as in this example, is very useful in fitting nonlinear models. The ability to keep some parameters fixed while optimizing over others can be used to simplify the model temporarily.

10.3 Some Details

This section covers some additional details and special cases in nonlinear fitting. Section 10.3.1 summarizes some of the special settings that can be used to control the iterative fitting. Section 10.3.2 looks at an example of the detailed numerical examination sometimes needed in nonlinear models. Section 10.3.3 considers the use of *weighted* nonlinear least-squares fitting.

10.3.1 Controlling the Fitting

Both `ms()` and `nls()` use several values to control characteristics of their optimization algorithms. The argument `control` is used to specify a list of control values to these functions. Any control parameters not specified are computed at default values by the fitting function. So, for example, to set the maximum number of iterations to 10 but leave all other control values at their defaults, we use:

```
myfit <- nls(myformula, control = list(maxiter=10))
```

The same mechanism works for either `nls()` or `ms()`.

To see all the possible controls, do `?nls.control` or `?ms.control`. Here are the three most common controls; they are supplied the same way for both fitting functions:

   maxiter: the maximum number of steps in the iteration (default 50);

tolerance: the tolerance for convergence of the iteration (the default depends on the function);

trace: control of trace printing during the iteration (default FALSE).

Either a logical value or a function can be given for trace. The trace, if done, is also stored in the component trace of the output structure. When trace is false, no summary of the iterations is printed or saved. Advanced users can define their own function to generate the trace in the form they wish to see it—the function trace.ms() illustrates how a trace function is called.

By default, the maximum number of iterations is 50. This limit is not often reached since the algorithm convergences quickly or reaches a point where the iterations cannot proceed. On rare occasions, the iterations may still appear to be progressing toward an optimum slowly but steadily when the limit is reached. You can try to restart the optimization with this limit increased. You can also decrease this limit if you believe the estimates must be reached sooner, or if you just can't afford 50 iterations.

Convergence is declared when the convergence criterion becomes less than the tolerance level. The default value is based on the relative precision of computations and is about 0.001 on typical current machines. Smaller values will require more iterations while larger values will result in convergence being declared earlier. The form of the convergence criterion differs between the two fitting functions. The nls() uses a relative offset criterion (Bates and Watts, 1988, Section 2.2) that measures the numerical imprecision in the parameter estimates compared to the statistical variability. The ms() algorithm will exit when any of several measures of progress in the optimization drop below chosen values. The tolerance value is used to set each of these to comparable values. You can also set the tolerances individually—see the detailed documentation for ms.control(). It would be advisable to leave these tolerances at their default settings until you see some undesirable behavior.

One more control variable, minscale, is common to both ms() and nls(). As described in Section 10.4, both a step size and a step direction are determined at each iteration. Since the approximation used to determine this step may not be valid over the entire extent of the step, a minimum step size is incorporated. Having determined a direction in which to look for an optimum, the algorithms try to make some improvement by taking a step in that direction. If the initial step does *not* produce a reduction in the fitting criterion, the algorithm reduces the scale of the step and tries again.

If the scale of the step becomes very small without the objective function decreasing, the model is probably incorrectly defined or poorly behaved. Typically, either the derivatives are not correct, with the result that the computed direction is not really pointing downhill, or there are some discontinuities in the function or its derivatives in this region. The algorithms will stop if the minimum scale factor is reached, and will indicate an error condition. The default value of the minimum scale factor is 0.001.

10.3.2 Examining the Model

Every chapter of this book has emphasized the importance of *looking* at the data and at the fitted model. The advice applies with special emphasis for nonlinear models. In earlier chapters, we supplied the data portion of the model, but the explicit combination of data with parameters was handled by the fitting software. We could reasonably hope that the *evaluation* of the model would produce correct results.

Now, however, the scope of models is much greater. The formula we supply is explicitly the computation to be used to combine parameters and data. Some careful examination of that formula often prevents frustrating problems during the fitting. Both empirical examination, often using plots, and mathematical thinking may be needed. Some questions to ask are:

- Can the model be expressed in terms of some intermediate functions that will make its behavior clearer?

- What will happen to the computations as particular elements take on extreme values (e.g., tending to $\pm\infty$)?

- Should the computations be reexpressed to avoid numerical problems or to speed up the computations?

In thinking about these questions, the origin of the model often helps. For example, many general nonlinear models arise when we need to generalize a model that could be handled by simpler methods (e.g., a linear model). The quantities arising in the simpler model are often important in understanding the generalized one.

Our first example was so simple that we might not expect any problems to arise. No such luck: the naive expression for the model can easily cause problems. We can begin by noting again that the likelihood only depends on the parameters through a linear predictor, via a function that we define on page 431:

```
lprob <- function(lp)log(1 + exp(lp)) - lp
```

where `lp` was `D * alpha` in our example. The computations depend on the trivial-looking univariate function whose graph is shown in Figure 10.4. Some of its mathematical properties are:

$$
\begin{aligned}
f(x) &= \log(1 + e^x) \\
&= x + f(-x) \\
&\rightarrow x \text{ as } x \rightarrow \infty \\
&\rightarrow 0 \text{ as } x \rightarrow -\infty
\end{aligned}
$$

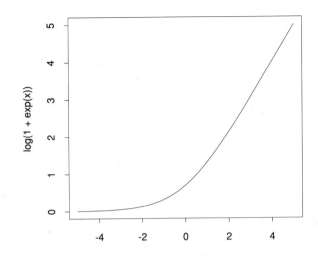

Figure 10.4: *Behavior of the function* $\log(1 + e^x)$, *which determines the log-likelihood in our table tennis example.*

This apparently mild-mannered function can easily cause problems if we compute it directly from its definition. With finite precision arithmetic, e^x will overflow while $f(x)$ is still moderate for positive x. The alternative form, $x + f(-x)$, allow us to restrict the exponential computations to the range $x < 0$. A second problem arises when x becomes large and negative: finite-precision computation of $1 + e^x$ will be exactly 1 when e^x is less than the relative precision of computations. The computed value for $f(x)$ will be zero before it should. Notice that the limit as $x \to -\infty$ is more precisely stated as

$$f(x)/e^x \to 1$$

so that $f(x)$ should behave like e^x for large negative x.

What is needed to avoid these problems is an explicit function to compute $\log(1 + e^x)$ accurately, by mapping all positive arguments onto negative ones, and by using an asymptotic computation in terms of e^x for large negative x. First, here is an S function to compute $\log(1 + \exp(x))$ for $x < 0$:

```
logexp.neg <- function(x) {
    y <- exp(x)
    small <- y < 4 * .Machine$double.eps
```

```
        y[small] <- y[small] * (1 - 0.5 * y[small])
        y[!small] <- log(1 + y[!small])
        y
}
```

Where `exp(x)` is at or near the relative precision of the machine, we used two terms from an expansion of $\log(1+y)$ for small y. Using `logexp.neg()`, we can now handle any arguments:

```
logexp <- function(x) {
    y <- x
    pos <- x > 0
    y[pos] <- x[pos] + logexp.neg( -x[pos])
    y[!pos] <- logexp.neg(x[!pos])
    y
}
```

We have not tried to be particularly efficient about this calculation; in practice, one might implement this as an algorithm in C or FORTRAN. The model on page 431 can now use the function

```
lprob <- function(x)logexp(x) - x
```

for safer computations.

When derivatives of models involving `logexp` are required, they typically enter in terms of the linear predictor; that is, the argument to `logexp(x)` comes from matrix multiplication of, say, `X` by a vector of parameters, `beta`. If `X` is supplied as an argument, the corresponding matrix of gradients can be returned. The derivative of $\log(1 + \exp(x))$ is

$$e^x/(1 + e^x) = 1/(1 + e^{-x})$$

with the left expression preferable for negative x and the right for positive x. This can be turned into an S function for the linear predictor:

```
glogexp <- function(x, X) {
    neg <- x < 0; pos <- !neg
    x  <- exp( ifelse(neg, x, -x) )
    x[neg] <- x[neg]/(1+x[neg])
    x[pos] <- 1/(1 + x[pos])
    x*X
}
```

Notice that we arranged that the argument to `exp()` would always be negative, to avoid possible overflow. Now we can provide gradients for our table tennis example:

```
lprob <- function(lp, X){
    value <- logexp(lp) - lp
```

```
      attr(value, "gradient") <- glogexp(lp,X) - X
      value
}
```

The analogous changes to computing second derivatives we leave as an exercise.

To see similar reasoning in a more realistically complicated situation, we return to the model (10.3) on page 425 for the soldering experiment. The contribution to the negative log-likelihood for observations with $y_i = 0$ included a term

$$\log(e^{-\tau \mathcal{S}_i} + e^{-e^{\mathcal{S}_i}})$$

This can be rewritten as

$$-e^{\mathcal{S}_i} + \log(1 + \exp(e^{\mathcal{S}_i} - \tau \mathcal{S}_i))$$

With this substitution, equation (10.3) can be rewritten in a computationally more convenient form:

$$\ell_i(\beta, \tau) = logexp(-\tau \mathcal{S}_i) + \begin{cases} e^{\mathcal{S}_i} - logexp(e^{\mathcal{S}_i} - \tau \mathcal{S}_i) & \text{if } y_i = 0 \\ y_i \mathcal{S}_i - e^{\mathcal{S}_i} & \text{if } y_i > 0 \end{cases}$$

using the `logexp` function defined above. The second use of this function particularly needs numerical care, since the argument to our favorite function now itself contains an exponentiation. We can rewrite the above as an S function:

```
zip <- function(y, X, beta, tau) {
    S <- X %*% beta
    tS <- tau * S
    eS <- exp(S)
    zero <- y < 0.5; pos <- !zero
    lkh <- logexp(-tS)
    lkh[pos] <- lkh[pos] + y[pos] * S[pos] -es[pos]
    lkh[zero] <- lkh[zero] + es[zero] - logexp(es[zero] - tS[zero])
    lkh
}
```

In terms of this function, the model formula given to `ms()` is

```
~ zip(defects, X, beta, tau)
```

assuming that the appropriate matrix of predictor variables is in X and that `defects` is the vector of defect counts.

Next, consider calculating derivatives as well. With a little bit of effort, you can verify that the derivatives of ℓ_i with respect to the elements of β satisfy

$$(\partial \ell_i / \partial \beta_j)/X_{ij} = -\tau g(-\tau \mathcal{S}_i) + \begin{cases} (e^{\mathcal{S}_i} - g(e^{\mathcal{S}_i} - \tau \mathcal{S}_i))(e^{\mathcal{S}_i} - \tau) & \text{if } y_i = 0 \\ y_i - e^{\mathcal{S}_i} & \text{if } y_i > 0 \end{cases}$$

where g stands for the derivative of the `logexp` function. The derivatives with respect to τ can be written:

$$\partial \ell_i / \partial \tau = -\mathcal{S}_i g(-\tau \mathcal{S}_i) + \begin{cases} \mathcal{S}_i g(e^{\mathcal{S}_i} - \tau \mathcal{S}_i) & \text{if } y_i = 0 \\ 0 & \text{if } y_i > 0 \end{cases}$$

Remember that, as always, the gradient computation should return a matrix with n rows and one column for each parameter. A function to compute both the likelihood and the gradient starts off just like `zip` above, and then implements the gradient calculations. Here is one implementation:

```
zip2 <- function(y, X, beta, tau) {
    S <- X %*% beta
    tS <- tau * S
    eS <- exp(S)
    zero <- y < 0.5; pos <- !zero
    lkh <- logexp(-tS)
    lkh[pos] <- lkh[pos] + y[pos] * S[pos] -es[pos]
    lkh[zero] <- lkh[zero] + es[zero] - logexp(es[zero] - tS[zero])
# now the gradients.  First the derivatives for beta
    g <-  X    #make g the right size
    g[pos,] <- y[pos] - es[pos]
    g0 <- glogexp(es[zero] - tS[zero])
    g[zero,] <- es[zero] - g0
    g <- (g + tau * glogexp(-tS)) * X
# now derivatives for tau
    gtau <- - S * glogexp(tS)
    gtau[zero] <- gtau[zero] + S[zero] * g0
# bind them together, as an attribute to lkh
    attr(lkh, "gradient") <- cbind(g,gtau)
    lkh
}
```

Similar work will produce expressions for the second derivatives.

10.3.3   Weighted Nonlinear Regression

As in linear regression, the sum-of-squares criterion in nonlinear regression can include weights. In linear regression, an optional argument supplies the weights. There is no need for a separate argument for nonlinear weighted regression because the form of the model is sufficiently general to include the weights. This generality is necessary if, as often occurs, the weights depend on the values of the parameters or the value of the predictor.

To use weights, the model is written without a response and with the "predictor" being the square root of the contribution to the weighted sum-of-squares. Suppose we wanted estimates for the Michaelis–Menten model applied to the treated enzyme data, with weights inversely proportional to the prediction. We could construct a function `weighted.MM` and use it as

```
> weighted.MM
function(resp, conc, Vm, K)
{
    pred <- (Vm * conc)/(K + conc)
    (resp - pred)/sqrt(pred)
}
> Pur.wt <- nls(~ weighted.MM(vel, conc, Vm, K), Treated,
+    list(Vm = 200, K = 0.05))
> Pur.wt
Residual sum-of-squares : 14.5969
parameters:
        Vm         K
 206.8338 0.05461018
formula:  ~ weighted.MM(vel, conc, Vm, K)
12 observations
```

We have used numerical derivatives here. Obtaining analytic derivatives would be delicate, even for this simple example.

As described in Carroll and Ruppert (1988), a weighted nonlinear regression with the weights given by a power of the predictor is very similar to taking a power transformation of both the predictor and the response. This "transform both sides" approach uses the Box-Cox form of power transformations

$$y^{(\lambda)} = \begin{cases} (y^\lambda - 1)/\lambda & \text{if} \quad \lambda \neq 0 \\ \log y & \text{if} \quad \lambda = 0 \end{cases}$$

and is implemented in the function `TBS`. For example, to fit the logarithm of the velocity to the logarithm of the predictor for the treated enzyme data we would use

```
> Pur.TBS <- nls(~ TBS(vel, Vm*conc/(K+conc), 0), Treated,
+    list(Vm=200, K=0.05))
> Pur.TBS
Residual sum-of-squares : 0.1676924
parameters:
        Vm         K
 203.5726 0.05284299
formula:  ~ TBS(vel, (Vm * conc)/(K + conc), 0)
12 observations
```

As can be seen, this produces results very similar to those from the weighted regression.

One advantage of the "transform both sides" approach is that the log-likelihood for the transformation parameter λ can be evaluated and used to choose the transformation. See Carroll and Ruppert (1988) for details.

10.4 Programming Details

Numerical methods for general optimization and for nonlinear least-squares have been the subject of much research. As we have been saying throughout the chapter, the problems are harder and more likely to produce numerical difficulties than those in any other chapter of the book. The numerical algorithms used here are, we believe, representative of the current state of the art. In this section, we give some information that may help users to get the most from the methods. As is the general philosophy in the book, informed users are encouraged to go beyond what we provide, to extend or replace the methods by others that may be more suited to their particular needs. In the case of nonlinear models, a warning label needs to be shown, however. These are indeed difficult numerical problems, and fiddling with the underlying algorithms is not recommended if you are not sure of what you are doing. If other numerical algorithms are to be used, they need to produce objects containing the information used by the summary and printing methods.

10.4.1 Optimization Algorithm

Like most methods for general optimization, the algorithm used here is based on a quadratic approximation to the objective function. When the model formula provides both first and second derivatives, this approximation is a local one using these derivatives (i.e., the algorithm is a version of Newton's method). When no derivatives or only first derivatives are supplied, the algorithm approximates the second derivative information, but in a method designed specifically for minimization, since the quadratic approximation is not a goal in itself. There are many such methods, the one used here being taken from the PORT subroutine library, and evolved from the published algorithm by Gay (1983).

Two distinctive numerical features of the algorithm are the particular method used to develop a quadratic approximation when second derivatives are absent and a "trust region," a running estimate of the size of the region around the current estimate in which a quadratic approximation to the objective function is likely to be trustworthy. These are discussed in the reference above and in the PORT library documentation, but are unlikely to affect S users.

The algorithm is capable of working with user models that return 0, 1, or 2 orders of derivatives. As discussed in Section 10.2.3, improvements in numerical accuracy and in efficiency often come when users supply analytical derivatives.

10.4.2 Nonlinear Least-Squares Algorithm

There are many algorithms for the problem of nonlinear least squares, even though it is clearly more specialized than general optimization. The nls() function uses a relatively simple approach—the Gauss–Newton algorithm with a step factor to ensure that the sum of squares decreases at each iteration. In this algorithm the residuals and the gradient are calculated at the current parameter values then a linear least-squares fit of the residual on the gradient gives the parameter increment. If applying the full parameter increment increases the sum-of-squares rather than decreasing it, the length of the increment is halved. The step factor is retained between iterations and started at twice the value that was successful on the last iteration except that it is not allowed to become greater than unity. In other words, if the last iteration required the increment to be reduced to one quarter of its original length before the sum of squares decreased, the next iteration starts at one half of the calculated increment.

If the gradient is not returned with the value of the formula, it is calculated using finite differences with forward differencing.

For partially linear models, the increment is calculated using the Golub-Pereyra method (Golub and Pereyra, 1973) as implemented by Bates and Lindstrom (1986).

The convergence criterion for both algorithms is the relative offset criterion described in Section 10.3.1.

Bibliographic Notes

The books by Bates and Watts (1988) and by Seber and Wild (1989) discuss nonlinear regression, including nonlinear least squares but also with some general treatment of likelihood and Bayesian inference. Both books give a discussion of computing algorithms; the method described in Bates and Watts (1988) is the basis for that used by nls().

For statistical theory for general nonlinear modeling, we are unaware of any single comprehensive treatment. What is needed is a set of general results relating the statistical model, based on likelihood function, Bayesian inference, or other models, to the data-based results that can be computed by an algorithm such as that used by ms(). The basic statistical results, due to Fisher and others, are classical, but applying them in practice and with due regard for issues of curvature, small-sample behavior, and many other questions, is decidedly difficult. With the much greater accessibility of computational methods, a good general statistical treatment would be a major contribution to the community. Meanwhile, books on general statistical inference provide some basic results; for example, Chapters 9–11 of Kalbfleisch (1979) or various results in Cox and Hinkley (1974), Lehmann (1986), or various other classic references. Unfortunately, these discussions are not usually from the viewpoint of actually fitting a model to data. The books on nonlinear regression, and even McCullagh and Nelder (1989) on generalized linear models, will have some relevant results, if the reader can make the necessary generalizations from the special cases discussed.

Computational issues for nonlinear models are discussed from a statistical viewpoint in Chapter 6 of Chambers (1977). More detail regarding optimization is given by a number of authors; for example, see Dennis and Schnabel (1983).

Appendix A

Classes and Methods: Object-oriented Programming in S

John M. Chambers

The version of S used in this book extends that described in $\boxed{\text{S}}$ in several ways, the most important being the use of *methods*. Throughout this book a phrase such as "the summary method for objects of class `"lm"`" means that, if `fuel.fit` is an object of this class, typing

```
> summary(fuel.fit)
```

invokes a function specially designed to carry out the computations of `summary()` on `lm` objects. You didn't call that function explicitly or give `summary()` an argument to tell it that `fuel.fit` was an `lm` object. Instead, a mechanism in the S evaluator figured out which method for `summary()` should be applied to `fuel.fit`, and arranged to call that method. This mechanism causes S evaluation to be *data driven* (another way of describing object-oriented programming).

From the view of the user who is taking the functions as they stand, the method mechanism should be invisible. Functions, operators, and assignments should adapt to the classes of objects introduced in this book, without explicit user action. If you don't expect to do much programming to specialize the functions we provide, then you don't need to read any further in this appendix. However, if you want to re-design some of the methods or to design new classes of objects for new applications,

455

you will need to know something about the mechanism. This appendix will describe how methods are organized, how classes of objects and their associated methods can be designed, and how the S evaluator arranges for data-driven computations to work. The description is far from the whole story but is intended to give enough information so that you can use classes and methods effectively in extending the modeling software.

A.1   Motivation

The scope of a programming language can be viewed at a fundamental level in terms of the computations that can be performed and the kinds of data on which they can act—in S terminology, the *functions* available and the *classes* of objects to which the functions can be applied. A language that hopes to grow and adapt to new applications faces the fundamental challenge of helping the designer to create the new software while preserving a simple view of the language for the user. The class/method mechanism is the core of our response to that challenge in S.

The challenge can be seen more clearly if we begin with the "primitive" view of the language, in terms of the functions and data classes that map directly onto the underlying implementation. For example, primitive objects in S are essentially atomic vectors of a few prescribed modes (numeric, logical, complex, and character string), plus recursive (list-like) objects made up from other objects used as elements and attributes. Most of the functions in $\boxed{\textsf{S}}$ are defined for such objects. The concern addressed by the class/method mechanism is to allow the language to deal with new classes of objects (and new functions as well) while preserving a simple and natural view of the language for ordinary users.

Even as simple a data structure as the *category* ($\boxed{\textsf{S}}$, page 136) illustrates the challenge. Conceptually, categories represent repeated values, each of which comes from a finite set of levels—for example, "Male" and "Female", or "Low", "Medium", and "High". While the values could be represented as character strings, it is essential that there is some known set of possible levels, so representing the category as a character vector would be misleading. Categories in S code the values as integers, and retain the levels as an attribute. This implementation does not correspond to the concept as well as it should. Numeric operations are meaningless conceptually for categories, but numeric expressions involving categories in S *will* produce a result. Sometimes, as in the second example, we would like to include the notion that the levels are *ordered*, but not explicitly numeric, while continuing to inherit the other properties of categories. Working with just the primitive objects, there is no simple way to add the concept of ordered categories to the software.

In this book we talk about factor objects, rather than categories. Factors are a class of objects that implement the same notion as categories, only often in a better way, because of the class/method mechanism. Section 3.2.1 discusses factors and outlines their properties. We will use factors as an example throughout this appendix.

Our goal is to incorporate new classes of objects into the language in a seamless way, with functions such as printing, plotting, and subsetting behaving "correctly" on the new objects, without any explicit action on the user's part. At the same time, the job of the applications programmer should be made as simple as possible when designing new classes of objects and providing software to use them. Object-oriented programming systems pursue this goal by allowing programmers to define *methods* that replace generic functions when the argument to the function belongs to a particular class of objects. This is the kind of mechanism presented here: designers of new classes of objects will implement methods for some of the generic functions in S to handle the new class. Users will continue to see only the generic functions, so their view of the language will not be made more complicated. At the same time, the designer does not need to rewrite the generic functions or to worry about ensuring that the new methods will be invoked. In addition, the notion of *inheritance*—that new classes can automatically inherit methods written for existing classes—further reduces the amount of programming for the designer.

A.2 Background

This section is addressed to those interested in programming languages in general or to those familiar with S who would like to see a general defense of adding methods and classes to the language. Others can skip ahead to Section A.3.

The mechanism for classes and methods in S has much in common with other object-oriented languages, but differs in a number of respects related to the nature of S. The S language follows, in an informal way, three main programming paradigms:

1. object-oriented programming;

2. functional languages;

3. interfaces.

The first of these paradigms is the topic of this appendix, but a few words about the other two will clarify the context.

Functional programming languages use a model in which the central activity is the evaluation of a function call (to use S terminology—"expansion" rather than evaluation is the common term). The language operates by reading the user's function calls and evaluating them. The result of the evaluation is to present the user with some computations (printed, or often in the case of S, plotted). Languages based on this paradigm aim for a simplicity and clarity that makes programming in them more straightforward and less error-prone than with "conventional" languages such as FORTRAN or C.

The paradigm of interfaces appears less frequently and less explicitly in computer science, but it is key to S and is also part of the background philosophy of many approaches to modern computing systems. A language like S wants to make use of a wide variety of computational techniques (numerical methods, text processing, symbolic computations, etc.). Good implementations of all of these within the language itself would be prohibitively difficult. Instead, we try wherever possible to define an *interface* to some other language or system in order to incorporate some of that system's abilities into S. An interface, as used in S, is a communication model that defines some simplified version of what the other system can do and a mapping from S expressions and objects into that model. For example, most text manipulation in S is handled by an interface to the UNIX shell. The model and the mapping in this case are based on the model that UNIX commands only read and write streams of bytes, possibly broken up into lines by newline characters.

These two paradigms influence the class/method mechanism in S. The functional language paradigm focuses our attention on generic functions and on the evaluation process as the center of the mechanism. The use of interfaces provided a strong incentive to adopt the class/method mechanism and continues to test out the strength of the mechanism.

Object-oriented programming replaces the idea of executing a program, in the traditional sense of a single set of machine instructions, with actions that take place as the result of passing *messages* between objects. The messages can be thought of as requests to the object—for example, that the object print itself. The action occurring in response to the message typically depends on the *class* to which an object belongs. The definition of the class determines for which messages the object has a *method*. A method implements, in some programming language, the definition of what should happen in response to the message. S shares an emphasis on general objects, and allows users to define classes of objects. Classes as discussed in $\boxed{\text{S}}$, Chapter 8, are informal. A convention on the essential data structure for the class (in terms of what components or attributes are expected) is shared among a collection of functions written to create and use objects from the class.

While this informal approach was adequate for many applications, it has some disadvantages. The user had to be conscious of the class structure in order to call the specific functions for that class. For example, to print a summary of an object of class "lsfit" one called a function for that class, `ls.summary()`. For a different class, there would be a different function. The only alternative was to write a single function that understood all the relevant classes of objects. This happened to some extent with the function `print()`, but no general `summary()` function existed.

In the class/method paradigm, there is a generic `summary()` function, which will be called by the user regardless of what class of object is to be summarized. The generic function, however, does hardly anything itself. Instead, it invokes a mechanism that:

- finds the class, if any, of the relevant argument to `summary()`;

- finds a special `summary` method for that class, if one exists;

- evaluates the method in place of the call to the generic.

The user need only remember how the generic function is called. Applying the function to different classes of objects will produce the results designed for those objects, automatically.

This is the mechanism provided now in S. Some of the features are:

- Users of generic functions can expect the functions to adapt to new classes of objects without any action on the user's part.

- Designers of new classes of objects can redefine generic functions by writing a *method*: a new function with the same arguments as the generic function.

- Classes and methods can use a general form of *inheritance*, the ability of one class of objects to inherit methods from one or more other classes. The designer of a new class supplies only methods that need to be different from those inherited.

- Methods can invoke inherited methods in a simple, general way, simplifying the modification of methods to new classes.

- Groups of generic S functions (e.g., operators like arithmetic) often map into an interface to one C routine. If all these functions adapt in essentially the same way to a new class of objects, a single *group method* can be written for all of them.

- Methods for operators are invoked when either or both of the operands unambiguously identify a method.

- Methods can be written for some key operations in the language that are not precisely functions (e.g., replacement and permanent assignment).

- Methods in S are functions that can be called explicitly. In particular, a user can override the standard method for an object and force it to be treated as an object from another class.

The use of methods allows programming in S to have much more of the style of object-oriented programming systems. However, methods in S differ from such systems in some interesting ways, partly from the influence of the other two paradigms. The last feature above is an example. Because all functions in S are objects and because function calls are the central, essentially the only, activity in an S session, methods are not restricted to automatic invocation, but can be used like any other function.

A.3   The Mechanism

Two conventions define the mechanism for methods in S. First, objects in S can have a special attribute, `"class"`. The class of an object can be extracted or set by the function `class()`:

```
> class(myobject) <- "factor"
> class(myobject)
 [1] "factor"
```

For the moment, we think of the class as either a single character string or NULL; this will be extended when we talk about classes that inherit methods from other classes.

Second, a mechanism in the S evaluator will automatically find an appropriate method and evaluate a call to it. The special function `UseMethod()` invokes this mechanism:

```
UseMethod("print", x)
```

with the following effect. The evaluator looks for an S function object to act as a method for the generic function `print()` for the object `x`. If `x` has class `"factor"`, a function named `print.factor` will be used as the method. The function call currently being evaluated is replaced by the corresponding call to `print.factor()`. The second argument to `UseMethod()` is normally omitted, in which case it is taken to be the first argument in the definition of the current function. More detail of how this mechanism works is covered in Section A.6.

The class/method mechanism in S is uniform and general, but not obligatory. Functions that look for methods are called *generic*, in the sense that they define only generically what their effect should be, leaving further specifications to be done by methods. Generic functions perform some standard computation, which we expect to be adapted to different classes of objects by other functions, the methods. The body of a standard generic function typically consists of a call to `UseMethod()`:

```
print <- function(x, ...)
    UseMethod("print")
```

The first argument to `print()` is the object that determines the method. The "..." argument is there so that methods may have additional arguments, specific to the particular method. Generics are perfectly free to have other arguments as well. It might make sense to do so if we asserted that *any* method for this generic must deal with the argument. In practice, nearly all generic functions look like the above.

In the case of basic S functions, for which we want to make the default computations particularly efficient, the interface to the method mechanism will be invoked directly from C code, and the body of the generic function will be a call to the `.Internal()` interface. Although these functions do not call `UseMethod()`, they mostly make use of the same mechanism. The main exceptions are the operators (arithmetic, comparison, or logical), which allow either or both operands to define the method, provided they do so consistently (see Section A.7).

There is a second special function, `NextMethod()`, which uses the method mechanism to simplify handling inherited methods. Section A.5 describes inheritance. We will discuss some important special cases, such as operators, subsets, and assignments, in Sections A.7 and A.8. Section A.10 lists those internal functions that take methods.

All objects implicitly inherit from class `"default"`. Corresponding to the generic function `print()` will be a default method, `print.default()`. Objects without a specific class or with a class inheriting no other print method will be printed using `print.default()`. For generic functions invoking methods via the `.Internal()` interface, the default method is also contained in the internal code.

A.4   An Example of Designing a Class

The essential programming exercise involving the method mechanism is to design or refine the behavior of a class of objects. How should users think about the objects in the class? How should generic functions behave for them?

As an example, we will consider the implementation of the class `"factor"`. The class/method mechanism allows a more consistent and natural implementation of such objects. What would otherwise be a very substantial job of reprogramming many S functions can be done quite simply by writing some appropriate methods so factors behave sensibly when used with common generic functions.

Designers of a class of objects need to consider the internal or *private* structure for the objects that will best implement the conceptual or *public* view. In the case of factors, we give the object an attribute, `levels`, that represents the set of possible values. The object itself then consists of a vector of integer values between 1 and `length(levels(x))`. We also allow elements of the object to be missing (`NA`). This implementation defines the private view of factors. As designers of the class, and implementers of the most basic methods for them, we work with this view. General users, however, should see only the public view—the view of factors as repeated values from the set of levels. Even designers of new methods later on may be able to work entirely or largely with the public view. In any case, methods must be written with a clear understanding of which view is being used. We will illustrate the important practical consequences of the public/private distinction later on.

Besides methods, a *generator* function must exist to generate objects from the new class. The function `factor()` will do this:

```
factor <- function(x, levels = sort(unique(x)),
  labels = as.character(levels)) {
    y <- match(x, levels)
```

```
    names(y) <- names(x)
    levels(y) <- labels
    class(y) <- "factor"
    y
}
```

In **S**, Section 8.1, a convention was set out that each class should also have functions to test membership in the class and to coerce to that class. For factors, these functions would be `is.factor()` and `as.factor()`. Such functions are still useful, but the method mechanism makes them less important. Methods are designed to produce suitable results without explicitly coercing objects to a particular class, and the inheritance mechanism to be described later generalizes the notion of testing membership. The function `inherits()` replaces the testing functions for all classes in most applications.

```
inherits(x, "factor")
```

returns TRUE if the class of x includes "factor".

For our first method, let's arrange to print the object. The method will be a function object named `print.factor`. Its first argument will be the object to be printed.

What *should* a printing method for factors do? We want users to think of the data as values from the `levels` set, and a natural way to do that is to print the vector computed as

```
levels(x)[x]
```

This is a vector whose elements are the levels of the corresponding elements of x. It is a character vector, but to emphasize the levels as a set, we will print it without quotes, using the `quote=F` argument to `print()`. One more detail: We allow NA in the data, so before printing we should turn any NA's into an additional level. Adding some code to achieve this, we get our method:

```
print.factor <- function(x, ...) {
    class(x) <- NULL
    l <- levels(x)
    if(any(is.na(x))) {
        l <- c(l, "NA")
        x[is.na(x)] <- length(l)
    }
    x <- l[x]
    print(x, quote = F)
    invisible(x)
}
```

The class of the object was set to NULL at the beginning of the method because this is a *private* method. The print.factor() definition depends on the implementation of factors. When it applies functions such as levels() and "["() to x, these are intended to be the default methods, regardless of whether the object x might have inherited some method for those functions. In contrast, *public* code in a method expects most of the computations to be done by another method, with just a little pre- or post-processing. The other method may be for the same class or for an inherited class, but the public method does not take explicit control of which method is used.

We'll improve print.factor() a little in later discussion, but the above is quite respectable. As an example of its use,

```
> sex <- factor( sample(c("Male","Female"), 25, T))
> sex
 [1] Male   Female Female Male   Female Female Female Female Male
[10] Male   Female Male   Female Male   Female Male   Female Male
[19] Female Male   Male   Female Female Female Male
```

In later sections, we will show methods for other generic functions, but first we look at a new class, ordered factors, as an example of inheritance.

A.5 Inheritance

A powerful tool in object-oriented programming is the ability of objects to inherit methods from one or more other classes. This greatly simplifies defining new classes that are adapted from existing classes. Only methods related to the new features need be written. In S, inheritance is easily incorporated. The class attribute of an object can be of any length. When it is of length greater than 1, the object can inherit methods from any of the classes included. For example,

```
> class(x)
[1] "ordered" "factor"
```

says that methods for either class "ordered" or class "factor" can be used for object x. The search for methods will proceed sequentially through all the elements of the class attribute. In this sense, "ordered" can be thought of as the principal class of x, and often we will speak of such an object as belonging to class "ordered" even though it inherits from other classes as well. All classes implicitly inherit from the class "default", which need not appear in the class attribute.

Let's define the class of *ordered factors* to be just like factors, except that the levels are now assumed to be ordered increasingly as given. A function to generate objects from this class could be as follows:

```
ordered <- function(x, levels = sort(unique(x)),
  labels = as.character(levels)){
   x <- factor(x, levels, labels)
   class(x) <- c("ordered", class(x))
   x
}
```

Generic functions applied to ordered factors will look first for methods for class
"ordered", then for class "factor", and finally for "default".

As an example, consider a character vector, say rating.text, containing character strings "Low", "Medium", and "High", which we will convert to an ordered factor:

```
> rating.text
 [1] "Medium" "High"   "Low"    "High"   "Medium" "Medium" "High"    "Low"
 [9] "Low"    "High"   "Medium" "High"   "High"   "High"   "Medium" "Low"
[17] "High"   "Low"    "High"   "Low"
> ratings <- ordered(rating.text, levels = c("Low","Medium","High"))
> ratings
 [1] Medium High   Low    High   Medium Medium High   Low    Low    High
[11] Medium High   High   High   Medium Low    High   Low    High   Low
```

The levels argument is needed in the call to ordered() to establish the correct
ordering. Because of inheritance, print(ratings) uses the method print.factor()
if the method print.ordered() is not found.

The function NextMethod(), which exploits inheritance, is a key tool in writing
methods. The call

```
NextMethod("print")
```

looks for an inherited method for the generic function "print"; that is, it searches for
a function whose name is "print." concatenated with one of the classes following the
current class, "ordered", in the class attribute of the object. It calls this function
and returns the value as the value of NextMethod(). All classes implicitly inherit
from the class "default", so that NextMethod() invokes the default method if no
other inherited method is found. Section A.6 gives a more precise definition of the
inheritance mechanism.

Suppose we decided to show the ordering of the levels after printing the data for
our new class. The essential printing is done by the inherited method, after which
print.ordered() adds a line at the end to show the levels:

```
print.ordered<- function(x, ...) {
    NextMethod("print")
    cat("\n", paste(levels(x),collapse=" < "), "\n")
    invisible(x)
}
```

Our previous example now prints as follows:

```
> ratings
 [1] Medium High    Low     High    Medium Medium High    Low     Low     High
[11] Medium High    High    High    Medium Low     High    Low     High    Low

Low < Medium < High
```

The function `NextMethod()` is useful even in cases not involving inheritance, as a way of invoking the default method for the generic function. For example, we should write a method for extracting subsets from `factor` objects, to ensure that the extracted object continues to belong to class `"factor"`. The generic function named `"["` extracts subsets. Its definition is

```
"[" <- function(x, ..., drop = T)
  .Internal(x[..., drop = drop], "S_extract", T, 1)
```

If there were no method `"[.factor"()`, the behavior would be as follows:

```
> sex[1:5]
[1] 2 1 1 2 1
attr(, "levels"):
[1] "Female" "Male"
```

The internal calculations remove the class of the object. Clearly we need a method, although only a simple one:

```
"[.factor" <- function(x, i){
    y <- NextMethod("[")
    class(y) <- class(x)
    y
}
```

Now subsets of factors retain their class nature:

```
> sex[1:5]
[1] Male    Female Female Male    Female
```

The style of `"[.factor"()` is a common way of writing methods: use the inherited method and then set some attributes of the result (here just the class), before returning it. The use of `NextMethod()` is crucial. Recursive use of the generic function directly can cause an infinite loop—the object in question is still a factor, so the generic function will invoke the method once again, and so on, until S complains when the maximum level of nesting of expressions is reached. In more complicated computations, the method may need to construct a new object, rather than counting on the inherited method for the current object (the method `Ops.factor()` on page 473 is an example).

The definition of `"[.factor"()` illustrates one other point: the argument lists for methods. When `UseMethod()` has found a method, it re-matches the actual call to the generic with the definition of the method. Similarly, `NextMethod()` matches arguments with the same order and names as the call to the generic. Methods do not need to have the same arguments as the generic. In most examples, the method should allow optional, named arguments to be passed down, if it plans to call `NextMethod()`. This means that most such methods should have "..." as an argument. In the case of `"[.factor"()`, however, we took the opposite approach. We decided that only one subscript argument made sense for factors, although other objects might meaningfully have more arguments. The definition here then produces a clean error message if an extra argument is included:

```
> sex[1,5]
Error in call to [.factor(): Argument number 3 in call not matched
```

whereas, if the second argument were "..." rather than "i", the message is:

```
> sex[1,5]
Error in x[..1, ..2]: No dim attribute for array subset
```

which is less helpful. This is a tradeoff between allowing more general use of the method and providing informative error messages; the relative importance of each should be examined case-by-case.

The object determining the choice of method is not modified when the methods are invoked, either directly or through inheritance. When `print.factor()` is called in printing `ratings`, for example, the class of x will be `c("ordered", "factor")`. This is why `"[.factor"()` sets the class by

```
class(y) <- class(x)
```

rather than just setting it to `"factor"`. The distinction is essential if other classes are to inherit the method `"[.factor"()` and retain their own class attribute. Special objects introduced into the frame provide full information on the actual methods and classes used—see Section A.6.

A related point applies to the design of classes. Asserting that one class inherits from another asserts that an inherited method *will* work; for example, the objects of class `"ordered"` must have all the information used to print objects of class `"factor"` if the `NextMethod()` call is to work in the definition of `print.ordered()`. So while S does not enforce any rules about which classes can inherit from which other classes, the designer needs to ensure that inheriting classes really do make sense when viewed by the inherited methods as objects of another class.

The distinction between *public* and *private* views of classes applies to these examples: `print.factor()` was a private method, depending on the `levels` attribute; `print.ordered()` was a public method, just invoking the inherited method and modifying its value. Similarly, `"[.factor"()` is a public method: its only assertion is that a subset of the factor object should still be considered a factor. Public methods have the advantage of being less dependent on implementation details, and are usually simpler to understand. Private methods tend to be more efficient, if carefully designed; in any case, some private methods will be necessary. Private methods should set the class of the object to NULL, as we did in `print.factor()`, to prevent accidental infinite loops and, more generally, to ensure that computations are done by the default methods.

The distinction is sometimes subtle and not always clear-cut. For example, another possible implementation of `"[.factor"()` would be:

```
"[.factor" <- function(x, i){
    oldclass <- class(x)
    class(x) <- NULL
    y <- x[i]
    class(y) <- oldclass
    y
}
```

This is somewhere between a public and private view. It does not use any explicit attributes of the `factor` object, but notice that, since it sets the class to NULL, the default method will always be used for `x[i]`. Therefore, this implementation would prevent anyone from defining a class from which `factor` inherits, with a method for `"["()`. Generally, the previous implementation is more flexible, although for factors the distinction makes little practical difference.

A.6   The Frames for Methods

Methods invoked through `UseMethod()` or through a `.Internal()` interface behave as if the call to the generic had, instead, been a call to the method. When an inherited method is invoked by a call to `NextMethod()`, it behaves as if called from the previous method, with that method's arguments. The specific way this works will be described below. The evaluator also arranges for some special objects to be inserted in the frame of the method; these define precisely the class, method, etc., being used. See page 470.

Arguments to Methods

When `UseMethod()` is called, the frame in the evaluator for the call to the generic function becomes the frame for the call to the method:

- The arguments in the call to the generic are re-matched to the formal arguments of the method, using the standard S rules for argument matching ($\boxed{S}$, page 354). The method will see argument matches as it would if the user's call had been directly to the method.

- Any other objects in the frame remain the same. In particular, the first argument, which defined the choice of method, will have been evaluated.

- The special objects defined on page 470 will be in the frame.

Most generic functions will have only two arguments, the object itself and "...", as illustrated on page 460. Methods can have other arguments to control aspects of the computation peculiar to them. As an example, consider the print method for data frames. Basically, the method just turns the data frame into a matrix and prints the result. Here is a slightly simplified version:

```
print.data.frame <- function(x, ..., quote = F, right = T)
{
    print(as.matrix(x), ..., quote = quote, right = right)
    invisible(x)
}
```

The method has arguments `quote=` and `right=` not found in the generic. These control whether quotes should be put around strings and whether character columns should be right-justified. They are also arguments to the `print.default()` method, which will eventually do the actual printing of the matrix. Having the arguments in this method allows their default values to be set differently from the defaults in other print methods, while still allowing the user to set the arguments explicitly:

```
> print(catalyst, quote=T)
    Temp  Conc Cat Yield
1   "160" "20" "A" "60"
2   "180" "20" "A" "72"
3   "160" "40" "A" "54"
4   "180" "40" "A" "68"
5   "160" "20" "B" "52"
6   "180" "20" "B" "83"
7   "160" "40" "B" "45"
8   "180" "40" "B" "80"
```

The existence of "..." in the generic and in the methods means that other arguments can be passed down that may be meaningful to later, inherited methods. For example, the argument `digits=` is meaningful to `print.default()` and other methods, but not to `print.data.frame()`. Constructing the methods as above will allow users to pass this argument down to the methods that understand it. The extra arguments to the method above came after the "...", forcing the arguments to be named in the call. This is a reasonable strategy; the natural order of such arguments is unclear, given that the user may combine optional arguments from more than one method.

Turning now to methods invoked as a result of a call to `NextMethod()`, these behave as if they had been called from the previous method with a special call. The arguments in the call to the inherited method are the same in number, order, and actual argument names as those in the call to the current method (and, therefore, in the call to the generic). The expressions for the arguments, however, are the names of the corresponding formal arguments of the current method. Suppose, for example, that the expression `print(ratings)` has invoked the method `print.ordered()`. When this method invokes `NextMethod()`, this is equivalent to a call to `print.factor()` of the form

```
print.factor(x)
```

where x is here the x in the frame of `print.ordered()`. If several arguments match the formal argument "`...`", those arguments are represented in the call to the inherited method by special names "`..1`", "`..2`", etc. The evaluator recognizes these names and treats them appropriately (see page 476 for an example).

This rather subtle definition exists to ensure that the semantics of function calls in S carry over as cleanly as possible to the use of methods (compare $\boxed{\mathbf{S}}$, page 354). In particular:

- Arguments are passed down from the current method to the inherited method with their current values at the time `NextMethod()` is called.

- Lazy evaluation continues in effect; unevaluated arguments stay unevaluated.

- Missing arguments remain missing in the inherited method.

- Arguments passed *through* the "`...`" formal argument arrive with the correct argument name.

- Objects in the frame that do not correspond to actual arguments in the call will not be passed to the inherited method.

The inheritance process is essentially transparent so far as the arguments go.

If the object driving the choice of method, x in the case of `print()`, has been modified, the modified version is the one seen by inherited methods. However, modifying this object does *not* alter the choice of which method is invoked next. The object `.Class` always determines the choice of method. If the method wants to change x and use the class of the new version to control inheritance, it should invoke the generic function, just as `print.data.frame()` did in the above example. It is possible to make changes to `.Class` instead, but this is not recommended unless you understand the inheritance mechanism thoroughly.

The rules above describe how `NextMethod()` constructs the new call, by default. If the current method wants to alter this call, the changes are provided as additional arguments to `NextMethod()`. Consider `print.factor()` again. It called `print()` rather than `NextMethod()` in order to change the default value of the `quote=` argument. Another approach would be to insert the argument into a call to `NextMethod()`:

```
NextMethod("print", quote = quote)
```

replacing the call to print(). In this case there is no particular difference, but if x still retained its original class, the call to print() would have caused an infinite loop. Any arguments can be given in this way, including the object itself. Named arguments override any correspondingly named arguments in the inherited call, and unnamed arguments are inserted at the beginning of the call. Generally, unnamed arguments should be used only to replace the object itself, but remember that doing so does *not* affect the choice of the next method.

Special Objects in the Frame

The method mechanism adds to the evaluation frame a complete picture of the current situation, in four special objects:

.Class: The class attribute corresponding to the current method. NextMethod() adds the current .Class to the inheriting .Class as the attribute previous. See the example below.

.Generic: The name of the generic function.

.Method: The name of the method being used, as a character vector. This object has a special form when methods are defined for operators: see the discussion in Section A.7 below.

.Group: The name of the group, in the case that the interface to methods comes through one of the internal interfaces in Table A.1 on page 472.

These objects are maintained and used by the S evaluator, but they can be used also in writing methods. To illustrate, suppose we trace print.factor() with the browser:

```
> trace(print.factor,browser)
> sex
browser: print.factor(sex)
b> ?
1: .Class
2: .Method
3: .Generic
4: .Group
5: x
b> .Class
[1] "factor"
b> .Method
[1] "print.factor"
b> .Generic
```

```
[1] "print"
b> .Group
[1] ""
```

Similarly, suppose we also trace `print.ordered()`, and then print `ratings`:

```
> trace(print.ordered,browser)
> ratings
browser: print.ordered(ratings)
b> .Class
[1] "ordered" "factor"
b> .Method
[1] "print.ordered"
b> 0
browser: print.factor(x)
b> .Class
[1] "factor"
attr(, "previous"):
[1] "ordered" "factor"
```

Typing 0 returns from the `browser()` trace in `print.ordered()`. `NextMethod()` is then called. It constructs a call to `print.factor()`, which we are also tracing. We print `.Class` again. The class is `"factor"`, as when we were printing `sex`, but now with an attribute `"previous"`, containing the same object that was in `.Class` in the previous method, `print.ordered()`.

Each time `NextMethod()` moves down through the inheritance pattern, it shifts the `.Class` along to the inherited class, and attaches the previous value as the `"previous"` attribute of the new one. This extra information in `.Class` provides a mechanism for working back through the entire network of methods involved in the current computation, should we want to do so.

A.7 Group Methods; Methods for Operators

Functions using the `.Internal()` interface generate calls to C routines; for example,

```
> exp
function(x)
.Internal(exp(x), "do_math", T, 108)
```

calls the C routine `do_math`. Usually one routine handles a number of related S functions; for example, there is one internal interface for all the usual operators (arithmetic, comparison, logical), either in binary or unary form, and one for all the "mathematical" functions that transform objects element by element, including functions such as `exp()`, as well as less obviously mathematical functions, such as `round()`. Table A.1 shows the functions in this group and in various other groups

Group	Functions

Math	`atan(x, y); cumsum(x); abs(x); acos(x);`
	 `acosh(x); asin(x); asinh(x); atanh(x);`
	 `ceiling(x); cos(x); cosh(x); exp(x);`
	 `floor(x); log(x); log10(x);`
	 `round(x, digits); signif(x, digits);`
	 `sin(x); sinh(x); tan(x); tanh(x);`
	 `gamma(x); lgamma(x); trunc(x)`

Summary	`all(x); any(x); max(x); min(x); prod(x);` `range(x); sum(x)`

Ops	`e1 + e2; e1 - e2; e1 * e2; e1 / e2; e1 ^ e2;`
	 `e1 < e2; e1 > e2; e1 <= e2; e1 >= e2; e1 != e2;`
	 `e1 == e2; e1 %% e2; e1 %/% e2; e1 & e2;`
	 `e1
		e2; -e1; !x`

Table A.1: *The groups of functions for which methods can be written*

of functions. Methods often will be identical, or nearly so, for all the functions in a group. One can take advantage of this by writing one *group* method, rather than a separate method for each function in the group. All the functions in the group will obey the method for objects from the new class, with only one piece of code to be written, saving substantially on work and clutter when implementing a method for all the 26 functions in the `Math` group, for example. The catch, of course, is that the new method must work correctly for all the functions in the group. The `NextMethod()` function and the `.Generic` object are the key tools in writing such methods.

Group methods are functions whose name consists of the *group* name, not the name of the individual function, followed by ".", followed by the name of the class. The function `Math.factor()`, then, provides the method for all the functions in the `Math` group applied to factors. Since we want to discourage the user from thinking of the levels as numbers, the definition of this method is simple:

```
Math.factor <- function(x,...)
stop("A factor is not a numeric object")
```

Trivial? Yes, but it prevents returning a meaningless (indeed, wrong) answer that would have resulted if the generic definition had been used.

It is possible to have methods for individual functions included in the group as well. The individual method is always chosen in preference to the group method. Group methods that can treat all the included generic functions the same way are the most convenient, but the method is free to do something special depending on the particular function. The value of the special object ".Generic" gives the name of the generic function. This could be an argument to the `switch()` function to enumerate special cases.

Methods for the `Math` group and the `Summary` group tend to be simple, if they are needed at all. The generic functions are meaningful only for numeric or, occasionally, logical data. If a method is needed at all, it will typically just produce an error message or do some simple conversion of the object to the appropriate mode. Writing methods for the `Ops` group tends to be more challenging. This group includes all the arithmetic, comparison, and logical operators. One is fairly likely to want a method for this group, so that operators will work when the operands are one object from a special class and one plain object, or when the two operands are compatible objects from the same special class.

Notice that S already supplies methods, implicitly, for arrays and time-series. A matrix and a vector can be operands to an operator, with the result being a matrix. Two matrices can be operands if their dimensions match. Similar but more liberal definitions apply to time series ($\boxed{\text{S}}$, page 296). New classes of objects ought to have similar methods where they make sense.

An interesting feature here is that the traditional object-oriented view is rather clumsy, because the *two* operands ought to be treated equally in deciding what should happen. Taken literally this suggests having to define a new class of objects for each *pair* of original classes. Then a method would be defined for the various operator "messages" for each class pair, either explicitly or by inheritance. Implementing such a scheme, however, would be very clumsy and in practice a symmetric, pure approach seems never to be taken. The focus in S on the function as the primary arbiter of what method to be used seems in this situation to be natural.

For the `Ops` group of functions in Table A.1, the internal interface invokes a special method if the two operands, taken together, suggest a single method—specifically, if one operand corresponds to a method that dominates that of the other operand, or if they both correspond to the same method. Otherwise, the default definition of the operator is used. Either a group definition or an individual definition of a method dominates if the other operand has no corresponding method, and an individual definition dominates a group definition.

With the factor class of objects, if a group method, `Ops.factor()` has been defined (we will define one below), then any operation involving two factors will be handed over to this method. So will any operation in which either the left or the right operand is a factor, and the other operand either has no class attribute or else is a class for which no special method is defined. Notice that the special method is responsible for doing some further analysis to determine which operand is the factor and to check that two factor operands are compatible:

```
Ops.factor <- function(e1, e2) {
        ok <- switch(.Generic, "=="=, "!=" = T, F)
        if(!ok) stop(paste('"',.Generic,
```

```
            '" not meaningful for factors', sep=""))
    nas <- is.na(e1) | is.na(e2)
    if(nchar(.Method[1])) {
            l1 <- levels(e1)
            e1 <- l1[e1]
    }
    if(nchar(.Method[2])) {
            l2 <- levels(e2)
            e2 <- l2[e2]
    }
    if(all(nchar(.Method)) && (length(l1) != length(l2) ||
        !all(sort(l2) == sort(l1))))
            stop("Level sets of factors are different")
    value <- NextMethod(.Generic)
    value[nas] <- NA
    value
}
```

This method works by converting each factor operand to the corresponding character vector, as defined by the levels. The call to switch() at the beginning of the method checks that the operator is meaningful, by examining .Generic. Only equality comparisons are allowed. If this check was omitted, arithmetic operators would be caught automatically, but comparisons, such as >, would be interpreted on the levels as character strings. This does not make sense—the levels set is explicitly unordered—so we catch it here.

The .Method object (page 470) shows which operands are factors. The first and second elements of .Method will be non-empty strings if the corresponding operand inherited from class "factor". In the case that both operands inherited, we implement a check that the two objects are compatible, in the sense that they use the same level set. The particular test there ignores, as it should, the order of appearance of the levels in the level set. When Ops.factor() is called on one factor operand and one character vector, the factor operand is converted to a character vector. Comparisons will be in this form:

```
> ratings=="Low"
 [1] F F T F F F F T T F F F F F F T F T F T
```

Without the special method, this would have ended up comparing the numbers used to code the levels with the string "Low".

Note the use of .Generic in the call to NextMethod() to avoid explicit mention of the individual function. If it was necessary to invoke the generic function directly, the simplest approach would be:

```
get(.Generic, mode = "function")(e1, e2)
```

An example of this is the method `Ops.ordered()`. It does *not* want to invoke the next method, which would be `Ops.factor()`, since that method is incompatible. Instead, it constructs some objects without any class and explicitly invokes the generic function on those.

Factors, unlike character vectors, can contain missing values, so the definition of `Ops.factor()` allows for them according to the standard S rule that an element of the result is `NA` wherever either operand is `NA`. The missing values are computed at the beginning of the function by the expression

```
nas <- is.na(e1) | is.na(e2)
```

and inserted into the result at the end by

```
value[nas] <- NA
```

Print `Ops.ordered`, the corresponding method for ordered factors, to see a similar general style, using the numeric code, however, rather than the levels. This method implements operations that are meaningful for ordered, but non-numeric, values.

A.8 Replacement Methods

Many S functions can appear on the left side of an assignment arrow, indicating replacement of some subset or attribute of the object that is the first argument of the function. These too are good candidates for special methods. S already provides a mechanism by which users can write their own replacement functions ($\boxed{\mathbf{S}}$, page 217). Replacements of the form

```
f(x) <- value
```

are evaluated by S as the expression

```
x <- "f<-"(x, value)
```

so that the user need only define the function `"f<-"()`. This mechanism combined with our mechanism for methods allows methods to be generated for replacement operations. To create a replacement method for function `f()` for objects inheriting, say, from class `"factor"`, we write a method function `"f<-.factor"()`. Parenthetically, the S evaluator actually intercepts replacements using `"["()`, `"[["()`, `"$"()`, `dim()`, `dimnames()`, `levels()`, and `tsp()` without calling a function. Nevertheless, replacement methods can be written for these, just as if a replacement function were being called, and definitions for `"[<-"()`, etc. are provided for reference.

To show an example of replacement methods, and a somewhat more ambitious example of methods, we consider a method for replacements using `[]` for data frames. We can write a replacement method for data frames, by defining a function `"[<-.data.frame"()`. The goal is to replace data as if the object were a matrix:

```
mydata[6, 1:2] <- 0
```

sets the first two variables of the sixth observation to 0. Here is a simplified version,
ignoring NA's, matrix-like variables, character subscripts, and various possible errors.
It illustrates some useful points, nevertheless:

```
"[<-.data.frame"<-
function(x, ..., value)
{
    cl <- class(x); class(x) <- NULL
    rows <- attr(x, "row.names")
    has.1 <- !missing(..1); has.2 <- !missing(..2)
    if(!has.1) {
        nrows <- length(rows)
        if(length(value) < nrows)
            value <- value[rep(1:length(value), length = nrows)]
        if(has.2)
            x[..2] <- list(value)
        else x[] <- list(value)
        class(x) <- cl
        return(x)
    }
    iseq <- seq(along = rows)[..1]
    if(has.2) jseq <- seq(along = x)[..2]
    else jseq <- seq(along = x)
    n <- length(iseq); p <- length(jseq)
    m <- length(value)
    if(m < p) value <- value[rep(1:m, length = p)]
    for(j in 1:p) {
        jj <- jseq[j]
        x[[jj]][iseq] <- value[[j]]
    }
    class(x) <- cl
    x
}
```

The first if() disposes of the special case that only columns are being replaced,
since this reduces to an ordinary replacement in a list. The remainder of the code
goes through all the columns selected, replacing the appropriate elements or rows.
The for loop applies the replacement to each variable in turn. Finally, the class
attribute is replaced and the entire data frame returned as the value of the function
call. Any replacement method, like any user-defined replacement function, should
return the new value for the entire object.

A.9 Assignment Methods

One other piece of S evaluation, in addition to function calls, may be relevant in designing new classes of data. The process of assignment associates a name with an S object, either in one of the frames active during the S session or as a permanent S object. Currently, methods are accepted for permanent assignment, by writing a method for the generic function "`<<-`"(). Methods for temporary assignments could have been allowed as well, but our feeling is that the overhead would be excessive.

To redefine permanent assignment, create a function "`<<-.`", followed by the name of the class. This method will be called just before a permanent assignment is committed ($\boxed{S}$, page 121). The arguments to the method are x (the name) and `value` (the S object to be assigned that name). As an example, we consider briefly a class of objects designed to support large amounts of data. Suppose we had defined a class `"extern"` to allow S to refer indirectly to objects, perhaps stored in a different format. The public view of external objects hides this indirect reference. Methods allow access to the data and other operations on it without reading the entire object into memory, as would happen for ordinary S objects. The private view of the object is defined by two attributes: the `class` attribute, with value `"extern"`, and the `where` attribute, whose value is a character string identifying the external data, say as a file name.

We will not try to describe how external objects would be handled in detail, but will use them to illustrate assignment methods. This class needs a method for assignment because the appropriate permanent file must be created when the assignment of an external object is to be committed. The other methods defined for `extern` objects will create new objects as the result of functions such as arithmetic operators. These objects will have special `where` names, referring to files that will be removed automatically at the end of the expression. Suppose permanent files for extern objects reside in the `.Ext` subdirectory of the working directory, in which they have the same name as the `extern` object that refers to them. The method for permanent assignment sets up the appropriate file and returns an `extern` object for assignment with the correct `where` attribute:

```
"<<-.extern" <- function(x, value) {
    perm.file <- paste(search()[1], ".Ext", x, sep="/")
    cur.file <- attr(value, "where")
    if(is.perm(cur.file)) {
      if(cur.file != perm.file)
        unix( paste("cp", cur.file, perm.file))
    } else unix( paste("mv", cur.file, perm.file)
    attr(value, "where") <- perm.file
    NextMethod()
}
```

The method essentially does two things: it puts the `where` file in the right place, by using either the `mv` or `cp` command; and it sets the `where` attribute in the object to the correct file name. Calling `NextMethod()` completes the standard assignment with the modified object.

External objects are related to a number of interesting applications of the classes/methods mechanism, including interfaces between S and database management software and the definition of classes of S databases. These applications are outside the scope of the present book, however, and we will not pursue them any further here.

A.10   Generic Functions

Generic functions that call `UseMethod()` are easy to recognize, of course, and new generic functions can be written in this form at any time. A new generic function, say `precis()`, usually looks like this:

```
precis <- function(x, ...)
    UseMethod("precis")
```

The inclusion of the name of the generic in the call to `UseMethod()` ensures that the generic function will be self-defining, so that the function object can be used anywhere in S without losing the identity of the related class.

For most generic functions, a default method will also be written:

```
precis.default <- function(x, ...) {
    if(is.null(class(x))) cat("Mode:", mode(x),
        "Length:", length(x), "\n")
    else cat("Class:", class(x),"\n")
}
```

It is not obligatory that the body of the generic consist only of the call to `UseMethod()`, although the keep-it-simple motto encourages it. The effect of `UseMethod()` is to replace the body of the generic function with the body of the method, after re-matching the arguments. See page 460.

The inclusion of "..." in the generic is recommended so methods can include optional arguments not meaningful to the generic. When `UseMethod()` invokes the method, the actual arguments to the generic are re-matched to the arguments of the method.

Generic functions working through the `.Internal` interface can't be recognized by looking at the definition of the generic. Table A.2 lists all the functions using the `.Internal()` interface for which methods can be written. There must *not* be a default method for these functions. The default is provided by the internal code. The list in Table A.2 is larger than that in Table A.1 because group methods are not allowed in all cases. The decision, somewhat arbitrary, was that some internal

x[..., drop]	x[[..., drop]]	x$name	e1 %% e2
e1 %/% e2	e1 * e2	e1 + e2	e1 - e2
e1 / e2	e1 ^e2	e1 < e2	e1 <= e2
e1 == e2	e1 > e2	e1 >= e2	e1 & e2
e1 && e2	e1 | e2	e1 || e2	e1!=e2
-e1	!x		
abs(x)	acos(x)	acosh(x)	all(x)
any(x)	as.*anything*(x)	as.vector(x, mode)	asin(x)
asinh(x)	atan(x,y)	atanh(x)	attr(x, which)
attributes(x)	ceiling(x)	cos(x)	cosh(x)
cumsum(x)	dim(x)	dimnames(x)	exp(x)
floor(x)	gamma(x)	lgamma(x)	length(x)
levels(x)	log(x)	match(x)	max(x)
min(x)	mode(x)	names(x)	prod(x)
range(x)	round(x, digits)	signif(x, digits)	sin(x)
sinh(x)	storage.mode(x)	sum(x)	tan(x)
tanh(x)	trunc(x)	tsp(x)	

Table A.2: *Functions in S for which internal code will detect special methods.*

interface routines handled a group of functions too diverse or too unlikely to be candidates for methods to justify the extra overhead in checking each time for a group method.

A.11 Comment

Methods provide a powerful tool for extending S to handle novel classes of objects. We have concentrated here on illustrating the technique and have kept the methods as simple as possible. In practical applications, designing the methods should be the most carefully thought-out part of the project. The best strategy to make methods correspond to the meaning of a class of objects in serious applications can be challenging and not entirely unambiguous. These strategic questions, although requiring care and sometimes introducing subtle issues, are not disadvantages. Rather, they illustrate the substantive needs that can be addressed directly with a rich software environment.

Having now persevered to the end, you may be interested to look back at some of the methods that have arisen throughout the book. They should illustrate the concepts behind the design of the various classes of objects, and in particular how those classes are interconnected to form our overall approach to statistical models.

Bibliographic Notes

There is an enormous literature on programming languages, and a very large one even on the subtopic of object-oriented programming. None of it is directly needed to use the facilities described here, since the approach in S is not directly derived from any one other approach. The styles of Lisp-based object-oriented programming and that of C++ represent two divergent approaches, described respectively in the books by Keene (1989) and Stroustrup (1986). Our approach is closer in some respects to the former, although, as noted, the other properties of S tend to produce a different use of methods and classes. A statistical system built directly on Lisp and having its own approach to object-oriented programming is LISP-STAT, described in Tierney (1990).

Functional programming is also widely discussed. A thorough treatment is given by Reade (1989). The book by Gelernter and Jagannathan (1990) discusses a wide range of programming languages, including both object-oriented and functional, attempting to model them all in a consistent way.

Appendix B

S Functions and Classes

This appendix contains detailed documentation for a selected subset of the functions, methods and classes of objects used in the book. Online documentation is available for these, and for all the other functions discussed in the book, by using the "?" operator. Where a generic function implies most of the information needed (e.g., `deviance`), documentation for the methods is omitted here. Conversely, if one method dominates the use in all models (e.g., `add1.lm`, `alias.lm`), only that method is included here. Finally, several methods are included if they are important and substantially different (e.g, `plot.factor`, `plot.gam`, and others for `plot`). If you want online documentation, you should not have to worry about such distinctions. Typing

    ?plot(myfit)

for example, shows you the methods for `plot()` applicable to the object `myfit`. To see all the `plot()` methods for any object whatever, do

    ?methods(plot)

The online documentation will be up to date with your version of S; therefore, it may reflect changes since this book was published.

481

| ? | On-line Information on Functions, Objects, and Calls | ? |

```
?
?name
?object
?name(object, ...)
?methods(name)
```

ARGUMENTS

name: a name or a character string giving the name of a function or operator. If omitted, documentation on ? is given (this documentation).

object: the name of an S object. Documentation will be offered on all the classes of objects from which the object inherits.

name(object, ...): a proposed call, typically to a generic function, with the first argument being some (existing) S object. Documentation will be offered on the function name itself and on all methods for name that might be used when the call is actually evaluated. However, the call is *not* evaluated: this use of ? is usually to decide what would happen *if* some proposed computation were done.

methods(name): all possible methods for function name will be presented, based on the functions available on the current search list.

In the cases where documentation is offered on all classes or methods, the options are presented to the user via the menu() function. All the possibilities (as a character vector) are returned (invisibly) as the value of ?. Not all the proposed documentation need exist: ? does not check for the existence of the documentation when it constructs the menu.

SEE ALSO

help, menu

EXAMPLE

```
?plot     # help on plot function
?myfit    # documentation for all the classes of object myfit
?"+"      # addition (and other arithmetic) Note the need for quotes
?plot(myfit)    # tell me about the plot methods for myfit
?">"(obj, 0)    # tell me about the ">" methods for obj
```

| **add1.lm** | Compute an Anova Object by Adding Terms | **add1.lm** |

```
add1.lm(object, scope, scale, keep, x)
```

ARGUMENTS

object: an `lm` object, or any object that inherits from class `lm`. In particular, a `glm` object is also appropriate for a *chi-squared* analysis based on the score-test.

scope: a `formula` object describing the terms to be added. This argument is required, and is parsed to produce a set of terms that may be added to the model on their own without breaking the hierarchy rules. The scope can also be a character vector of term labels. Any "." in `scope` is interpreted relative to the formula implied by the `object` argument.

scale: the multiplier of the *df* term in the `Cp` statistic. If not supplied, `scale` is estimated by the residual variance of `object`, or else in the case of a `glm` object the dispersion parameter.

keep: a character vector of names of components that should be saved for each augmented model. Only names from the set `coefficients`, `fitted.values`, `residuals`, `x.residuals`, `effects`, `R` are allowed. `keep=T` implies the complete set. `x.residuals` for a given term is the `X` matrix corresponding to that term, adjusted for all the terms in the model `object`. The other components are as in `object`. The default behavior is not to keep anything.

x: a model matrix that includes all the terms in `object` as well as all those to be added. This is an optional argument, used, for example, by `step.glm()`, and saves recomputing the model matrix every time.

VALUE

Using the `"R"` component of `object`, as well as the corresponding `qr` object, each of the superset models corresponding to `object` plus a term as specified in `scope` are fitted. An `anova` object is constructed, consisting of the term labels, the degrees of freedom, the residual sum of squares, and the `Cp` statistic for each superset model. If `keep` is missing, this is what is returned. If `keep` is present, a list with components `"anova"` and `"keep"` is returned. In this case, the `"keep"` component is a matrix of mode `"list"`, with a column for each superset model, and a row for each component kept.

add1() handles weighted `lm` objects, in particular `glm` objects. The weighted residual sum of squares is a Pearson chi-square statistic based on the weights of the model `object`, and a one-step iteration towards the super-set model. This results in a *score test* for the inclusion of each term. The function add1() is used as a primitive in `step.glm()`.

This function is a method for the generic function add1() for class "`lm`". It can be invoked by calling add1(x) for an object x of the appropriate class, or directly by calling add1.lm(), regardless of the class of the object.

SEE ALSO

drop1, anova, step, step.glm, step.gam

EXAMPLE

```
add1(lm.object, ~ .^2)
    # consider all interactions of terms in lm.object for inclusion
add1(glm.ob, ~ . -Age + poly(Age,2) + log(Age) + sqrt(Age))
    # try some candidate transformations for Age.
```

alias.lm	Alias Pattern for `lm` Objects	**alias.lm**

```
alias.lm(object, complete=T, partial=T, pattern=T, ...)
```

ARGUMENTS

object: a fitted model, inheriting from `lm`.

complete:

partial: flags indicating whether information for complete and partial aliasing should be included in the result.

pattern: should the resulting alias matrices be simplified by calling the `pattern()` function.

VALUE

a list potentially containing components for complete aliasing and for partial aliasing; each is included only if both requested and found to exist in the model. The component for complete aliasing is a matrix with columns corresponding to effects that are linearly dependent on the rows (i.e., effects that are completely aliased with the estimable effects). Partial aliasing is essentially the correlation of the estimable effects, with the diagonal elements set to zero.

This function is a method for the generic function `alias()` for class `"lm"`. It can be invoked by calling `alias(x)` for an object `x` of the appropriate class, or directly by calling `alias.lm()`, regardless of the class of the object.

---

anova	Compute an Anova Table	**anova**

```
anova(object, ..., test = "none")
anova(object)
```

ARGUMENTS

object: a model object, such as those produced by `lm()`, `glm()`, `aov()`, `loess()`, etc.

...: optional additional model objects; `anova()` will behave differently if additional models are provided.

test: the type of test statistic to be included in the table. The default is `"none"`, and other choices depend on the method. Typical choices are `"F"`, `"Chi"`, or `"Cp"`.

VALUE

an `anova` object. This class of objects inherits from the class `"data frame"`, and consequently suitable methods exist for printing, subsetting, etc. An additional `"heading"` attribute is a character vector that is printed at the top of the table.

If called with a single object as an argument, `anova` produces a table with rows corresponding to each of the terms in the object, plus an additional row for the residuals. The method for `aov` objects is similar to `summary()`.

When two or more objects are used in the call, a similar table is produced showing the effects of the pairwise differences between the models, considered sequentially from first to last.

SEE ALSO

`anova.lm`, `anova.glm`, `anova.gam`, `anova.aov`, `anova.loess`

EXAMPLE

```
> anova(glm.object)
Analysis of Deviance Table
Binomial model
Response: Kyphosis
Terms added sequentially (first to last)
```

```
              Df Deviance Resid. Df Resid. Dev
       NULL                    80        83.23
  bs(Start, 5)  5    23.49     75        59.75
      Number   1     1.73      74        58.02
anova(gas.null, gas.alternative)
Model 1: loess(formula = NOx ~ E, span = 1)
Model 2: loess(formula = NOx ~ E, span = 2/3)
Analysis of Variance Table
Model   Enp       RSS     Test   F Value     Pr(F)
1       3.5     0.5197  1 vs. 2    10.14    0.0009
2       5.5     0.3404
```

---

aov	Fit an Analysis of Variance Model	**aov**

```
aov(formula, data, projections = F, ...)
```

ARGUMENTS

formula: the formula for the model.

data: if supplied, a data frame in which the variables named in the formula are to be found. If data is omitted, the current search list is used; for example, a data frame may have been attached or variables may be objects in the working database.

projections: if TRUE, the fitted model will include a matrix of the projections onto the terms in the model. This matrix will be the component "projections" of the returned fit, or of the fits for the error strata in the case of multiple error strata. See proj() for the description of the projections. This adds substantially to the size of the returned object (the matrix has as many rows as observations and as many columns as terms), but if you plan to use the projections, it is more efficient to compute them during the fit rather than by calling proj() later.

...: other arguments can be supplied that are meaningful to lm(). In particular, weights, subsetting of the data frame and treatment of missing values can be supplied. See lm.

VALUE

an object describing the fit. There are two cases: if there is no Error term in the model, the object is of class "aov". This class inherits from the class of linear models (class "lm"). See aov.object and lm.object. The formula may optionally specify special blocking or error structure if it includes a term that calls the special function Error(). For example,

```
aov(response ~ time * concentration + Error(blocks)
```

specifies that factor `blocks` defines an error stratum. The resulting model will include two error strata, `blocks` and `Within`. In the case of multiple error strata, `aov` fits a separate model for each stratum. Specifically, the response is projected onto each term in the error model, and these projections are then used to fit separate models. The object returned by `aov` has class `aovlist` and is a list of `aov` objects of the form above, one for each stratum. In addition, the `"aov.list"` object has an attribute `"call"` containing the call.

EXAMPLE

```
gunaov <- aov(Rounds ~ Method + Physique/Team, gun)
```

aov.genyates	Anova for Balanced Designs	**aov.genyates**

```
aov.genyates(formula, data, onedf=F)
```

ARGUMENTS

`formula`: the formula for the model. The formula must not include an `Error` term.

`data`: if supplied, a data frame in which the objects named in the formula are to be found. If `data` is omitted, the current search list is used; frequently, a data frame will have been attached.

`onedf`: logical expression of length 1. When `onedf==T`, the function returns single degree of freedom projections in component `proj`. When `onedf==F` (the default), the function collapses the single degree of freedom projections into multi-degree of freedom projections. Each column of the collapsed result represents one term of the analysis of variance table. The sum of squares of each column is the sum of squares for the corresponding term in the model.

VALUE

an object describing the fit. It will be of class `aov`. In addition, the fit contains the projection matrix:

`proj`: an orthogonal matrix, identical to the result of the `proj.lm` function applied to the `aov` structure. Thus the two expressions:

```
proj.lm(aov(formula, data, qr=T, onedf=F))
aov.genyates(formula, data, onedf=F)$proj
```

yield the same result.

SEE ALSO

proj, proj.lm, aov

EXAMPLE

> aov.genyates(Yield ~ Temp * Conc * Cat, catalyst)

---

aov.object	Analysis of Variance (aov) Object	**aov.object**

This class of objects is returned from `aov()` and other functions to represent an analysis of variance. Class `aov` inherits from class `lm`. With a matrix response, the object inherits from `c("maov", "aov")` so that methods can use the matrix nature of the response, fitted values, etc. Objects of this class have methods for the functions `print()`, `summary()`, and `predict()` functions, among others. In addition, any function that can be used on an `lm` object can be used on an `aov` object. The components of a legitimate `aov` object are those of an `lm` object. See `lm.object` for the list. The residuals, fitted values, coefficients, and effects should be extracted by the generic functions of the same name, rather than by the "**$**" operator.

For a multivariate response, the coefficients, effects, fitted values and residuals are all matrices whose columns correspond to the response variables.

If the model formula contained an `Error` specification for fitting in strata, the object is of class `aovlist`, and each component of the list is an `aov` object. The first component of the list is for the degenerate "intercept" stratum. The object in this case also has attributes describing the overall model; in particular, the `call` and the `terms` object correspond to the components of the same name in an `lm` object.

---

aovlist.object	See `aov.object`	**aovlist.object**

---

as.data.frame	See `data.frame`	**as.data.frame**

---

as.factor	See `factor`	**as.factor**

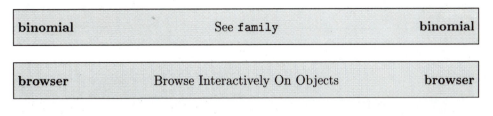

binomial	See `family`	binomial

browser	Browse Interactively On Objects	**browser**

```
browser(object, ...)
```

ARGUMENTS

object: Some S object.

VALUE

typically the value in a `return()` expression typed to the browser.

The methods for this function are used for the interaction provided in looking at S objects. Nearly always, the objects will be list-like or in some sense composed of components that are themselves useful to be examined. The various methods amount, generally, to constructing a suitable list object which then becomes the argument to `browser.default()`.

This is a generic function. Functions with names beginning in "`browser.`" will be methods for this function.

SEE ALSO

`trace, restart, debugger, traceback`

EXAMPLE

```
browser(treeobject) #examine a tree-based fitted model
browser() # examine the current evaluation frame
```

---

browser.default Browse Interactively in a Function's Frame **browser.default**

```
browser.default(frame=, catch=TRUE, parent=, message=,
          prompt="b> ", readonly=F)
```

ARGUMENTS

frame: either a list or a number, the latter meaning the corresponding frame in the evaluator. This is particularly useful when using the browser interactively: `traceback()` returns the active frame numbers, and one of these can be given to `browser` to examine data in another frame. By default the frame of the function calling `browser` is used. Therefore in its usual use, adding the expression `browser()` to a function allows you to see what the function has done so far. If `frame` is a list, 'new.frame() is called to make an evaluation-frame copy of it.

catch: logical; should errors and interrupts be caught in the browser? If `TRUE`, then the browser will be restarted after such errors, but see the quit signal comment below. If `FALSE`, any errors will return to the S prompt level.

parent: optional frame to be used as the parent frame of `frame`. Defaults to `sys.parent(2)` if `frame` is missing and to `sys.parent(1)` if `frame` is specified.

message: optional text to be printed instead of the standard browser message.

prompt: character string to be printed to prompt for input. This allows changing the prompt to distinguish between several versions of `browser` that may be in effect at the same time. If `frame` is specified numerically, the default prompt includes the frame number, e.g. `"b(5)> "`.

readonly: optional flag. If `FALSE`, and if `frame` is missing or numeric, assignments will cause changes in the corresponding evaluation frame that persist after the return from `browser()`.

VALUE

the value returned in a `return` expression typed by the user; if you return by giving a response 0 to the prompt, the value is `NULL`.

When the browser is invoked, you will be prompted for input. The input can be any expression; this will be evaluated in the `frame`. Three kinds of expressions are special. The response ? will get you a list of menu-selectable items (the elements of the frame). A numeric response is taken to be such a selection. A `return` expression returns from the browser with this value. The expression `substitute(x)` is useful to see the actual argument that was given, corresponding to the formal argument `x`.

The quit signal (usually the character control-backslash) will exit from the browser, and from the whole expression that generated the call to the browser, returning to the S prompt level. (Don't type two control-backslash characters, or one if `catch=FALSE`; either action will terminate your session with S!)

This function is a method for the generic function `browser()` for class `"default"`. It can be invoked by calling `browser(x)` for an object `x` of the appropriate class, or directly by calling `browser.default()`, regardless of the class of the object.

SEE ALSO

> `trace, restart, debugger, traceback`

EXAMPLE

```
trace(foo, browser)  #call browser on entering foo()
options(interrupt=browser)  #invoke browser upon interrupts
myfun <- function(x, y) {
    # lots of computing
    browser() #now check things just before the moment of truth
    .C("myroutine", x, y, w)
}
```

bs	Generate a Basis for Polynomial Splines	**bs**

```
bs(x, df, knots, degree=3, intercept=FALSE)
```

ARGUMENTS

x: the predictor variable.

df: degrees of freedom; one can specify `df` rather than `knots`; `bs()` then chooses `df-degree-1` knots at suitable quantiles of `x`.

knots: the *internal* breakpoints that define the spline; the range of the data provide the boundary knots. The default is `NULL`, which results in a basis for ordinary polynomial regression. Typical values are the mean or median for one knot, quantiles for more knots.

degree: degree of the piecewise polynomial—default is 3 for cubic splines.

intercept: if `TRUE`, an intercept is included in the basis; default is `FALSE`.

VALUE

a matrix of dimension `length(x)` * `df`, where either `df` was supplied or if knots were supplied, `df = length(knots) + 3 + intercept`.

`bs()` is based on the function `spline.des()` written by Douglas Bates. It generates a basis matrix for representing the family of piecewise polynomials with the specified interior knots and degree, evaluated at the values of `x`. A primary use is in modeling formulas to directly specify a piecewise polynomial term in a model.

SEE ALSO

`ns, poly, lo, s, smooth.spline`

EXAMPLE

`lm(y ~ bs(age, 4) + bs(income, 4)) # an additive model`

---

burl.tree	View all Splits for Nodes of a Tree Object	**burl.tree**

`burl.tree(tree, nodes, ...)`

ARGUMENTS

`tree:` fitted model object of class `tree`. This is assumed to be the result of some function that produces an object with the same named components as that returned by the `tree()` function.

`nodes:` an integer vector containing indices (node numbers) of all nodes to be examined. If missing, users select nodes as described below.

VALUE

the primary purpose of `burl.tree()` is its graphical side effect: for each node selected or specified, a plot of the change in deviance at each possible split, on each available predictor. For continuous predictors, a high density plot displays the change in deviance for each cut-point. For factor predictors, a scatterplot displays the change in deviance against an encoding of the subset split; the plotting symbol is the left-hand split. For the last node specified or selected, `burl.tree()` returns a named (by predictor variable) list containing the following details of the competition for the best split at that node.

ARGUMENTS

`x:` vector of cutpoints or sequence numbers (subset splits).

y: vector of deviance change if node is split at x.

cutleft: character vector of left-hand splits.

cutright: character vector of right-hand splits.

numl: the number of observations in the left-hand split at each x.

GRAPHICAL INTERACTION

This function checks that the user is in *split-screen mode.* A dendrogram of tree is expected to be visible on the current screen, and a graphics input device (e.g., a mouse) is required. Clicking (the selection button) on a node results in the additional screens being filled with the information described above. This process may be repeated any number of times. Warnings result from selecting leaf nodes. Clicking the exit button will stop the burling process and return the list described above for the last node selected. See .Device and split.screen for specific details on graphic input and *split-screen mode.*

Graphical parameters (see par()) may also be supplied as arguments to this function. In addition, the high-level graphics control arguments described under plot.default() and the arguments to title() may be supplied to this function.

EXAMPLE

```
z <- tree(Mileage~Weight + Type)
tree.screens()
plot(z)
burl.tree(z)
```

C	Factor with Chosen Contrasts	**C**

```
C(object, contr, how.many)
```

ARGUMENTS

object: a factor or ordered factor.

contr: what contrasts to use. May be one of four standard names (helmert, poly, treatment, or sum), a function, or a matrix with as many rows as there are levels to the factor.

how.many: optionally, the number of contrasts to be assigned to the factor, if fewer than k-1, where k is the number of levels. Note that setting this in a model formula is an assertion that the coefficients for the remaining factors are either known to be negligible or else should be aliased with other coefficients.

VALUE

a factor, with the contrasts attribute set as above. May be used inline in a model formula or to create a new factor.

EXAMPLE

```
# use treatment contrasts for factor Cat
aov(Yield ~ Cont* C(Cat, treatment), catalyst)
# only fit linear and quadratic effects
aov(Defects ~ C(Reliability, poly, 2) * Type * Plant)
```

coefficients	Extract Coefficients, etc. from a Model	coefficients

```
coefficients(object)
residuals(object)
fitted.values(object)
```

ARGUMENTS

object: any object representing a fitted model, or, by default any object with a
component named by the name of the extractor function.

VALUE

the coefficients, residuals, or fitted values defined by the model in `object`.
While for some models this will be identical to the component of the object
with the same name, you are encouraged to use the extractor functions,
since these will call the appropriate method for this class of object. For
example, residuals from generalized linear models come in four flavors,
and the typically most useful one is *not* the component.

NOTE

As a special inducement to use the extractor function rather than the
component, three abbreviated versions of these functions exist; namely,
`coef()`, `resid()`, and `fitted()`.

SEE ALSO

`predict, effects`

EXAMPLE

`residuals(kyph.fit)`

---

contrasts	The Contrast Matrix for a Factor	**contrasts**

```
contrasts(x)
```

ARGUMENTS

   x: a factor or ordered factor.

VALUE

   a matrix, with as many rows as there are levels for x, say k, and at most
   k-1 columns. When x is used in a model, such as through lm() or aov(),
   the portion of the model matrix assigned to x will be the result of matrix-
   multiplying the dummy matrix for the levels of x by contrasts(x). If x
   has an attribute "contrasts", this is the value of contrasts(x); otherwise,
   the standard contrasts are computed and returned. These are given by
   calling one of two functions, as named by options("contrast"). The first
   function is the default for factor objects and the second the default for
   ordered factor objects.

EXAMPLE

```
contrasts(f) <- contrasts(f)[,1:3] #only the first 3 contrasts
```

---

coplot	Conditioning Plot	**coplot**

```
coplot(formula, data, given.values, panel = points, rows, columns,
      show.given = TRUE, add = FALSE, xlab, ylab, xlim, ylim, ...)
```

ARGUMENTS

formula: formula defining the response and the predictors involved in the plotting.
   This is an S expression of the form:

   y ~ x | g1

   or

   y ~ x | g1 * g2

   where y is the response, x is the predictor against which y is plotted on the
   dependence panels, and g1 and g2 are given predictors. These variables
   may specify numeric vectors or factors. The formula may be given literally,
   or it may be an expression that evaluates to a formula.

data: data frame in which the formula will be evaluated. If missing, evaluation will take place as if the formula were evaluated in the frame of the function calling `coplot`.

given.values: a numeric vector, character vector, or two-column matrix that specifies the given values when there is one given predictor, or a list of two such objects when there are two. If missing, reasonable things happen.

panel: a user-supplied function of x and y that determines the method of plotting on the dependence panels.

rows: for the case of one given predictor, the number of rows of the matrix of dependence panels. If missing, the following is the default: let k be the number of given values; if `columns` is missing, then

```
rows <- ceiling(sqrt(k))
```

else

```
rows <- ceiling(k/columns)
```

This argument is not used if there are two given predictors.

columns: for the case of one given predictor, the number of columns of the matrix of dependence panels. If missing, the following is the default: let k be the number of given values; if `rows` is missing,

```
columns <- ceiling(k/ceiling(sqrt(k)))
```

else

```
columns <- ceiling(k/rows)
```

This argument is not used if there are two given predictors.

show.given: if `FALSE`, given panels are not included.

add: if `TRUE`, add to the current plot.

Graphical parameters (see `par()`) may also be supplied as arguments to this function. Graphical parameters (see par) may also be supplied as arguments to this function. The arguments `xlab` and `ylab` are as for other graphics functions, except that the former is a character vector of labels for predictors. If the elements have names, they are matched to the names in formula; if not, the elements are assigned, in order, according to the order in which they appear in formula. If missing, the names of the predictors in formula are used. The arguments `xlim` and `ylim` are axis limits as in other graphics functions.

For an example of the output of `coplot()`, see Figure 3.8 on page 78.

SEE ALSO

```
co.intervals, panel.smooth
```

EXAMPLE

```
# the following makes a coplot of NOx against C given E
# with smoothings of the scatterplots on the dependence panels:
E.intervals <- co.intervals(ethanol$E, 16, 0.25)
coplot(NOx ~ C | E, given.values = E.intervals, data = ethanol,
    panel = function(x, y) panel.smooth(x, y, span = 1,
    degree = 1))
```

data.frame	Construct a Data Frame Object	**data.frame**

```
data.frame( ..., row.names, check.rows = F, check.names = T)
as.data.frame(object)
```

ARGUMENTS

... : objects to be included in the data frame. These can be vectors (numeric, character, or logical), factors, numeric matrices, lists, or other data frames. Matrices, lists, and data frames provide as many variables to the new data frame as they have columns, elements, or variables, respectively. Numeric vectors and factors are included as is, and non-numeric vectors are coerced to be factors, whose levels are the unique values appearing in the vector. Making any of the above the argument in a call to the function I() prevents the expansion or conversion.

row.names: optional argument to provide the `row.names` attribute. If included, can either provide an explicit set of row names or indicate that one of the variables should be used as the row names. In the latter case, `row.names` can either be a numeric index for the variable or the name that the variable would have in the data frame. The indicated variable will be dropped as a variable and used for the row names. By default, **data.frame** tries to construct the row names from the `dimnames` attribute of a matrix argument, from the `row.names` argument of a previous data frame, or, if none of these produces row names, by using the row numbers. However the row names are constructed, they are required to be unique. Note that arguments `row.names`, `check.rows`, and `check.names`, if supplied, must be given by name.

check.rows: flag; if TRUE, the rows are checked for consistency. If several arguments imply row names, the function will check that these names are consistent.

Generally only useful in computations that claim to have selected the same rows from several parallel sources of data.

`check.names`: flag; if `TRUE`, the variable names will be made into legal S object names, by replacing illegal characters, like blanks, parentheses, or commas by ".". (type `?make.name` for details).

`object`: an object to be coerced to be a data frame. If not a data frame already, this object is likely to be a matrix or a list.

VALUE

a data frame, consisting of all the variables supplied in the arguments. The variables are required to have the same number of observations. All the variables should have names; in the case of list arguments this is required, and in other cases `data.frame` will construct default names but issue a warning. The elements of the data frame are the variables. In addition, the data frame will always have an attribute `"row.names"` containing the row names.

EXAMPLE

```
# two lists, taking one component as row names
data.frame(car.specs, car.report[-1], row.names = "Model")
```

---

data.frame.object   Data Frame Objects   **data.frame.object**

Data frames are objects that combine the behavior of data, in the sense that they can be addressed by rows (meaning observations) and columns (meaning variables), with the behavior of lists or frames in S, in the sense that the variables can be used like individual objects—for example, by attaching the object to the search list, by setting it up as a frame in the evaluator or the browser, or by passing it to a model-fitting function along with a formula using the variable names in the data frame.

Many matrix-like computations are defined as methods for data frames, notably, subsets and the `dim` and `dimnames` attributes. However, data frames are not matrices; most importantly, any object can become a variable in the data frame, so long as it is addressable by the observations. In practice, this means that the variables should be one of vectors, matrices, or some other class of objects that can itself be treated as either a vector or matrix (in particular, can be subset like a vector or matrix). If the variable is vector-like, it should have length equal to the number of rows; if matrix-like, it should have the same number of rows as the data frame.

The definition of the dimension and the dimnames of a data frame is done differently from that of a matrix. Every data frame is required to have an attribute `"row.names"` whose length is, by definition, the number of rows of the data frame. The number of columns is by definition the number of variables; that is, the length of the data frame as a list. The dimnames list is equivalent to

```
list(row.names(x), names(x))
```

Both the row names and the names are required to be there and to have no duplicate values.

ATTRIBUTES

The following attributes must be included and behave as follows.

row.names: a vector of length equal to the number of observations (and therefore equal to either the length or the number of rows of every variable). There must be no duplicate values. Where no explicit row names are supplied in creating the data frame, `1:nrows(x)` will be used.

names: the names must exist, be of full length, and be unique.

SEE ALSO

> data.frame, design, design.object, pframe.object

---

data.matrix   Convert a Data Frame into a Numeric Matrix   **data.matrix**

```
data.matrix(frame)
```

ARGUMENTS

frame: a data frame, or else a frame that inherits from class `"data.frame"` (design or model frame).

VALUE

a numeric matrix containing the numeric information in `frame`. The matrix has a column for each numeric vector, a set of columns for each matrix, and a column for each factor in `frame`. Factors are first transformed to numeric values using `codes()`. If factors are present in `frame`, the matrix returned has an attribute called `"column.levels"`, a list with an element for each column of the matrix. The elements of this list are either NULL or else contain the levels of the factors prior to conversion by `codes()`.

SEE ALSO

> codes, as.matrix, model.matrix

---

deriv	Symbolic Partial Derivatives of Expressions	**deriv**

```
deriv(expr, namevec, function.arg, tag=".expr")
```

ARGUMENTS

expr: expression to be differentiated, typically a formula, in which case the expression returned computes the right side of the $\sim$ and its derivatives.

namevec: character vector of names of parameters.

function.arg: optional argument vector or prototype for a function. If present, the returned value is in the form of a function instead of a multiple statement expression. When `function.arg` is given as a function prototype, the function arguments can have defaults.

tag: base of the names to be given to intermediate results. Default `".expr"`.

VALUE

a multiple-statement expression or a function definition. When evaluated, these statements return the value of the original expression along with an attribute called `"gradient"`. This is the gradient matrix of the expression value with respect to the named parameters.

If `function.arg` is a character vector, the result is a function with the arguments named in `function.arg`. If `function.arg` is a function, the result is a function with the same arguments and default values.

While generating this sequence of expressions, the function attempts to eliminate repeated calculation of common subexpressions. Sometimes user assistance is needed, as in the example below. To improve readability, expressions that are used only once are folded back into the expression where they are used. Since parentheses are always added when such expressions are folded in, there may be redundant parentheses in the final expressions.

The symbolic differentiation and the simplification of the result are highly recursive. Even for relatively simple expressions, S can reach its limit on the number of nested expressions and give an error message. The remedy is to increase the value of the option `expressions` when this happens.

EXAMPLE

```
# value and gradient of the Michaelis-Menten model
> deriv(~ Vm*conc/(K+conc),c("Vm","K"))
expression({
      .expr1 <- Vm * conc
      .expr2 <- K + conc
      .value <- .expr1/.expr2
      .grad <- array(0, c(length(.value), 2),
            list(NULL, c("Vm", "K")))
      .grad[, "Vm"] <- conc/.expr2
      .grad[, "K"] <-  - (.expr1/(.expr2^2))
      attr(.value, "gradient") <- .grad
      .value
}
)
# to obtain a function as the result
> deriv(~ Vm*conc/(K+conc),c("Vm","K"),
+   function(Vm, K, conc = 1:10) NULL)
function(Vm, K, conc = 1:10)
{
      .expr1 <- Vm * conc
      .expr2 <- K + conc
      .value <- .expr1/.expr2
      .grad <- array(0, c(length(.value), 2),
              list(NULL, c("Vm", "K")))
      .grad[, "Vm"] <- conc/.expr2
      .grad[, "K"] <-  - (.expr1/(.expr2^2))
      attr(.value, "gradient") <- .grad
      .value
}
```

design	Generate a Design Object	**design**

```
design(..., factor.names)
```

ARGUMENTS

   ...: objects that can be interpreted as factors in a design: vectors, data frames, matrices, or factors themselves. Each object or column of a data frame or matrix will be considered as a template for a factor. Numeric vectors and matrices will be converted to factors, with the unique values of the vector or the column of the matrix defining the levels.

factor.names: optional vector of names for the factors. If omitted, names will be constructed. In the case that the argument(s) are matrices with `dimnames` for the columns, these dimnames will be used. Otherwise, the standard factor names are used.

VALUE

an object of class `design`, inheriting from the class `data.frame`. This function should be compared and contrasted with `data.frame()`, which does not force all variables to be factors, and with `data.matrix()`, which in a sense performs the inverse operation to `design()`, by converting factors to numeric variables.

SEE ALSO

`fac.design, oa.design`

EXAMPLE

```
# dmat is a numeric matrix with appropriate levels in the
# rows; myfac is a factor defined on the same observations.
mydesign <- design(dmat, myfac)
```

---

design.object   Design Objects   **design.object**

Designs inherit from data frames. By virtue of having class `"design"`, they are treated differently by some generic functions, notably `plot()`. The assumption is that a design object starts life as a data frame, all of whose variables are factors. Then, one or more quantitative variables are added. Unless told otherwise, methods for designs tend to assume that the first quantitative variable found in the design should be the response.

ATTRIBUTES

Designs need have no attributes other than those for data frames. They may have special information indicating that they were produced as fractional factorial designs or as orthogonal array designs (see `fac.design()` or `oa.design()`). The important function `factor.names()` returns or sets the names for the factors and for their levels. It does not use an explicit attribute of this name, however. Instead, it uses the names of the design object and the levels of each factor variable.

SEE ALSO

`data.frame.object`, `data.frame`, `design`, `fac.design`, `oa.design`

---

design.table   Arrange Response as a Multiway Array   **design.table**

```
design.table(design, response)
```

ARGUMENTS

`design:` a data frame representing a design, perhaps with a response included.

`response:` a response variable. If omitted, the first non-factor included in the design is used.

VALUE

a multiway array, whose elements are formed from the response. Suppose `d` is a numeric vector whose elements are the number of levels in each of the factors in the design. Then the dimension attribute of the array is `d` if there are no replicated values in the design, or `c(nrep, d)` otherwise, where `nrep` is the maximum number of replications. The dimnames attribute of the array is the factor names of `design`, with an initial element of `1:nrep` if there are replications.

Note that rearranging the response as a multiway array makes more sense for complete designs. The function will work fine for fractional designs, but the resulting array will be mostly `NA`s, and hard to look at. Such designs are usually easier to look at just by printing them, but if you have some functions that work on multiway arrays with `NA`s, the value of `design.table` can be used whether the design is complete or not.

---

deviance   Extract the Deviance from a Fitted Model Object   **deviance**

```
deviance(object)
```

ARGUMENTS

object: a fitted model object, typically of class `glm` or `gam`, although others are possible.

VALUE

the deviance of the fitted model is returned. For `glm`, `gam`, and `tree` models, the `"deviance"` is a component of `object`, in which case `deviance()` is a simple extractor function. For other models, the `gaussian` family is assumed and the weighted residual sum of squares is returned.

---

drop1.lm   Compute an Anova Object by Dropping Terms   **drop1.lm**

```
drop1(object, scope, scale, keep)
```

ARGUMENTS

object: an `lm` object, or any object that inherits from class `lm`. In particular, a `glm` object is also appropriate for a *chi-squared* analysis based on the score test.

scope: an optional `formula` object describing the terms to be dropped. Typically this argument is omitted, in which case all possible terms are dropped (without breaking hierarchy rules). The scope can also be a character vector of term labels. If the argument is supplied as a `formula`, any "." is interpreted relative to the formula implied by the `object` argument.

scale: the multiplier of the df term in the `Cp` statistic. If not supplied, `scale` is estimated by the residual variance of `object`, or else in the case of a `glm` object the dispersion parameter.

keep: a character vector of names of components that should be saved for each
        subset model. Only names from the set `coefficients`, `fitted.values`,
        `residuals`, `x.residuals`, `effects`, `R` are allowed. `keep=T` implies the com-
        plete set. `x.residuals` for a given term is the `X` matrix corresponding to
        that term, adjusted for all the other terms in the model `object`. The
        other components are as in `object`. The default behavior is not to keep
        anything.

VALUE

using the `"R"` component of `object`, each of the subset models corresponding to the terms specified in `scope` is computed. An `anova` object is constructed, consisting of the term labels, the degrees of freedom, the residual
sum of squares, and the Cp statistic for each subset model. If `keep` is missing, this is what is returned. If `keep` is present, a list with components
`"anova"` and `"keep"` is returned. In this case, the `"keep"` component is a
matrix of mode `"list"`, with a column for each subset model, and a row
for each component kept.

`drop1()` handles weighted `lm` objects, including `glm` objects. The weighted
residual sum of squares is a Pearson chi-square statistic based on the
weights of the full model, and a one-step iteration towards the subset
model. This results in a *score test* for the removal of each term. The
function `drop1()` is used as a primitive in `step.glm()`.

This function is a method for the generic function `drop1()` for class `"lm"`.
It can be invoked by calling `drop1(x)` for an object `x` of the appropriate
class, or directly by calling `drop1.lm()`, regardless of the class of the object.

SEE ALSO

        `add1`, `step`, `step.glm`, `step.gam`

EXAMPLE

```
drop1(lm.ob)
drop1(lm.ob, ~ . - Age) # drop all terms except Age
drop1(lm.ob, keep=T)
```

fac.design	Generate Factorial Designs	**fac.design**

```
fac.design(levels, factor.names, replications = 1,
    row.names, fraction)
```

ARGUMENTS

`levels`: vector of the number of levels for each factor in the design.

`factor.names`: optional factor names attribute. This may be a character vector, giving the names of the factors, or a list. If it is a list, the names attribute of the list is the names of the factors, and the elements of the list (which need not be of mode character) give the levels of the corresponding factor. If factor names are not given, they default to `std.factor.names(length(levels))`; namely, `"A"`, `"B"`, etc. If a factor's levels are not named, the levels are set to the factor name (possibly abbreviated) followed by level numbers.

`replications`: the number of times the complete design should be replicated.

`row.names`: optional names to use for the rows of the design. Defaults to `1:nrows`.

`fraction`: optional definition for the fraction desired in a fractional factorial design. This may either be a numerical fraction (e.g, 1/4 for a quarter replicate), or a model formula giving one or more defining contrasts, as in the example below. See `fractionate()` for details. Fractional factorials are provided only for two-level factors.

VALUE

a design corresponding to the factors specified. The `design` object is a data frame, with variables in the frame corresponding to each of the factors requested in the design.

SEE ALSO

`design`, `oa.design`, `fractionate`.

EXAMPLE

```
# a 1/4 replicate of a 2^5 design,
> fac.design(rep(2,5), names = fnames,
+    fraction =  ~ A:C:D + A:B:E )
   react acidcon acidamt reactim reactem
1   4mod    dil    2mol      2     low
2   5mol    dil   2.5mol     2     low
```

```
3   5mol      con     2mol      4      low
4   4mod      con    2.5mol     4      low
5   5mol      con     2mol      2      high
6   4mod      con    2.5mol     2      high
7   4mod      dil     2mol      4      high
8   5mol      dil    2.5mol     4      high
Fraction: y. ~ A:B:C + B:D:E
```

factor	Create Factor Object	**factor**

```
factor(x, levels, labels)
is.factor(x)
as.factor(x)
```

ARGUMENTS

x: data, to be thought of as taking values on a finite set (the `levels`). Missing values (`NA`s) are allowed.

levels: optional vector of levels for the factor. Any data value that does not match a value in `levels` will be `NA` in the factor.

labels: optional vector of values to use as labels for the levels of the factor, in place of the levels set.

VALUE

object of class `"factor"`, representing values taken from the finite set given by `levels()`. It is important that this object is *not* numeric; in particular, comparisons and other operations behave as if they operated on values from the levels set, which is always of mode character. `NA` can appear, indicating that the corresponding value is undefined. The expression `na.include(f)` returns a factor like `f`, but with `NA`s made into a level.

`is.factor` returns `TRUE` if `x` is a factor object, `FALSE` otherwise.

`as.factor` returns `x`, if `x` is a factor, `factor(x)` otherwise.

SEE ALSO

ordered, na.include.

EXAMPLE

```
factor(occupation)  # "doctor", "lawyer", etc.
# make readable labels
occ <- factor(occupation,level=c("d","l"),
    label=c("Doctor","Lawyer"))
# turn factor into character vector
as.vector(factor)
colors <- factor(color,c("red","green","blue"))
table(colors)  #table counting occurrences of colors
```

---

factor.names   Extract or Set Factor and Level Names   **factor.names**

```
factor.names(design)
factor.names(design) <- values
```

ARGUMENTS

design: a design, typically to be used with the analysis of variance and/or other
functions for designed experiments.

values: either a list, similar to the returned value of `factor.names()`, or a vector,
to be used as the names of the factors. In the second case, the factor levels
will default as below.

VALUE

`factor.names()` returns a list, whose **names** attribute contains the names
of the factors in the design, and whose elements are the levels for the
corresponding factors. Defaults will be produced wherever necessary: the
factor names default to "A", '"B", etc., and the levels to abbreviated factor
names, with "1", etc. pasted on.

When used on the left side of an assignment, `factor.names` takes **values**
and coerces them to the form just described, using the default rules.

EXAMPLE

```
> factor.names(design.1)
$Temperature:
[1] 160 180
$Concentration:
[1] 20 40
$Catalyst:
[1] "cat A" "cat B
> factor.names(design2) <- c("Glass","Phosphor")
```

```
> factor.names(design2)
$Glass:
[1] "G1" "G2" "G3"
$Phosphor:
[1] "P1" "P2"
```

family	Generate a Family Object	**family**

```
family(object)
binomial(link = logit)
gaussian()
Gamma(link = inverse)
inverse.gaussian()
poisson(link = log)
quasi(link = identity, variance = constant)
```

ARGUMENTS

link: the choices of link functions are `logit`, `probit`, `cloglog`, `identity`, `inverse`, `log`, `"1/mu^2"`, and `sqrt`. Not all links are suitable for all families. The following table summarizes the suitable pairings:

	binomial	gaussian	Gamma	inverse.gaussian	poisson	quasi
logit	*					*
probit	*					*
cloglog	*					*
identity		*	*		*	*
inverse			*			*
log			*		*	*
1/mu^2				*		*
sqrt					*	*

The function `power()` can also be used to generate a *power* link function object for use with `quasi()`; `power()` takes an argument `lambda`.

variance: the choices of variance functions are `constant`, `mu(1-mu)`, `mu`, `mu^2`, and `mu^3`. This argument may be used only with `quasi()`; each of the other families implies a variance function.

object: any object from which a family object can be extracted. Typically a fitted model object, with a default of `gaussian`.

VALUE

a family object, which is a list of functions and expressions used by `glm()` and `gam()` in their iteratively reweighted least-squares algorithms. Each of the names, except for `quasi` and the family extractor function `family()`, are associated with a member of the exponential family of distributions. As such, they have a fixed variance function. There is typically a choice of link functions, with the default corresponding to the *canonical* link for that family. The `quasi` name represents *Quasi-likelihood* and need not correspond to any particular distribution; rather `quasi()` can be used to combine any available link and variance function.

Users can construct their own families, as long as they have compatible components having the same names as those, for example, of `binomial()`. The easiest way is to use `quasi()` with home-made `link` and `variance` objects; otherwise `make.family()` can be used, or else direct construction of the family object. When passed as an argument to `glm()` or `gam()` with the default link, the empty parentheses () can be omitted. There is a print method for the class `"family"`.

SEE ALSO

`family.object, glm, gam, robust, power`

EXAMPLE

```
binomial(link = probit) # generate binomial family with probit link
glm(formula, family = binomial)
robust(gaussian) # create a robust version of the binomial family
gam(formula, family = robust(quasi(link = power(2))))  # the works!
```

family.object	A Family of GLM Models	**family.object**

This class of objects is returned by one of the *family* functions. See
`family` for the choices. It is a list of functions and expressions that define
the IRLS iterations for fitting `glm` and `gam` models. These `family` objects
allow a great deal of flexibility in the use of `glm()` and `gam()`. In particular,
they allow construction of robust fitting algorithms and composite link
functions. There is a `print()` method for `family` objects, that produces a
simple summary without any details; use `unclass(family.object)` to see
the contents.

COMPONENTS

The following components, with a corresponding functionality, are required for a `family` object.

family: a character vector giving the family name, and the names of the link and
variance functions.

link: a function with argument `mu` that transforms from the scale of the mean
to the scale of the linear or additive predictor `eta`.

inverse: a function with argument `eta`, the inverse of the link.

deriv: a function with argument `mu`, the derivative of the link function.

initialize: an expression to initialize the values of the fitted values `mu` in the
body of `glm()` or `gam()`. Other values can also be initialized, such as the
prior weights `w`, or the maximum number of iterations `maxit`, to name
two. Modifying these expressions should be done with some care, and is
only recommended for experienced users. Other variables local to `glm()` or
`gam()` can be initialized as well; see `binomial()$initialize` for an example.
The initialize expression can also be used to transform a response variable
having specialized structure into the required vector response `y`. Once
again the binomial serves as an example.

variance: a function with argument `mu`, the variance function.

deviance: the deviance function has four arguments:

```
deviance(mu, y, w, residuals = F)
```

and returns the deviance, a quantity similar to the residual sum of squares
for a Gaussian least squares model. If `residuals=T`, `deviance()` returns a

vector of deviance residuals, whose weighted sum of squares is the deviance.

weight: an expression for updating the iterative weights. For the `binomial` family, this expression is `w*mu*(1 - mu)`, for the `gaussian` it is `w`, where `w` are the prior weights.

---

fitted.values	See `coefficients`	**fitted.values**

formula	Define or Extract a Model Formula	**formula**

```
formula(object)
```

ARGUMENTS

object: either a formula expression (a call to the $\sim$ operator), or an object that defines such an expression, such as a fitted model or a `terms` object.

VALUE

an object of class `"formula"`, essentially just the call to $\sim$.

This is a generic function. Functions with names beginning in `"formula."` will be methods for this function.

SEE ALSO

```
formula.object
```

EXAMPLE

```
sqrt(skips) ~ . #or, equivalently
formula(sqrt(skips) ~ .)
formula(fuel.fit) # find the formula
```

---

formula.object	Model Formula Objects	**formula.object**

This class of objects represents the structural models in all model-fitting functions, and is used also in a number of other functions, particularly for plots. Formulas are their own value; that is, they represent an expression calling the operator $\sim$, but evaluating this expression just returns the expression itself. The purpose of formula objects is to supply the essential information to fit models, produce plots, etc., in a readable form that can be passed around, stored in other objects, and manipulated to determine the terms and response of a model. Names in the formula will eventually be interpreted as objects, often as variables in a data frame. This interpretation, however, only takes place when the related subexpressions have been removed from the formula object.

Useful generic functions for formulas include: `terms()`, `update()`, and `plot()`.

---

fractionate	Produce a Fractional Factorial Design	**fractionate**

```
fractionate(design, fraction)
```

ARGUMENTS

`design`: a design object; that is, a data frame containing factors.

`fraction`: the fraction desired. This is either a numeric fraction (e.g., 1/2, 1/4), or a formula containing the terms to be used as defining contrasts. If numeric, `fractionate()` will choose a fraction according to a set of defining contrasts representing an attempt to allow estimation of as many low-order effects as possible. (The fractions are in the object `dimdc.list` if you want to look at them.) If a formula is supplied, its terms should generally be simple high-order interactions—that is, factor names linked by ":". Each such term defines one interaction (combination of factors). The design will be divided in half according to whether the corresponding contrast variable has value +1 or -1. The term can appear in the formula with sign either "+" or "-", and the positive or negative half will be chosen accordingly. Notice that negative and positive terms do not cancel in this use of formulas.

VALUE

a new design object, containing the rows of the original design specified by `fraction`. The function only works for 2^k designs. Defining fractions for factors with three or more levels is more complicated, and we have not attempted to do so. The design has an attribute `"fraction"` containing the defining contrast(s).

EXAMPLE

```
# a 1/4 replicate from the full 2^5 design in full.design
# specifying the -1 fraction for the first contrast
> davies.design <- fractionate(full.design,
+     fraction = ~ -A:C:D + A:B:E)
```

gam	Fit a Generalized Additive Model	**gam**

```
gam(formula, family = gaussian, data, weights, subset, na.action,
                start, control, trace=F, model=F, x=F, y=T, ...)
```

ARGUMENTS

formula: a formula expression as for other regression models, of the form `response` $\sim$ `predictors`. See the documentation of `lm()` and `formula` for details. Nonparametric smoothing terms are indicated by `s()` for smoothing splines or `lo()` for `loess` smooth terms. See the documentation for `s` and `lo` for their arguments. Additional smoothers can be added by creating the appropriate interface. Interactions with nonparametric smooth terms are not fully supported, but will not produce errors; they will simply produce the usual parametric interaction.

family: a family object—a list of functions and expressions for defining the `link` and `variance` functions, initialization, and iterative weights. Families supported are `gaussian`, `binomial`, `poisson`, `Gamma`, `inverse.gaussian` and `quasi`. Functions such as `binomial()` produce a family object, but can be given without the parentheses. Family functions can take arguments, as in `binomial(link=probit)`.

data: an optional data frame in which to interpret the variables occurring in the formula.

weights, subset, na.action: the optional weights for the fitting criterion, subset of the observations to be used in the fit, and function to be used to handle any `NA`s in the data. These are interpreted as in the `lm()` function.

`start`: an optional vector of initial values on the scale of the additive predictor.

`control`: a list of iteration and algorithmic constants. See `gam.control()` for their names and default values. These can also be set as arguments to `gam()` itself.

`...`: all the optional arguments to `lm()` can be given to `gam()`, including `weights`, `subset` and `na.action`.

VALUE

an object of class `gam` is returned, which inherits from both `glm` and `lm`. Can be examined by `print()`, `summary()`, `plot()`, and `anova()`. Components can be extracted using extractor functions `predict()`, `fitted()`, `residuals()`, `deviance()`, `formula()`, and `family()`. Can be modified using `update()`. It has all the components of a `glm` object, with a few more. Other generic functions that have methods for `gam` objects are `step()` and `preplot()`. Use `gam.object` for more details.

The model is fit using the *local scoring* algorithm, which iteratively fits weighted additive models by *backfitting.* The backfitting algorithm is a Gauss-Seidel method for fitting additive models, by iteratively smoothing partial residuals. The algorithm separates the parametric from the nonparametric part of the fit, and fits the parametric part using weighted linear least squares within the backfitting algorithm. Although nonparametric smooth terms `lo()` and `s()` can be mixed in a formula, it is more efficient computationally to use a single smoothing method for all the smooth terms in an additive model. In this case the entire local scoring algorithm is performed in FORTRAN.

SEE ALSO

`gam.object`, `glm`, `family`

EXAMPLE

```
gam(kyphosis ~ s(age,4) + Number, family = binomial)
gam(ozone^(1/3) ~ lo(rad) + lo(wind, temp))
gam(kyphosis ~ poly(Age,2) + s(Start), data=kyph.data,
    subset = Number>10)
```

gam.object	Generalized Additive Model Object	**gam.object**

This class of objects is returned by the `gam()` function to represent a fitted generalized additive model. Class `gam` inherits from class `glm`, since the parametric part a `gam.object` is fit by weighted least-squares; the object returned has all the components of a `glm`. Objects of this class have methods for the functions `print()`, `plot()`, `summary()`, `anova()`, `predict()`, `fitted()`, and `step()`, among others.

COMPONENTS

The following components must be included in a legitimate `gam` object. The residuals, fitted values, coefficients and effects should be extracted by the generic functions of the same name, rather than by the "$" operator. The `family()` function returns the entire family object used in the fitting, and `deviance()` can be used to extract the deviance of the fit.

`coefficients`: the coefficients of the parametric part of the `additive.predictors`, which multiply the columns of the model matrix. The names of the coefficients are the names of the single-degree-of-freedom effects (the columns of the model matrix). If the model is overdetermined there will be missing values in the coefficients corresponding to inestimable coefficients.

`additive.predictors`: the additive fit, given by the product of the model matrix and the coefficients, plus the columns of the `"smooth"` component.

`fitted.values`: the fitted mean values, obtained by transforming the component `additive.predictors` using the inverse link function.

`smooth, nl.df, nl.chisq, var`: these four characterize the nonparametric aspect of the fit. `smooth` is a matrix of smooth terms, with a column corresponding to each smooth term in the model; if no smooth terms are in the `gam` model, all these components will be missing. Each column corresponds to the strictly nonparametric part of the term, while the parametric part is obtained from the model matrix. `nl.df` is a vector giving the approximate degrees of freedom for each column of `smooth`. For smoothing splines specified by `s(x)`, the approximate `df` will be the trace of the implicit smoother matrix minus 2. `nl.chisq` is a vector containing a type of score test for the removal of each of the columns of `smooth`. `var` is a matrix like `smooth`, containing the approximate pointwise variances for the columns of `smooth`.

residuals: the residuals from the final weighted additive fit; also known as *working residuals*, these are typically not interpretable without rescaling by the weights.

deviance: up to a constant, minus twice the maximized log-likelihood. Similar to the residual sum of squares.

null.deviance: the deviance corresponding to the model with no predictors.

iter: the number of *local scoring* iterations used to compute the estimates.

family: a three-element character vector giving the name of the family, the link, and the variance function; mainly for printing purposes.

weights: the iterative weights from the final IRLS fit

The object will also have the components of an `lm` object: `coefficients`, `residuals`, `fitted.values`, `call`, `terms` and some others involving the numerical fit. See `lm.object`.

Gamma	See `family`	**Gamma**

gaussian	See `family`	**gaussian**

glm	Fit a Generalized Linear Model	**glm**

```
glm(formula, family = gaussian, data, weights, subset, na.action,
        start, control, trace=F, model=F, x=F, y=T, ...)
```

ARGUMENTS

formula: a formula expression as for other regression models, of the form `response` ~ `predictors`. See the documentation of `lm()` and `formula` for details.

family: a family object—a list of functions and expressions for defining the `link` and `variance` functions, initialization, and iterative weights. Families supported are `gaussian`, `binomial`, `poisson`, `Gamma`, `inverse.gaussian`, and `quasi`. Functions such as `binomial()` produce a family object, but can be given without the parentheses. Family functions can take arguments, as in `binomial(link=probit)`. See `family`.

data: an optional data frame in which to interpret the variables occurring in the formula.

weights, subset, na.action: the optional weights for the fitting criterion, subset of the observations to be used in the fit, and function to be used to handle any NAs in the data. These are interpreted as in the lm() function.

start: a vector of initial values on the scale of the linear predictor.

control: a list of iteration and algorithmic constants. See glm.control() for their names and default values. These can also be set as arguments to glm() itself.

trace: if TRUE, details of the iterations are printed. Can also be set in the control argument.

model: if TRUE, the model.frame is returned. If this argument is itself a model.frame, then the formula and data arguments are ignored, and model is used to define the model.

x: if TRUE, the model.matrix is returned.

y: if TRUE, the response variable is returned (default is TRUE).

qr: if TRUE, the QR decomposition of the model.matrix is returned.

...: all the optional arguments to lm() can be provided to glm(), including weights, subset, and na.action. Note that weights refers to original prior weights, not the iterative weights used in fitting. See lm for documentation of these arguments.

VALUE

an object of class glm is returned, which inherits from lm. Can be examined by print(), summary(), plot(), and anova(). Components can be extracted using predict(), fitted(), residuals(), deviance(), formula(), and family(). Can be modified using update(). It has all the components of an lm object, with a few more. Other generic functions that have methods for glm objects are drop1(), add1(), step(), and preplot(). See glm.object for further details.

The model is fit using *iterative reweighted least squares* (IRLS). The working response and iterative weights are computed using the functions contained in the family object. glm models can also be fit using the function gam(). The workhorse of glm() is the function glm.fit(), which expects an x and y argument rather than a formula.

SEE ALSO

glm.object, gam, family, glm.fit

EXAMPLE

```
glm(Count ~ ., data = solder, family = poisson)
glm(Kyphosis ~ poly(Age, 2) + (number > 10)*Start,
    family = binomial)
glm(ozone^(1/3) ~ bs(rad, 5) + poly(wind, temp, degree = 2))
```

glm.object	Generalized Linear Model Object	**glm.object**

This class of objects is returned by the `glm()` function to represent a fitted generalized linear model. Class `glm` inherits from class `lm`, since it is fit by iterative reweighted least squares; the object returned has all the components of a weighted least squares object. The class of `gam` objects, on the other hand, inherit from class `glm`. Objects of class `glm` have methods for the functions `print()`, `plot()`, `summary()`, `anova()`, `predict()`, `fitted()`, `drop1()`, `add1()`, and `step()`, among others.

COMPONENTS

The following components must be included in a legitimate `glm` object. The residuals, fitted values, coefficients and effects should be extracted by the generic functions of the same name, rather than by the "$" operator. The `family()` function returns the entire family object used in the fitting, and `deviance()` can be used to extract the deviance of the fit.

coefficients: the coefficients of the `linear.predictors`, which multiply the columns of the model matrix. The names of the coefficients are the names of the single-degree-of-freedom effects (the columns of the model matrix). If the model is overdetermined, there will be missing values in the coefficients corresponding to inestimable coefficients.

linear.predictors: the linear fit, given by the product of the model matrix and the coefficients; also the `fitted.values` from the final weighted least-squares fit.

fitted.values: the fitted mean values, obtained by transforming `linear.predictors` using the inverse link function.

residuals: the residuals from the final weighted least-squares fit; also known as *working* residuals, these are typically not interpretable without rescaling by the weights.

deviance: up to a constant, minus twice the maximized log-likelihood. Similar to the residual sum of squares.

`null.deviance:` the deviance corresponding to the model with no predictors.

`iter:` the number of IRLS iterations used to compute the estimates.

`family:` a three-element character vector giving the name of the family, the link, and the variance function; mainly for printing purposes.

`weights:` the iterative weights from the final IRLS fit.

The object will also have the components of an `lm` object: `coefficients`, `residuals`, `fitted.values`, `call`, `terms`, and some others involving the numerical fit. See `lm.object`.

---

inherits	Test Inheritance of an Object	**inherits**

```
inherits(x, what, which=F)
```

ARGUMENTS

`x:` any object, possibly but not necessarily having a class attribute.

`what:` a character vector of possible classes.

`which:` option; if `TRUE`, the returned value specifies which classes matched the object; otherwise, the value is a single logical, suitable for use in an `if` expression test.

VALUE

`TRUE` if any of the classes in the class attribute of `x` match (exactly) any of the strings in `what`.

EXAMPLE

```
# the definition of as.factor()
function(x) if(inherits(x,"factor")) x else factor(x)
```

interaction	Compute the Interaction of Several Factors	**interaction**

```
interaction(design, drop = F)
interaction(..., drop = F)
```

ARGUMENTS

design:

...: the arguments to `interaction` can be either a data frame containing all the factors to be used *or* all the individual factors. It will not understand a combination of factors and designs as arguments; you have to pick one form or the other.

drop: if `TRUE` the levels of the new factor not represented in the data are dropped.

VALUE

a new factor, whose levels are all possible combinations of the factors supplied as arguments. If `drop = T`, only the levels represented in the new factor are retained.

EXAMPLE

```
> attach(catalyst)
> Temp
[1] 160 180 160 180 160 180 160 180
> Conc
[1] 20 20 40 40 20 20 40 40
> interaction(Temp, Conc)
[1] 160.20 180.20 160.40 180.40 160.20 180.20 160.40 180.40
```

---

interaction.plot   Plots of the Response for Pairs of Factors   **interaction.plot**

```
interaction.plot(x.factor, trace.factor, response, ..., fun = mean,
    trace.label = deparse(substitute(trace.factor)))
```

ARGUMENTS

`x.factor`: factor to be plotted on the x-axis It may be a factor in a design object.

`trace.factor`: factor whose levels will be separate traces. It may be a factor in a design object.

`response`: vector containing the response. It may be contained in a data frame.

`...`: optional specification of graphical parameters, including parameters for matplot, to be applied before doing the plot (and reset after the plot is finished).

`fun`: a function or the name of a function. It should be a summary function returning one number on each call.

`trace.label`: heading given to factor plotted as traces.

VALUE

a plot will be created showing the requested function of responses for each level of the `x.factor` at each level of the `trace.factor`. By default, lines for each value of the `trace.factor` are drawn in different styles so that they may be more easily distinguished. Note: Ignore warning messages: missing values generated coercing to double.

For an example of the output of `interaction.plot()`, see Figure 5.5 on page 168.

EXAMPLE

```
> attach(catalyst)
> interaction.plot(x.factor = Conc, trace.factor = Cat, Yield)
> detach()
```

---

inverse.gaussian   See `family`   **inverse.gaussian**

---

lm	Fit Linear Least-Squares Regression Model	lm

```
lm(formula, data, weights, subset,
   na.action, method="qr", model=F, x=F, y=F, ...)
```

ARGUMENTS

formula: a formula object, with the response on the left of a $\sim$ operator, and the terms, separated by "+" operators, on the right. This argument is passed around *unevaluated*; that is, the variables mentioned in the formula will be defined when the model frame is computed, not when lm() is initially called. In particular, if data is given, all these names should generally be defined as variables in that data frame.

data: an optional data.frame in which to interpret the variables named in the formula, or in the subset and the weights argument.

weights: optional weights; if supplied, the algorithm fits to minimize the sum of the weights multiplied into the squared residuals. The weights must be nonnegative and it is strongly recommended that they be strictly positive, since zero weights are ambiguous, compared to use of the subset argument.

subset: optional expression saying that only a subset of the rows of the data should be used in the fit. This argument, like the terms in formula, is evaluated in the context of the data frame, if present. The specific action of the argument is as follows: the model frame, including weights and subset, is computed on *all* the rows, and then the appropriate subset is extracted. A variety of special cases make such an interpretation desirable (e.g., the use of lag() or other functions that may need more than the data used in the fit to be fully defined). On the other hand, if you meant the subset to avoid computing undefined values or to escape warning messages, you may be surprised. For example,

```
lm(y ~ log(x), mydata, subset = x > 0)
```

will still generate warnings from log(). If this is a problem, do the subsetting on the data frame directly:

```
lm(y ~ log(x), mydata[,mydata$x > 0])
```

na.action: a missing-data filter function, applied to the model.frame, after any subset argument has been used.

method: the least-squares fitting method to be used; the default is `"qr"`. The method `"model.frame"` simply returns the model frame.

model, x, y, qr: flags to control what is returned. If these are `TRUE`, then the model frame, the model matrix, the response, and/or the QR decomposition will be returned as components of the fitted model, with the same names as the flag arguments.

... : additional arguments for the fitting routines. The most likely one is `singular.ok=T`, which instructs the fitting to continue in the presence of over-determined models (the default method recognizes this, but if new fitting methods are written, they don't have to do so).

VALUE

an object representing the fit. Generic functions such as `print()` and `summary()` have methods to show the results of the fit. See `lm.object` for the components of the fit, but the functions `residuals()`, `coefficients()`, and `effects()` should be used rather than extracting the components directly, since these functions take correct account of special circumstances, such as overdetermined models. The response may be a single numeric variable or a matrix. In the latter case, coefficients, residuals, and effects will also be matrices, with columns corresponding to the response variables. In either case, the object inherits from class `"lm"`. For multivariate response, the first element of the class is `"mlm"`.

EXAMPLE

```
lm(Fuel ~ . , fuel.frame)
```

lm.object	Linear Least-Squares Model Object	lm.object

This class of objects is returned from the `lm()` function to represent a fitted linear model. Class `lm` is also inherited by other fitted models, when the fitting computation is based eventually on linear least-squares. Examples include `aov`, `glm`, and `gam` objects. If the response variable is a matrix, the class of the object is `c("mlm", "lm")` so that methods can use the matrix nature of the response, fitted values, etc. Objects of this class have methods for the functions `print()`, `plot()`, `summary()`, and `predict()`, among others. In addition, the function `kappa()` can be used to estimate how ill-determined the model was, and the function `qqnorm()` applied to the residuals is a good test of the distributional assumptions.

COMPONENTS

The following components must be included in a legitimate `lm` object. The residuals, fitted values, coefficients, and effects should be extracted by the generic functions of the same name, rather than by the "$" operator. For pure `lm` objects this is less critical than for some of the inheritor classes.

`coefficients`: the coefficients of the least-squares fit of the response to the columns of the model matrix. The names of the coefficients are the names of the single-degree-of-freedom effects (the columns of the model matrix). If the model was overdetermined and `singular.ok` was true, there will be missing values in the coefficients corresponding to inestimable coefficients.

`residuals, fitted.values`: the residuals and fitted values from the fit.

`effects`: orthogonal, single-degree-of-freedom effects. Using the `"qr"` method, there will be as many of these as observations. The first `rank` of them correspond to degrees of freedom in the model and are named accordingly.

`R`: the triangular factor of the decomposition. For `method=qr`, this is determined by the orthogonal decomposition of the model matrix. For other methods, it may be computed by other calculations, but note that summary methods for `lm` objects assume the existence of this component. If it is not computed, the methods will fail.

`rank`: the computed rank (number of linearly independent columns in the model matrix). If the rank is less than the dimension of `R`, columns of `R` will have been pivoted, and missing values inserted in the coefficients. The upper-left `rank` rows and columns of `R` are the nonsingular part of the fit, and the remaining columns of the first `rank` rows give the aliasing information (see `alias()`).

assign: the list of assignments of coefficients (and effects) to the terms in the model. The names of this list are the names of the terms. The ith element of the list is the vector saying which coefficients correspond to the ith term. It may be of length 0 if there were no estimable effects for the term.

terms: an object of mode `expression` and class `term` summarizing the formula. Used by various methods, but typically not of direct relevance to users.

call: an image of the call that produced the object, but with the arguments all named and with the actual formula included as the formula argument.

df.residual: the number of degrees of freedom for residuals.

qr: optionally, the `qr` decomposition object. See `qr` for its structure. Depends on using method `"qr"`.

model: optionally the model frame, if `model=T`.

x: optionally the model matrix, if `x=T`.

y: optionally the response, if `y=T`.

For a multivariate response, the object returned has class `"mlm"`, and the coefficients, effects, fitted values, and residuals are all matrices whose columns correspond to the response variables.

lo	Specify a loess fit in a GAM formula	**lo**

```
lo(..., span=0.5, degree=1)
```

ARGUMENTS

...: the unspecified ... can be a comma-separated list of numeric vectors, numeric matrix, or expressions that evaluate to either of these. If it is a list of vectors, they must all have the same length.

span: the number of observations in a neighborhood. This is the smoothing parameter for a `loess` fit.

degree: the degree of local polynomial to be fit; can be 1 or 2.

VALUE

a numeric matrix is returned. The simplest case is when there is a single argument to `lo()` and `degree=1`; a one-column matrix is returned, consisting of a normalized version of the vector. If `degree=2` in this case, a two-column matrix is returned, consisting of a 2d-degree orthogonal-polynomial basis. Similarly, if there are two arguments, or the single argument is a two-column matrix, either a two-column matrix is returned if `degree=1`, or a five-column matrix consisting of powers and products up to degree 2. Any dimensional argument is allowed, but typically one or two vectors are used in practice. The matrix is endowed with a number of attributes; the matrix itself is used in the construction of the model matrix, while the attributes are needed for the backfitting algorithms `all.wam()` or `lo.wam()` (weighted additive model). Local-linear curve or surface fits reproduce linear responses, while local-quadratic fits reproduce quadratic curves or surfaces. These parts of the `loess()` fit are computed exactly together with the other parametric linear parts of the model.

Note that `lo()` itself does no smoothing; it simply sets things up for `gam()`.

SEE ALSO

s, bs, ns, poly, loess

EXAMPLE

```
y ~ Age + lo(Start, span=.5)
    # fit Start using a loess smooth with a span of 0.5.
y ~ lo(Age) + lo(Start, Number)
y ~ lo(Age, 0.5) # the argument name for span is not needed.
```

loess	Fit a Local Regression Model	**loess**

```
loess(formula, data, subset, na.action, model = FALSE, weights,
      family = c("gaussian", "symmetric"), normalize = TRUE,
      span = 3/4, enp.target, degree = 2, drop.square, parametric,
      control = loess.control(), ...)
```

ARGUMENTS

formula: a formula object, with the response on the left of a ∼ operator, and the terms, separated by "*" operators, on the right. This argument is passed around *unevaluated* , that is, the variables mentioned in the formula will be defined when the model frame is computed, not when `loess()` is initially called. In particular, if `data` is given, all these names should generally be defined as variables in that data frame, and in no case should you expect that names of local variables in the function calling `loess()` can appear in the formula and be matched to those local variables.

data: an optional `data.frame` in which to interpret the variables named in the formula, the `subset` and the `weights` argument.

subset: optional expression saying that only a subset of the rows of the data should be used in the fit. This argument, like the terms in `formula`, is evaluated in the context of the data frame, and should typically only involve variables in that frame. The specific action of the argument is as follows: the model frame, including weights and subset, is computed on *all* the rows, and then the appropriate subset is extracted. There are a variety of special cases that make such an interpretation desirable (e.g, the use of `lag()` or other functions that may need more than the data used in the fit to be fully defined). On the other hand, if you meant the subset to avoid computing undefined values or to escape warning messages, you may be surprised. For example,

```
loess(y ∼ log(x), mydata, subset = x > 0)
```

will still generate warnings from `log()`. If this is a problem, do the subsetting on the data frame directly:

```
loess(y ∼ log(x), mydata[,mydata$x > 0])
```

na.action: a missing-data filter function, applied to the `model.frame`, after any subset argument has been used.

model: if `TRUE`, the model frame is returned.

weights: optional expression for weights to be given to individual observations in the sum of squared residuals that forms the local fitting criterion. By default, an unweighted fit is carried out. If supplied, `weights` is treated as an expression to be evaluated in the same data frame as the model formula. It should evaluate to a non-negative numeric vector. If the different observations have nonequal variances, `weights` should be inversely proportional to the variances.

family: the assumed distribution of the errors. The values are `"gaussian"` or `"symmetric"`. The first value is the default. If the second value is specified, a robust fitting procedure is used.

normalize: logical that determines if numeric predictors should be normalized. If `TRUE`, the standard normalization is used. If `FALSE`, no normalization is carried out.

span: smoothing parameter.

enp.target: another way to specify the amount of smoothing. An approximation is used to compute a value of `span` that will yield approximately `enp.target` equivalent number of parameters.

degree: overall degree of locally-fitted polynomial. 1 is local-linear fitting and 2 is local-quadratic fitting.

drop.square: for cases with `degree` equal to 2 and with two or more numeric predictors, this argument specifies those numeric predictors whose squares should be dropped from the set of fitting variables. The argument can be a character vector of the predictor names given in `formula`, or a numeric vector of indices that gives positions as determined by the order of specification of the predictor names in `formula`, or a logical vector of length equal to the number of predictor names in formula.

parametric: for two or more numeric predictors, this argument specifies those variables that should be conditionally parametric. The method of specification is the same as for `drop.square`.

control: a list that controls the methods of computation in the loess fitting. The list can be created by the function `loess.control()`, whose documentation describes the computational options.

...: arguments of the function `loess.control()` can also be specified directly in the call to `loess` without using the argument `control`.

VALUE

> an object of class `"loess"` representing the fitted model. See the documentation for `loess.object` for more information on the components.

SEE ALSO

> `specs.loess, pointwise, loess.control`.

EXAMPLE

```
> attach(ethanol)
> loess(NOx ~ C * E, span = 1/2, degree = 2, parametric = "C",
+ drop.square = "C")
Call:
loess(formula = NOx ~ C * E, span = 1/2, degree = 2,
      parametric = "C", drop.square = "C")

Number of Observations:        88
Equivalent Number of Parameters: 9.2
Residual Standard Error:       0.1842
Multiple R-squared:            0.98
Residuals:
    min    1st Q  median   3rd Q     max
 -0.5236 -0.0973 0.01386 0.07345 0.5584
```

loess.object	*Loess Model Object*	**loess.object**

This class of objects is returned from the `loess()` function to represent a fitted local regression model. Objects of this class have methods for the functions `print()`, `plot()`, `preplot()`, `predict()`, and `anova()` functions, among others.

COMPONENTS

`fitted.values`: surface evaluated at the observed values of the predictors.

`residuals`: response minus fitted values.

`terms`: an object of mode `expression` and class `term` summarizing the formula. Used by various methods, but typically not of direct relevance to users.

`call`: an image of the call that produced the object, but with the arguments all named and with the actual formula included as the formula argument.

model: the model frame, which is present only if the argument `model` is `TRUE`. The model frame contains the data—after transformation, subsetting, and treating missing values—to which the local regression model is fitted.

The remaining components of the `loess` object are lists: `surface`, `errors`, `control`, `inference`, and `predictors`. The first two contain all of the information about the specification of the local regression model, apart from the selection of the data used in the fit. The third contains the computational options. The contents of these three components can be inspected by using `specs.loess`. The component `inference` contains information that is used by other local regression functions to carry out inferences. The component `predictors` contains information about the predictors.

SEE ALSO

    loess, predict.loess, plot.loess, pointwise, anova.loess, specs.loess,
    preplot.loess, loess.control

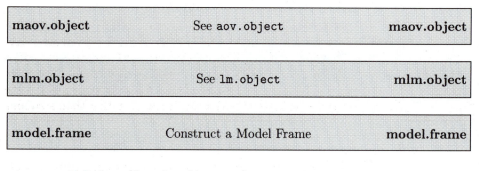

| **maov.object** | See `aov.object` | **maov.object** |

| **mlm.object** | See `lm.object` | **mlm.object** |

| **model.frame** | Construct a Model Frame | **model.frame** |

    model.frame(formula, data, ...)

ARGUMENTS

formula: the formula or other object defining what terms should be included in the model frame. Besides being a formula object, this can be a fitted model of various kinds, in which case the formula used in fitting the model defines the terms.

data: optional data frame from which the model frame is to be constructed.

...: other arguments to the model fitting functions, such as `weights=`, `subset=`, `na.action=` are passed on to 'model.frame().

Typically, `model.frame()` is called less often by users than by functions that are either fitting a model or summarizing one. The default method for `model.frame()` constructs the model frame from the terms (usually inferred from the formula), the data if any, and any special expressions such as subsets, weights, or whatever the particular fitting method needs.

VALUE

a data frame representing all the terms in the model (precisely, all those terms of order 1; i.e., main effects), plus the response if any, and any special extra variables (such as weight arguments to fitting functions). One such argument is handled specially—namely, `subset=`. If this argument is present, it is used to compute a subset of the rows of the data. It is this subset that is returned. The returned data frame has an attribute `terms` containing the terms object defined by the formula. The response and any extra variables other than `subset` are stored in the data frame. They should be retrieved from the frame by using

```
model.extract(fr, response) # for response
model.extract(fr, weights) # for weights=
```

and so on for whatever names were used in the arguments to `model.frame()`. Other than `subset`, the names of such extras are arbitrary; they only need to evaluate to a legitimate variable for the data frame (e.g., a numeric vector, matrix, or factor). The names of such variables are specially coded in the model frame so as not to conflict with variable names occurring in the terms. You should always use `model.extract()`, which shares the knowledge of the coded names with `model.frame()`, rather than assuming a specific coding.

NOTE

Model frames are more typically produced as a side-effect of fitting a model rather than directly by calling `model.frame()`. Functions like `lm()` take an option `model=T`, that produces the model frame as a component of the fit.

EXAMPLE

```
model.frame(fuel.fit)
model.frame(sqrt(skips) ~ ., solder)
```

| **ms** | Fit a Nonlinear Model by Minimum Sums | **ms** |

ms(formula, data, start, control, trace)

ARGUMENTS

formula: the nonlinear model formula. There will be no left side to the $\sim$ expression. Unlike formulas for linear models, nonlinear formulas include the parameters to be estimated. The right side of the formula is essentially an arbitrary S expression. When evaluated, it should return values to be minimized.

data: a data frame in which to do the computations. In addition to the usual data variables, the data frame may contain parameters (set, typically, by using the param() assignment for objects of class "pframe") that establish initial values for the fit, or are used for any other purpose.

start: optional starting values for the iteration. If start is omitted, the model-fitting will look for starting values as ordinary objects with the names of the parameters. Note that whenever the names of the parameters are not supplied explicitly, the assumption is that any names occurring in formula that are not variables in the data frame are parameters.

If start is supplied, it can be either a list or a numeric vector. The list is the most general and is recommended for unambiguous specification of the parameters. In either case, names(start) gives the names of the parameters. Notice that the list form allows the individual parameter names to refer to subsets of the parameters of arbitrary length. If a numeric starting vector is supplied the named parameters must each be of length 1.

control: optional list of control values to be used in the iteration. See ms.control() for the possible control parameters and their default settings.

trace: should a trace function be called after each step of the iteration? Default FALSE. Otherwise, trace can be either TRUE or the name of a function to use as a tracer. The standard tracer function is trace.ms(). Also available, by trace="browser.ms", is an invocation of the interactive browser, in a frame containing all the fitting information. See the definition of these functions for the calling sequence to any do-it-yourself tracer function. The use of special trace functions with ms() should be distinguished from the standard S tracing. The latter is simpler and usually the best way to track the modeling. Tracing through `ms()` allows access in S to the internal flags of the FORTRAN minimization algorithm. If you don't need to look at that information, you can usually trace a function you have written to compute the model information. Often tracing on exit, for example,

```
trace(mymodel, exit = browser)
```

is a good way to look at your function `mymodel()` just before it returns the next model values.

VALUE

an object of class `"ms"` with the the final parameters, function and derivative values, and some internal information about the fit. See `ms.object`.

SEE ALSO

`ms.control, nls`

EXAMPLE

```
fit.alpha <- ms( ~ lprob(D * alpha), pingpong)
```

---

ms.object	Nonlinear Fitting Object	**ms.object**

This object is returned by the function `ms()` to represent the result of fitting a nonlinear model by general minimization.

COMPONENTS

The object contains the final parameter values, corresponding function and gradient values, and final values for the flags generated internally in the minimization algorithm. If the model was defined in terms of n contributions from n observations, as in the case of minimizing the negative log-likelihood, the function value and derivatives will also be returned on a per-observation form for use in plots, etc.

`parameters`: the final values of the parameters in the estimation.

`formula`: the formula used for the estimation.

`call`: an image of the call to `ms()`, but with all the arguments explicitly named, so that the `data` component of the call will always give the `data` argument, and so on.

pieces, slopes, curves: these are the contributions of the N observations to, respectively, the value, the gradients and the hessian of the objective function. The first is a vector of length N, the other two are matrices with N rows. The slopes component is only returned if derivatives are computed and the curves component only if second derivatives are computed.

scale: the scaling vector used by the optimization algorithm.

opt.parameters, flags: these are the floating point and integer parameters used and generated by the underlying FORTRAN algorithm. You hope you don't need to know about them, but if you do, see the documentation for the algorithm dmnf in the Port library.

SEE ALSO

    nls.object

---

nls	Nonlinear Least Squares	**nls**

    nls(formula, data, start, control, algorithm)

ARGUMENTS

formula: the nonlinear regression model as a formula.

data: a data frame in which to do the computations. In addition to the usual data variables, the data frame may contain parameters (set, typically, by using the param() assignment for objects of class "pframe") that establish initial values for the fit, or are used for any other purpose.

start: optional starting values for the iteration. If start is omitted, the model-fitting will look for starting values as ordinary objects with the names of the parameters. Note that whenever the names of the parameters are not supplied explicitly, the assumption is that any names occurring in formula that are not variables in the data frame are parameters. On the whole, setting up the parameters in the data frame is often simpler, particularly if you want to experiment interactively with different starting values.

If start is supplied, it can be either a list or a numeric vector. The list is the most general and is recommended for unambiguous specification of the parameters. In either case, names(start) gives the names of the parameters. Notice that the list form allows the individual parameter names to refer to subsets of the parameters of arbitrary length. If a numeric starting vector is supplied the named parameters must each be of length 1.

control: optional list of control values to be used in the iteration, including the maximum number of iterations, tolerance for convergence, possible tracing, and scaling factors. For the complete list of the available control options and their default settings, see the documentation for `nls.control`.

algorithm: which algorithm to use. The default algorithm is a Gauss-Newton algorithm. If algorithm is `"plinear"` the Golub-Pereyra algorithm for partially linear least-squares models is used.

For the default algorithm the left side of `formula` is the response to be fitted. The right side should evaluate to a numeric vector of the same length as the response. If the value of the right side has an attribute called `"gradient"` this should be a matrix with the number of rows equal to the length of the response and one column for each of the parameters. The skelton of functions to provide this can be formed using `nl.deriv`. When there are linear parameters in the model as well as nonlinear parameters, the `"plinear"` algorithm can be used. The right side of the `formula` should evaluate to the derivative matrix for the linear parameters, conditional on the nonlinear parameters. This matrix can be given instead as a vector whose length is a multiple of the length of the left side. If the `"gradient"` attribute is included, it should be an array of dimension the number of observations by number of linear parameters by number of nonlinear parameters.

VALUE

an object inheriting from class `"nls"`, containing the parameters, residuals, fitted values, and derivatives of the model at the end of the iteration.

EXAMPLE

```
# fitting Michaelis and Menten's original data
> conc <- c(0.3330, 0.1670, 0.0833, 0.0416,
+      0.0208, 0.0104, 0.0052)
> vel <- c(3.636, 3.636, 3.236, 2.666, 2.114, 1.466, 0.866)
> Micmen <- data.frame(conc=conc, vel=vel)
> param(Micmen,"K") <- 0.02; param(Micmen,"Vm") <- 3.7
> fit <- nls(vel~Vm*conc/(K+conc),Micmen)
```

---

nls.object	Nonlinear Least-Squares Object	**nls.object**

This is an object inheriting from class `"nls"` with the following components:

parameters: the final value of the parameters in the estimation.

formula: the formula used for the estimation.

call: an image of the call to `nls()`, but with all the arguments explicitly named, so that the `data` component of the call will always give the `data` argument, and so on.

pnames: parameter names

residuals: the final value of the residuals.

fitted.values: the final value of the right side of `formula`.

data: a copy of the `data` argument with the final value of the parameters.

R: the upper-triangular R matrix from a QR decomposition of the gradient matrix at the final value of the parameters.

---

ns	Generate a Basis Matrix for Natural Cubic Splines	**ns**

```
ns(x, df, knots, intercept=F)
```

ARGUMENTS

x: the predictor variable.

df: degrees of freedom. One can supply `df` rather than knots; `ns()` then chooses `df-1-intercept` knots at suitably chosen quantiles of `x`.

knots: breakpoints that define the spline. The default is no knots; together with the natural boundary conditions this results in a basis for linear regression on `x`. Typical values are the mean or median for one knot, quantiles for more knots.

intercept: if `TRUE`, an intercept is included in the basis; default is `FALSE`.

VALUE
> a matrix of dimension `length(x) * df` where either `df` was supplied or if `knots` were supplied, `df = length(knots) + 1 + intercept`.
>
> `bs()` is based on the function `spline.des()` written by Douglas Bates. It generates a basis matrix for representing the family of piecewise-cubic splines with the specified sequence of interior knots, and the natural boundary conditions. These enforce the constraint that the function is linear beyond the boundary knots, which are taken to be at the extremes of the data. A primary use is in modeling formula to directly specify a natural spline term in a model.

SEE ALSO
> `bs, poly, lo, s`

EXAMPLE
> ```
> lsfit(ns(x,5),y)
> lm(y ~ ns(age, 4) + ns(income, 4)) # an additive model
> ```

---

oa.design	Generate an Orthogonal Array Design	**oa.design**

```
oa.design(levels, factor.names, min.resid.df=0)
```

ARGUMENTS

`levels`: vector of the number of levels for each factor in the desired design. Currently only two or three levels are allowed.

`factor.names`: optional factor names attribute. This may be a character vector, giving the names of the factors, or a list. If it is a list, the names attribute of the list is the names of the factors, and the elements of the list (which need not be of mode character) give the levels of the corresponding factor. If factor names are not given, they default to `std.factor.names(length(levels))`—namely, `"A"`, `"B"`, etc. If a factor's levels are not named, the levels are set to the factor name (possibly abbreviated) followed by level numbers.

`min.resid.df`: minimum residual degrees of freedom requested for a main-effects-only model.

VALUE
> a design for the factors specified, generated by selecting some of the columns from one of a stored catalog of orthogonal array designs. The

design object is a data frame, with variables in the frame corresponding to each of the factors requested in the design. Three additional attributes are special to orthogonal array designs: "**generating.oa**" gives the name of the object that contains the complete orthogonal array design from which the result was generated; "**selected.columns**" says which columns of this object were used to produce the result; "**residual.df**" gives the number of residual degrees of freedom in the design, when only the main effects are fitted (you may want to check this value to see how many more residual degrees of freedom than are needed in your application are available). **oa.design** may not be able to find a design as requested. If so, an error stop is made.

SEE ALSO

fac.design, design, fractionate.

EXAMPLE

```
oa <- oa.design(c(2,3,3,3,3,3))
#produces an 18 run design with 6 degrees of freedom
#for error assuming only main effects are fit.
```

ordered	Create or Modify Ordered Factors	**ordered**

```
ordered(x, levels, labels)
ordered(x) <- levels
```

ARGUMENTS

    x: data to be made into an ordered factor.

levels: optional vector of levels for the factor. Any data value that does not match a value in **levels** is coded in the output vector as **NA**. The levels will be assumed ordered (low to high) in the order given. If omitted, the sorted unique values of **x** will be used.

labels: optional vector of values to use as labels for the levels of the factor.

VALUE

an ordered factor, i.e., an object of class c("ordered", "factor").

When **ordered()** is used on the left of an assignment, the levels of **x** will be taken to be ordered according to the argument on the right side of the assignment. Typically, **levels** in this case will consist of some permutation of the current levels of x. If values in `levels(x)` are missing from `levels`, any corresponding data values in x will become `NA`.

The assignment can also be applied to a data frame, in which case the right side is taken to apply to each of the variables in the data frame. The right side should be either a logical vector of length equal to the number of variables, or else a list of the same length. In the case of a list, each element acts like the right side of an assignment of the ordered attribute of the corresponding variable.

SEE ALSO

    factor, design, data.frame

EXAMPLE

    ratings <- ordered(ratings.text, c("Low","Med","High"))
    # reverse the ordering
    ordered(ratings) <- c("High","Med","Low")

---

pairs	Produce a Scatter Plot Matrix	**pairs**

    pairs(x, labels = names(x), panel = points, ...)

ARGUMENTS

  x: matrix-like object; pairs of columns will be plotted.

labels: optional character vector for labeling the variables in the plots. The strings `labels[1]`, `labels[2]`, etc. are the labels for the 1st, 2nd, etc., panel in the diagonal panels. If supplied, the label vector must have length equal to `ncol(x)`.

 panel: a user-supplied function of x and y that determines the method of plotting on the panels.

Graphical parameters (see `par()`) may also be supplied as arguments to this function.

This is a generic function. Functions with names beginning in `"pairs."` will be methods for this function.

For an example of the output of `pairs()`, see Figure 3.7 on page 77.

EXAMPLE

    pairs(ethanol)

panel.smooth Smoothing Scatterplots on Multipanel Displays **panel.smooth**

```
panel.smooth(x, y, span = 2/3, degree = 1, family = c("symmetric",
    "gaussian"), zero.line = FALSE, evaluation = 50, ...)
```

ARGUMENTS

x: refers to abscissas of points on a panel.

y: refers to ordinates of points on a panel.

span: smoothing parameter.

degree: overall degree of locally fitted polynomial. 1 is locally linear fitting and 2 is locally quadratic fitting.

family: the values are "gaussian" or "symmetric". In the first case, local fitting methods are used. In the second case, the default, local fitting is used together with a robustness feature that guards against distortion by outliers.

zero.line: if TRUE, the line y = 0 is drawn on the panel.

evaluation: number of values at which the loess curve is evaluated.

This function adds smooth curves to the scatterplots on multipanel displays made by graphical functions such as pairs() and coplot(). The smoothing method used is loess(). The fit is evaluated at evaluation equally spaced points from min(x) to max(x) and then graphed by connecting the successive plotting locations by line segments.

Graphical parameters (see par()) may also be supplied as arguments to this function.

SEE ALSO

coplot, loess

EXAMPLE

```
E.intervals <- co.intervals(E, 16, 0.25)
coplot(NOx ~ C | E, given = E.intervals, data = ethanol,
    panel = function(x, y) panel.smooth(x, y, span = 1, degree = 1))
```

parameters	Parameters in a Parametrized Data Frame.	parameters

```
parameters(x)
parameters(x) <- value
param(x, what)
param(x, what) <- value
```

ARGUMENTS

　　x: a data frame, specifically one inheriting from class "pframe".

　what: character string, the name of the parameter. Parameters must be addressed by name, like attributes, which they very much resemble.

　value: the value for the assignment. If assigning all the parameters, this should be a list.

VALUE

parameters() returns or sets all the parameters. param() returns or sets the specific parameter named in what. The parameters are arbitrary named quantities. When a pframe object is attached or is the data argument to a model-fitting function, the parameters become available for computations just like the variables (i.e., the components) of the data frame. However, they are otherwise unrestricted; in particular, they do not need to correspond to the set of observations (the rows) of the data frame. The names of the parameters must be unique and must not conflict with the names of the variables in the data frame.

EXAMPLE

```
param(myframe, "lambda") <- 1.5
```

---

partition.tree	Plot a Low-Dimensional Tree Object	**partition.tree**

```
partition.tree(tree, label, add = F)
```

ARGUMENTS

  tree: fitted model object of class `tree`. This is assumed to be the result of some function that produces an object with the same named components as that returned by the `tree()` function.

  label: a column name of `tree$frame` that defines how each partition will be labeled.

  add: logical; if `TRUE`, the partition or step function is added to the current plot.

VALUE

  this function is used for its graphical side effect, a plot of the relationship $y = f(x)$ (one-variable tree) or $y = f(x1, x2)$ (two-variable tree). The function stops if `tree` consists of more than two predictors or if the predictors are factors. For a single predictor, $y=f(x)$ is a step function. For two predictors, the prediction space is carved into partitions, displaying the fitted values in each.

---

pframe.object	Parametrized Data Frame Objects	**pframe.object**

  Parametrized frames (`"pframe"` objects) inherit from data frames. Their essential difference is that they contain special *parameters* , which are just a list of arbitrary named objects, kept separate from the variables in the data frame specifically so they will not be subject to the constraints that all variables apply to the same set of observations. When a `pframe` is attached to the search list, the constraint is dropped and the parameters become accessible by name, just as the variables are.

ATTRIBUTES

  In addition to the attributes of data frames:

parameters: the parameter objects as a list. The list must be named, and the names must be unique.

SEE ALSO

  `data.frame.object`, `data.frame`.

---

plot   Plot an Object   **plot**

```
plot(x, ...)
plot(x, y, ...)
```

ARGUMENTS

 x: an S object. The plot will either display the data defined by x alone, or will plot data supplied in x (usually on the horizontal axis) versus data supplied by a second argument, y (usually on the vertical axis).

   Graphical parameters (see par()) may also be supplied as arguments to this function. In addition, the high-level graphics control arguments described under plot.default() and the arguments to title() may be supplied to this function.

   This is a generic function. Functions with names beginning in "plot." will be methods for this function.

---

plot.data.frame Distributional Plots of Variables **plot.data.frame**

```
plot.data.frame(object, labels = dimnames(data)[[2]], ...)
```

ARGUMENTS

object: data frame to be plotted.

labels: character vector of labels for the variables. If the elements have names, they are matched to the names of the variables of the data frame; if not, the elements are assumed to be in the same order as the columns of the data frame.

   This function makes plots of the sample distributions of the variables of a data frame. The observations of numeric predictors are graphed by a quantile plot; that is, if x is a numeric variable, sort(x) is graphed against ppoints(x). For a factor, counts of occurrences of levels are graphed.

   Graphical parameters (see par()) may also be supplied as arguments to this function.

   This function is a method for the generic function plot() for objects of class "data.frame". It can be invoked by calling plot(x) for an object x of the appropriate class, or directly by calling `plot.data.frame()`, regardless of the class of the object.

For an example of the output of `plot.data.frame()`, see Figure 3.6 on page 76.

EXAMPLE

```
plot(ethanol)  # ethanol is a data frame
x <- data.frame(weight = rawdata[, 1], sex = rawdata[, 2])
plot(x, labels = c("Weights of Mongooses", "Sex of Mongooses"))
```

---

plot.design   Plot a Function of Each Level of Factors or Terms   **plot.design**

```
plot.design(x, y, fun = mean, ...)
```

ARGUMENTS

x: either a data frame containing the design factors and, optionally, the response, or a formula.

y: the response, if not given in x.

This function is a method for the generic function `plot()` for class `"design"`. It can be invoked by calling `plot(x)` for an object x of the appropriate class, or directly by calling `plot.design()`, regardless of the class of the object. There are two basic styles for calling this method. If x is a design or data frame, then y can select a response variable from the design or, if y is a formula, it can define both terms and response, relative to the given design. Conversely, if x is not a design, it is assumed to be a formula or terms object, or something else from which a model frame can be computed. Then the design factors and response are determined from that model frame.

fun: a function or the name of a function. It should be a summary function returning one number on each call. The plot shows this function of the response for each level of each factor.

The supplied function will be called once for each level of each factor in the design. A plot will show these summary values. The levels for a particular factor are shown along a vertical line, and the overall value of `fun()` for the response is drawn as a horizontal line.

Graphical parameters (see `par()`) may also be supplied as arguments to this function.

This function is a method for the generic function `plot()` for class `"design"`. It can be invoked by calling `plot(x)` for an object `x` of the appropriate class, or directly by calling `plot.design()`, regardless of the class of the object.

For an example of the output of `plot.design()`, see Figure 5.2 on page 164.

EXAMPLE

```
# draw trimmed means
plot.design(catalyst, fun = function(x) mean(x, trim = .05),
    col = 2)
#choose which factors to include in the plot
plot.design(Yield ~ Conc + Cat, data = catalyst)
```

plot.factor	Summary Plots by Factors	**plot.factor**

```
plot.factor(x, y, style="box", rotate, boxmeans, character)
```

ARGUMENTS

x, y: the first argument gives either factor(s) or a formula. In the former case `x` can be a single factor or a design, and `y` is expected to be the response. In the latter case, both the response and the factors are determined by `x`; if `y` is given, it is a data frame in which to evaluate the formula.

style: character string indicating style of plot. Possible values are `"box"` (default), `"fraction non0"`, `"shaded bar"`, and `"character"`. The names can be abbreviated (one character is enough). The four styles produce: boxplots; bar-plots of the fraction of observations not equal to 0; shaded bars with each shaded area representing the number of observations having a particular value; characters plotted at the values of `y`, the characters given by argument `character=`. Supplying argument `character=` implies the style.

rotate: if `TRUE`, *x*-axis labels will be rotated.

boxmeans: `TRUE` if you want the mean of the boxplots to be indicated. Applicable for `style="box"` only.

character: name of factor whose levels will be used as plotting characters.

Produces a set of plots, one for each factor. Levels of the factor are arranged along the *x*-axis. Depending on the value of `style`, a box, bar, shaded bar, or characters will be plotted at each level.

Graphical parameters (see `par()`) may also be supplied as arguments to this function.

This function is a method for the generic function `plot()` for class `"factor"`. It can be invoked by calling `plot(x)` for an object `x` of the appropriate class, or directly by calling `plot.factor()`, regardless of the class of the object.

For an example of the output of `plot.factor()`, see Figure 5.3 on page 166.

EXAMPLE

```
# do box plots of all factors.
> plot.factor(Yield ~ ., catalyst, character = catalyst$Temp,
    main = "Yield, points identified by levels of Temp")
# do character plot of interaction
> attach("solder.balance")
> Boards <- interaction(Solder, Mask, Opening)
> plot.factor(Boards, skips, character = Panel, rotate = T)
> detach()
```

---

plot.gam	Plot Components of a GAM Object	**plot.gam**

```
plot.gam(x, residuals, rug, se, scale, ask = F)
```

ARGUMENTS

   x: a gam object, or a `preplot.gam` object. The first thing `plot.gam()` does is check if `x` has a component called `preplot`; if not, it computes one using `preplot.gam()`. Either way, it is this `preplot.gam` object that is required for plotting a gam object.

residuals: if TRUE, partial deviance residuals are plotted along with the fitted terms—default is FALSE. If `residuals` is a vector with the same length as each fitted term in `x`, then these are taken to be the overall residuals to be used for constructing the partial residuals.

   rug: if TRUE (the default), a univariate histogram or `rugplot` is displayed along the base of each plot, showing the occurrence of each x-value; ties are broken by jittering.

   se: if TRUE, upper and lower pointwise twice-standard-error curves are included for each plot. The default is FALSE.

scale: a lower limit for the number of units covered by the limits on the y-axis for each plot. The default is `scale=0`, in which case each plot uses the range of the functions being plotted to create their `ylim`. By setting `scale` to be the maximum value of `diff(ylim)` for all the plots, then all subsequent plots will produced in the same vertical units. This is essential for comparing the importance of fitted terms in additive models.

ask: if `TRUE`, `plot.gam()` operates in interactive mode.

VALUE

a plot is produced for each of the terms in the object `x`. The function currently knows how to plot all main-effect functions of one or two predictors. So in particular, interactions are not plotted. An appropriate x-y plot is produced to display each of the terms, adorned with residuals, standard-error curves, and a rugplot, depending on the choice of options. The form of the plot is different, depending on whether the x-value for each plot is numeric, a factor, or a matrix.

When `ask=T`, rather than produce each plot sequentially, `plot.gam()` displays a menu listing all the terms that can be plotted, as well as switches for all the options. An additional switch called `browser` allows users to temporarily regain control, and is useful for setting `par()` options, or for adding additional information to the current plot.

A `preplot.gam` object is a list of precomputed terms. Each such term (also a `preplot.gam` object) is a list with components `x`, `y` and others—the basic ingredients needed for each term plot. These are in turn handed to the specialized plotting function `gplot()`, which has methods for different classes of the leading `x` argument. In particular, a different plot is produced if `x` is numeric, a category or factor, a matrix, or a list. Experienced users can extend this range by creating more `gplot()` methods for other classes.

Graphical parameters (see `par()`) may also be supplied as arguments to this function.

This function is a method for the generic function `plot()` for class `"gam"`. It can be invoked by calling `plot(x)` for an object `x` of the appropriate class, or directly by calling `plot.gam()`, regardless of the class of the object.

For an example of the output of `plot.gam()`, see Figure 7.7 on page 263.

SEE ALSO

`preplot, predict.gam, gplot`

EXAMPLE

```
plot(gamob, ask=T) # interactive version
plot(gamob, residuals=T, se=T, rug=F)
gamob$preplot <- preplot(gamob)
plot(gamob)
```

plot.loess	Graphical Display of Local Regression Surface	**plot.loess**

```
plot.loess(object, given = 6, evaluation = 50, confidence = 0,
    coverage = 0.99, ranges = NULL, which.plots = NULL, xlab,
    ylab, rows, columns, show.given = TRUE, ...)
plot.preplot.loess(object, xlab = object$xlab, ylab = object$ylab,
    which.plots = object$which.plots, rows, columns,
    show.given = TRUE, ...)
```

ARGUMENTS

object: a `loess` object or a `preplot.loess` object. The latter is created by the function `preplot.loess()` when the user wishes to save the evaluations of the surface that are carried out to make the coplot.

given: number of conditioning values for a numeric given variable.

evaluation: number of points at which the curve on each dependence panel is computed.

confidence: number of points at which the confidence intervals on each dependence panel are displayed.

coverage: the level of the confidence intervals expressed as a fraction.

ranges: the ranges of evaluations of numeric variables. If there are `k` numeric predictors, `ranges` should either be a list of minimum and maximum pairs of length `k`, or a 2 by `k` matrix. Matching is done the same way as for `xlab`. If missing, the range for a numeric predictor is the range of the predictor observations.

which.plots: a character vector of names of predictors that determines the coplots that are made. If missing, all coplots are made.

xlab: character vector of labels for predictors. If the elements have names, they are matched to the names in `formula`; if not, the elements are assigned, in order, according to the order in which the predictors appear in `formula`. If missing, names are taken from `terms`.

`ylab`: a label for the response. If missing, the name is taken from `terms`.

`rows`: for the case of one given predictor, the number of rows of the matrix of dependence panels. If missing, the following is the default: let `k` be the number of given values; if `columns` is missing, then

```
rows <- ceiling(sqrt(k))
```

else

```
rows <- ceiling(k/columns)
```

This argument is not used if there are two given predictors.

`columns`: for the case of one given predictor, the number of columns of the matrix of dependence panels. If missing, the following is the default: let `k` be the number of given values; if `rows` is missing,

```
columns <- ceiling(k/ceiling(sqrt(k)))
```

else

```
columns <- ceiling(k/rows)
```

This argument is not used if there are two given predictors.

`show.given`: if `FALSE`, given panels are not included.

Graphical parameters (see `par()`) may also be supplied as arguments to this function.

These functions graph the fitted surface of a local regression model for one, two or three predictors. For one predictor, a curve is graphed against the predictor. For two or three predictors, a coplot is made against each predictor, conditional on the others. Each dependence panel of a coplot shows a curve that is a slice through the surface and is based on an evaluation for `evaluaton` equally spaced values of the predictor ranging between values specified by `ranges`; in addition, confidence intervals at `confidence` equally spaced values over the same range are shown. Normally, the user will want to make all coplots, but coplots against just certain predictors can be made by using the argument `which.plots`.

Graphical parameters (see `par()`) may also be supplied as arguments to this function.

This function is a method for the generic function `plot()` for class `"loess"`. It can be invoked by calling `plot(x)` for an object `x` of the appropriate class, or directly by calling `plot.loess()`, regardless of the class of the object.

For an example of the output of `plot.loess()`, see Figure 8.18 on page 345.

SEE ALSO

> loess, predict.loess, pointwise, anova.loess, specs.loess, preplot.loess, loess.control.

EXAMPLE

> ```
> ethanol.model <- loess(NOx ~ C * E, data = ethanol, span = 1/2,
>     drop.square = "C", parametric = "C")
> plot(ethanol.model, confidence = 7, which.plots = "C",
>     coverage = .95)
> ```

| **plot.tree** | Plot a Tree Object | **plot.tree** |

> ```
> plot.tree(x, type = "", ...)
> ```

ARGUMENTS

> x: fitted model object of class **tree**. This is assumed to be the result of some function that produces an object with the same named components as that returned by the **tree()** function.

> type: if "u", uniform spacing of nodes is used; default is nonuniform spacing based on change of deviance of parent and children nodes. The **mkh** graphical parameter is changed so that functions that require node coordinates, such as **text()** and **identify()**, can *query* it.

VALUE

> this function is used for its graphical side effect, an unlabeled dendrogram of the tree object **x**. If assigned, the coordinates of the nodes in **x** are returned as a list with components **x** and **y**.

> Graphical parameters (see **par()**) may also be supplied as arguments to this function.

> This function is a method for the generic function **plot()** for class **"tree"**. It can be invoked by calling **plot(x)** for an object **x** of the appropriate class, or directly by calling **plot.tree()**, regardless of the class of the object.

> For an example of the output of **plot.tree()**, see Figure 9.13 on page 400.

EXAMPLE

> ```
> zauto <- tree(Mileage ~ Weight + Displacement)
> plot(zauto)
> text(zauto) # put some labels on the plot
> identify(zauto) # find out what car is where
> ```

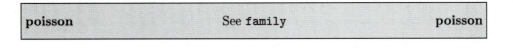

poisson	See `family`	**poisson**

poly	Generate a Basis for Polynomial Regression	**poly**

```
poly(x, 3)
poly(x, y, 2)
poly(...)
```

ARGUMENTS

...: the arguments to `poly` can be a comma-separated list of numeric vectors or matrices. If the final argument is atomic, positive, and integer-valued, it is taken to be the `degree` of the polynomial.

VALUE

a matrix of orthonormal polynomials is returned.

For a single vector argument and a trailing `degree` argument (first case above), a matrix of orthonormal polynomials of given degree is returned (the constant column is excluded). The orthogonality is with respect to the data.

For several arguments (vector, matrix, or both), each of the column vectors is used to generate orthogonal polynomials of the required `degree`. The columns will be a subset of the tensor product of the the orthogonal polynomials of given degree of each of the individual variables. The matrix has an attribute `degree` that is a vector giving the degree of each column.

SEE ALSO

`bs, ns, formula`

EXAMPLE

```
glm(Kyphosis ~ poly(Age, 3) + Start, family = binomial)
lm(NOx ~ poly(C, E, 4)
```

---

post.tree   PostScript Presentation Plot of a Tree Object   **post.tree**

```
post.tree(tree, pretty = 0, file, ...)
```

ARGUMENTS

   tree: fitted model object of class `tree`. This is assumed to be the result of some
          function that produces an object with the same named components as
          that returned by the `tree()` function.

 pretty: an integer denoting the extent to which factor levels in split labels will be
          abbreviated. The default (0) signifies no abbreviation. A NULL signifies
          using elements of `letters` to represent the different factor levels.

   file: ASCII file (`tree.ps` by default) to contain the output.

    ...: additional graphical arguments to control title and pointsize.

VALUE

          this function is used for its graphical side effect, a plot of `tree` in the
          PostScript page description language. The plot is different from the den-
          drogram produced by `plot()` and is intended for presentation. The edges
          connecting uniformly spaced nodes are labeled by left and right splits.
          Nodes are represented by ellipses (interior nodes) and rectangles (leaves)
          and labeled by `yval`. Under each node, either the within-node deviance is
          printed (regression trees) or the misclassification error rate (classification
          trees). The function is independent of the current graphics device. It
          creates a file which can be sent directly to a PostScript printer.

predict	Make Predictions from a Fitted Model Object	**predict**

```
predict(object, newdata, type, se.fit = F)
```

ARGUMENTS

object: a fitted model object, such as those produced by `lm()`, `glm()`, `loess()`, etc.

newdata: a data frame containing the values at which predictions are required. This argument can be missing, in which case predictions are made at the same values used to compute the object. Only those predictors, referred to in the right side of the formula in `object`, need be present by name in `newdata`. Some methods allow additional flexibility in the `newdata` argument; all allow a list rather than a data frame, and some, such as `predict.lm()`, allow an appropriate `model.matrix`.

type: type of predictions. The defaults differ for different methods, and for some only one type is sensible: predictions of the response. For objects with distinct terms, such as `lm`, `glm`, and `gam` objects, `type = "terms"` produces a matrix of predictions with a column for each term. For `"glm"` and `"gam"` models, the default is `type = "link"`, in which case predictions are on the scale of the linear or additive predictor, respectively.

se.fit: if `TRUE`, pointwise standard errors are computed along with the predictions.

VALUE

a vector or array of predictions, or a list consisting of the predictions and their standard errors if `se.fit = T`. A standard use of `predict()` is to simply extract the fitted values from a fit object, or in the case of generalized models, to extract the linear or additive predictor.

When standard errors are requested, the output of `predict()` is a list that includes components `"fit"`, `"se.fit"`, `"residual.scale"`, and `"df"`. The `"se.fit"` component consists of pointwise standard errors, and consequently has the same shape as the `"fit"` component. Although the computations of these standard errors differ for the different classes of models, they all have a similar flavor. The fitted values are linear either in the response or in some derived *pseudo-response;* the *raw* standard errors are the norm of these linear weights. The `"residual.scale"` component is the scale estimate used by `predict()` to scale the raw standard errors in computing the `"se.fit"` component; `"df"` is the degrees of freedom of this scale estimate. This allows rescaling of the `"se.fit"` component by other scale factors.

This is a generic function. Functions with names beginning in `"predict."` will be methods for this function.

WARNING

`predict()` can produce incorrect predictions when the `newdata` argument is used if the formula in `object` involves *data-dependent* transformations, such as `poly(Age, 3)` or `sqrt(Age - min(Age))`. The `predict.gam()` method overcomes this for the `gam`, `glm`, and `lm` classes. In other cases, this can be overcome by explicitly supplying the derived `matrix` for predictions, rather than a data frame.

SEE ALSO

`fitted, expand.grid`

EXAMPLE

```
#extract the fitted linear predictor from a glm object
predict(glmob)
predict(gamob, newdata, type="terms")
```

---

predict.gam   Make Predictions from a Fitted GAM Object   **predict.gam**

```
predict.gam(object, newdata, type, se.fit = F, terms)
```

ARGUMENTS

object: a fitted `gam` object, or one of its inheritants, such as a `glm` or `lm` object.

newdata: a data frame containing the values at which predictions are required. This argument can be missing, in which case predictions are made at the same values used to compute the object. Only those predictors, referred to in the right side of the formula in object need be present by name in `newdata`.

type: type of predictions, with choices `"link"` (the default), `"response"`, or `"terms"`. The default produces predictions on the scale of the additive predictors, and with `newdata` missing, `predict()` is simply an extractor function for this component of a `gam` object. If `"response"` is selected, the predictions are on the scale of the response, and are monotone transformations of the additive predictors, using the inverse link function. If `type="terms"` is selected, a matrix of predictions is produced, one column for each term in the model.

se.fit: if `TRUE`, pointwise standard errors are computed along with the predictions.

terms: if type="terms", the terms= argument can be used to specify which terms should be included; the default is labels(object).

VALUE

a vector or matrix of predictions, or a list consisting of the predictions and their standard errors if se.fit = T. If type="terms", a matrix of fitted terms is produced, with one column for each term in the model (or subset of these if the terms= argument is used). There is no column for the intercept, if present in the model, and each of the terms is centered so that their average over the original data is zero. The matrix of fitted terms has a "constant" attribute which, when added to the sum of these centered terms, gives the additive predictor. See the documentation of predict() for more details on the components returned.

This is a *safe* method of prediction for the classes gam, glm, and lm. Naive use of the generic predict() can produce incorrect predictions when the newdata argument is used, if the formula in object involves *data-dependent* transformations, such as poly(Age, 3) or sqrt(Age - min(Age)). These problems are overcome by predict.gam() by taking the following steps. A combined data frame is constructed containing the predictors in object, using both the data used in fitting object, as well as newdata. From this, a combined model frame and model matrix is constructed, and object is refitted using the top portion of both of these (belonging to the fitting set). The GAM iterations are not repeated; rather one final IRLS step is performed, using the working weights and response from the final iteration in the creation of the original object. In this way, it is guaranteed that any coefficients that are estimated can be applied to both the fitting or prediction portions of the model matrix.

This function is a method for the generic function predict() for class "gam". It can be invoked by calling predict(x) for an object x of the appropriate class, or directly by calling predict.gam(), regardless of the class of the object.

SEE ALSO

fitted, expand.grid

EXAMPLE

```
predict(gamob) # extract the additive predictors
predict(gamob, newdata, type="terms")
predict.gam(lmobjet, newdata) # safe prediction for lm object
```

```
predict.loess(object, newdata,  se.fit = FALSE)
```

ARGUMENTS

object: a `loess` object.

newdata: a data frame specifying the values of the predictors at which the evaluation is to be carried out. The default is to predict at the data used to fit `object`.

se.fit: if `TRUE`, estimates of the standard errors of the surface values and other statistical information are returned along with the surface values.

VALUE

if `se.fit=FALSE`, a vector or array of surface values evaluated at `newdata`. The evaluation is on the scale of the expression given on the left side of `formula`; for example, if the expression is `log(temperature)`, then the evaluation is on the log scale. If `se.fit=TRUE`, then a list is returned, with the following components.

fit: the evaluated loess surface at `newdata`.

se.fit: estimates of the standard errors of the surface values.

residual.scale: estimate of the scale of the residuals.

df: the degrees of freedom of the t-distribution used to compute pointwise confidence intervals for the evaluated surface. The function `pointwise()` can be used to compute such intervals.

For one predictor, `newdata` can be a vector rather than a data frame. For two or more predictors, the names of `newdata` must include the names of predictors used in `formula` as they appear on the database from which they come. For example, if the right side of `formula` is `log(E)*C`, then there must be names `C` and `E` in `newdata`. Note that the specification of `E` in this example is not on the transformed but rather on the original scale.

For two or more predictors, there are two data structures that can be given to `newdata`. The first is a plain old data frame; the result is a vector whose length is equal to the number of rows of `newdata`, and the element of the vector in position `i` is the evaluation of the surface at row `i` of `newdata`. A second data structure can be used when the evaluation points form a grid. In this case, `newdata` is the result of the function `expand.grid()`.

If se.fit=FALSE, the result of predict.loess() is a numeric array whose dimension is equal to the number of predictors; if se.fit=TRUE, then the components fit and se.fit are both such arrays.

The computations of predict.loess() that produce the component se.fit are much more costly than those that produce fit, so the number of points at which standard errors are computed should be modest compared to those at which we do evaluations. Often this means calling predict.loess() twice, once at a large number of points with se.fit equal to FALSE to get a thorough description of the surface, and once at a small number of points to get standard-error information.

Suppose the computation method for loess surfaces is interpolate, the default for the argument surface. Then the evaluation values of a numeric predictor must lie within the range of the values of the predictor used in the fit. The evaluation values for a predictor that is a factor must be one of the levels of the factor. For any evaluation point for which these conditions are not met, an NA is returned.

This function is a method for the generic function predict() for class "loess". It can be invoked by calling predict(x) for an object x of the appropriate class, or directly by calling predict.loess(), regardless of the class of the object.

SEE ALSO

loess, plot.loess, pointwise, anova.loess, specs.loess, preplot.loess, loess.control.

EXAMPLE

```
# Example 1 - evaluation at the 5 values of C and E in newdata
> ethanol.cp$call
loess(formula = NOx ~ C * E, span = 1/2, degree = 2,
    parametric = "C", drop.square = "C")
> predict(ethanol.cp, newdata)
[1] 0.2815825 2.5971411 3.0667178 3.2555778 1.0637788
# Example 2 - evaluation at 9 grid points
> C.marginal <- seq(min(C), max(C), length = 3)
> E.marginal <- seq(min(E), max(E), length = 3)
> CE.grid <- expand.grid(list(C = C.marginal, E = E.marginal))
> predict(ethanol.cp, CE.grid)
          E=0.5350 E=0.8835  E=1.2320
C= 7.50 -0.1039991 3.399360 0.6823181
C=12.75  0.2057837 3.850801 0.6481270
C=18.00  0.5155665 4.302243 0.6139359
# Example 3 - evaluate and compute estimates of standard errors
> gas.m$call
```

```
    loess(formula = NOx ~ E, span = 2/3, degree = 2)
    > predict(gas.m, newdata = seq(min(E), max(E),length = 5),
    +    se.fit = T)$se.fit
    [1] 0.2694392 0.1536510 0.1489403 0.1665470 0.3237732
```

predict.tree	Predictions from a Fitted Tree Object	**predict.tree**

```
    predict.tree(object, newdata, type = c("vector", "tree"))
```

ARGUMENTS

object: fitted model object of class `tree`. This is assumed to be the result of some function that produces an object with the same named components as that returned by the `tree()` function.

newdata: data frame containing the values at which predictions are required. The predictors referred to in the right side of `formula(object)` must be present by name in `newdata`. If missing, fitted values are returned.

type: character string denoting whether the predictions are returned as a vector (default) or as a tree object. If `"tree"`, a `tree` object is returned with new values for `frame$n` and `frame$yval/yprob`.

VALUE

vector of predicted responses obtained by dropping `newdata` down `tree`. For factor predictors, if an observation contains a level not used to grow the tree, it is left at the deepest possible node and `frame$yval` at that node is the prediction.

This function is a method for the generic function `predict()` for class `"tree"`. It can be invoked by calling `predict(x)` for an object x of the appropriate class, or directly by calling `predict.tree()`, regardless of the class of the object.

| **preplot** | Precompute a Plotting Object | **preplot** |

```
preplot(object, newdata)
```

ARGUMENTS

`object`: a fitted model object, such as those produced by `gam()` and `loess()`.

`newdata`: a data frame containing the values at which evaluations are required. This is often missing, in which case evaluations are made at the same values used to compute the object. Only those predictors referred to in the right side of the formula in `object` need be present by name in `newdata`.

VALUE

an object set up for plotting to describe the fit, optionally at the location specified in `newdata`. The purpose of calling `preplot()` is to precompute information to be used in the plot, without necessarily doing the plot at the same time. The object returned differs depending on the kind of fit (that is, on which method is being used), but in all cases can be given directly to the `plot()` function. This implies that a suitable plot method has been written for whatever class of object `preplot()` returns.

This is a generic function. Functions with names beginning in `"preplot."` will be methods for this function.

SEE ALSO

`predict`

EXAMPLE

```
preob <- preplot(gam.object, newdata)
plot(preob, resid=T, se=T)
```

proj	Projection Matrix	**proj**

```
proj(object, onedf=T)
```

ARGUMENTS

object: any object of class `lm` or any object that inherits from class `lm`. It will run faster if the object contains either a `qr` component or a `proj` component.

onedf: logical flag. When `TRUE` (the default for objects of class `lm`), the function returns a matrix of single-degree-of-freedom projections of the response variable onto the columns of the predictor matrix. The default method does not use this argument. When `FALSE`, an option available in `proj.lm` and the default for objects of class `aov`, the function collapses the single degree of freedom projections into multi-degree-of-freedom projections. Each column of the collapsed result represents one term of the analysis of variance table. The sum of squares of each column is the sum of squares for the corresponding term in the model formula, with degrees of freedom given by the `df` attribute of the result. The formula itself is returned in the `formula` attribute.

VALUE

matrix of orthogonal columns, one column for each column in an orthogonalized model matrix. In the default method, each column in the result is the projection of the response variable onto a column in the Q matrix from the QR decomposition of the model matrix. The sum across the columns gives the column of predicted values. The sum of the squared values in each column is the single-degree-of-freedom sum of squares for the corresponding column of the model matrix. The `onedf` attribute returns the value of the `onedf` argument. The method for `lm` objects appends a column of residuals. The method for `aovlist` objects returns a list of projection matrices, one for each stratum in the design.

This is a generic function. Functions with names beginning in "proj." will be methods for this function.

EXAMPLE

```
> lm.object <- lm(cost ~ age + type + car.age, claims)
> tmp <- proj(lm.object)
> gunaov.qr <- aov(Rounds ~ Method + Physique/Team, gun, qr = T)
> gunaov.proj <- proj(gunaov.qr)
```

prune.tree	Cost-complexity Pruning of Tree Object	**prune.tree**

```
prune.tree(tree, k, newdata)
```

ARGUMENTS

tree: fitted model object of class tree. This is assumed to be the result of some function that produces an object with the same named components as that returned by the tree() function.

k: cost-complexity parameter defining either a specific subtree of tree (k a scalar) or the sequence of subtrees minimizing the cost-complexity measure (k a vector). If missing, k is determined algorithmically.

newdata: a data frame containing the values at which predictions are required. The sequence of cost-complexity subtrees is evaluated on these data. If missing, the data used to grow the tree are used. The response as well as the predictors referred to in the right side of the formula in tree must be present by name in newdata. These data are dropped down each tree in the cost-complexity sequence and deviances calculated by comparing the supplied response to the prediction.

VALUE

if k is supplied and is a scalar, a tree object is returned that minimizes the cost-complexity measure for that k. Otherwise, an object of class tree.sequence is returned. A plot() method exists for objects of this class. It displays the value of the deviance for each subtree in the cost-complexity sequence. An additional axis displays the values of the cost-complexity parameter at each subtree. The object contains the following components:

size: number of terminal nodes in each tree in the cost-complexity pruning sequence.

deviance: total deviance of each tree in the cost-complexity pruning sequence.

k: the value of the cost-complexity pruning parameter of each tree in the sequence.

nodes: vector indicating the pruning order of the nodes that define the subtrees in the sequence. The first element of nodes is 0, indicating the full tree; the next element is the number of the node that is the root of the next subtree to be pruned, etc.

qqline	Produce a Median Line through a `qqnorm` Plot	**qqline**

```
qqline(x)
```

ARGUMENTS

x: typically a residual vector

This function fits and plots a line through the first and third quartile of the data, and the corresponding quantiles of the standard normal distribution.

EXAMPLE

```
# to check Gaussian distribution of the errors :
qqnorm(galaxy.m$residuals)
qqline(galaxy.m$residuals)
```

qqnorm.aov	Normal or Half-Normal Plots of Effects	**qqnorm.aov**

```
qqnorm(x, full=F, label, omit=NULL, ...)
```

ARGUMENTS

x: the result of fitting an analysis of variance model, including the cases of multivariate response and multiple-error strata.

full: TRUE for a full normal probability plot of the effects, FALSE for a half-normal plot of the absolute values of the effects.

label: should some of the points be labeled? Can be set to TRUE, in which case the user will be prompted to identify the points to label, or to a single number, n, in which case the n largest effects (in absolute value) will be labeled.

omit: those effects that should be omitted from the plot. By default, the first effect (assumed to be the intercept) is omitted. Either numeric or character data can be given. Character data are matched against the effect labels. The intercept is always omitted: if you really want to see it, set its name explicitly to anything other than "(Intercept)"—e.g.,

```
names(x$effects)[1] <- "*I*"
```

Graphical parameters (see `par()`) may also be supplied as arguments to this function.

VALUE

the points plotted, suitable to use as an argument to `identify()` to label interesting points after the fact. In the case of multiple responses or error strata, the points from the last plot are returned.

A quantile-quantile plot is produced, either of the `effects` component from the fit or of its absolute value, plotted against the appropriate quantiles of the normal distribution. In the case of multiple response or multiple strata, one plot will be produced for each separate set of effects. Subtitle labels will be composed in this case identifying the response and/or the stratum.

This function is a method for the generic function `qqnorm()` for class `"aov"`. It can be invoked by calling `qqnorm(x)` for an object `x` of the appropriate class, or directly by calling `qqnorm.aov()`, regardless of the class of the object.

For an example of the output of `qqnorm.aov()`, see Figure 5.6 on page 170.

quasi	See `family`	**quasi**

randomize	Random Ordering of the Runs of a Design	**randomize**

```
randomize(design, restrict)
```

ARGUMENTS

design: a design; that is, a data frame representing factors for an experimental design.

restrict: an optional vector specifying some factors (either numerically or by name) in the design matrix. If `restrict` is supplied, randomization will occur only *within* levels of the combination of the restricted factors. The runs (rows) of the design will be ordered by the restricted factors, and then randomly within this order.

VALUE

a permutation of the rows of the design, randomized in the sense above.

EXAMPLE

```
# a 3^2 factorial, randomized within levels of the first factor
> mydesign <- fac.design(rep(3,2))
> perm <- randomize(mydesign,"A")
 > perm
[1] 7 4 1 2 8 5 9 6 3
```

raov	Random-effects Analysis of Variance	**raov**

```
raov(formula, data, ...)
```

ARGUMENTS

formula: formula and optional data frame for analysis of variance model, or a previously computed analysis of variance fit. These arguments are interpreted in the standard form for analysis of variance. The usual optional arguments for aov() and lm() can be supplied to raov() also.

VALUE

a fitted anova model, similar to that returned by aov(), and containing in addition two components used for computing the estimated random-effects. The difference from standard anova models appears when the summary method is called: this method will compute and print the estimated variances for the random-effects model. Specifically, the ordinary mean-squares and the component ems.coef from the fitted anova model are used to estimate the variances for the usual random effects model for each relevant main factor or interaction in the model.

The analysis produced by raov() is valid only for the fully random model on a balanced design, with one error stratum. The function will check for balance, by calling replications().

EXAMPLE

```
# in the design pigment, Batch and Sample have
# attribute "random" set to TRUE
> praov <- raov(Moisture ~ Batch/Sample, pigment)
> summary(praov)
                     Df Sum of Sq Mean Sq
            Batch 14    1210.9  86.495
Sample %in% Batch 15     869.8  57.983
         Residuals 30      27.5   0.917
Random Effects -- estimated variances:
 Batch Sample %in% Batch Residuals
 7.128             28.533     0.917
```

| **read.table** | Create a Data Frame by Reading a Table | **read.table** |

```
read.table(file, header=F, sep, row.names, col.names)
```

ARGUMENTS *eagle.dat ← read.table ("a:eagle.dat", header = T)*

file: the name of the text file for the data. The file should contain one line per row of the table, with fields separated by the character in **sep**.

header: should the first line of the file be used as a header, specifying the names of the variables in the data frame?

sep: the field separator. If missing, any amount of white space can separate items.

row.names: optional specification of the row names for the data frame. If provided, it can give the actual row names, as a vector of length equal to the number of rows, or it can be a single number or character string. In the latter case, the argument indicates which variable in the data frame to use as row names (the variable will then be dropped from the frame). If **row.names** is missing, the function will use the first nonnumeric field with no duplicates as the row names. If no such field exists, the row names are **1:nrow(x)**. Row names, wherever they come from, must be unique.

col.names: optional names for the variables. If missing, the header information, if any, is used; if all else fails, **"V"** and the field number are be pasted together. Variable names, wherever they come from, must be unique. Variable names will be converted to syntactic names before assignment, but *not* if they came from an explicit **col.names** argument.

as.is: control over conversions to factor objects. By default, non-numeric fields are turned into factors, except if they are used as row names. If some or all fields should be left as is (typically producing character variables), set the corresponding element of **as.is** to **TRUE**. The argument will be replicated as needed to be of length equal to the number of fields; thus, **as.is=TRUE** leaves all fields unconverted.

VALUE

a data frame with as many rows as the file has lines (or one less if **header==T**) and as many variables as the file has fields (or one less if one variable was used for row names). Fields are initially read in as character data. If all the items in a field are numeric, the corresponding variable is numeric. Otherwise, it is a factor (unordered), except as controlled by the `as.is` argument. All lines must have the same number of fields (except the header, which can have one less if the first field is to be used for row names).

This function should be compared to `scan()`; `read.table()` tries much harder to interpret the input data automatically, figuring out the number of variables and whether fields are numeric. It also produces a more structured object as output. The price for this, aside from `read.table()` being somewhat slower, is that the input data must themselves be more regular and that `read.table()` decides what to do with each field, except for the use of the `as.is` argument. With `scan()`, input lines do not need to correspond to one complete set of fields, and the user decides what mode each field should have. Overall, `read.table()` will usually be the easy way to construct data frames from tables. If it doesn't do what you want, consider the functions `scan()`, `make.fields()`, or `count.fields()`, as well as text-editing tools and languages outside S.

---

replications	Number of Replications of Terms	**replications**

`replications(formula, data)`

ARGUMENTS

formula: a formula, terms object, or design. If a design, then `data` need not be given, and the formula `"~ ."` is implied.

data: a design or data frame.

VALUE

if the design is balanced—that is, for each term in the formula all levels are replicated the same number of times—then `replications` returns a vector of length equal to the number of terms, containing the number of replications for each term. Otherwise, the object returned is a list, with one element for each term. If the individual term is balanced, the corresponding element of the list will be a single number, the number of replications. Otherwise, the element will give the pattern of replications for the levels of the corresponding term. If the term is a main effect, the element will be a vector of length equal to the number of levels. If the term is a two-way interaction, the element will be a two-way array, and so on. The names or dimnames of the elements will identify the corresponding levels of the factor(s) in the term. See the example below. The value of `replications()` provides an easy test for overall balance or for balance of individual terms:

```
is.numeric(replications(formula, data)
sapply(replications(formula, data),
    function(x)length(x)==1 )
```

respectively test for complete or term-by-term balance. The second test is sensible only if the first one fails.

EXAMPLE

```
> reps <- replications(Yield ~ .^2, catalyst[-1, ])
> reps$"Conc:Cat"
   A B
20 1 2
40 2 2
```

residuals	See coefficients	**residuals**

robust	Generate a Robust Family Object	**robust**

```
robust(family = gaussian, scale=0, k=1.345, maxit=10)
```

ARGUMENTS

family: a family object, as produced by gaussian(), binomial(), poisson(), etc.

scale: a scale parameter can be supplied. If omitted, a robust scale is estimated based on a scale of 1 and the fit of the first glm() or gam() iteration.

k: a threshold used in the computation of the robustness weights.

maxit: the maximum number of iterations.

VALUE

a new family object, with its component functions suitably modified to perform a robust version of the glm() or gam() fit.

SEE ALSO

glm, gam, family, quasi, family.object.

EXAMPLE

```
# Fit a linear model robustly
glm(formula, family=robust)
# Fit a generalized additive model robustly
gam(formula, family=robust(binomial))
```

| **s** | Specify a Smoothing Spline Fit in a GAM Formula | **s** |

```
s(x, df=4, spar=0)
```

ARGUMENTS

  x: the univariate predictor, or expression, that evaluates to a numeric vector.

  df: the target equivalent degrees of freedom, used as a smoothing parameter.
       The real smoothing parameter (spar below) is found such that df=tr(S)-1,
       where S is the implicit smoother matrix. Values for df should be greater
       than 1, with 1 implying a linear fit.

  spar: can be used as smoothing parameter, with values larger than 0.

VALUE

       the vector x is returned, endowed with a number of attributes. The vector
       itself is used in the construction of the model matrix, while the attributes
       are needed for the backfitting algorithms all.wam() or s.wam() (weighted
       additive model). Since smoothing splines reproduces linear fits, the linear
       part will be efficiently computed with the other parametric linear parts of
       the model.

       Note that s() itself does no smoothing; it simply sets things up for gam().

SEE ALSO

       lo, smooth.spline, bs, ns, poly

EXAMPLE

```
# fit Start using a smoothing spline with 4 df.
y ~ Age + s(Start, 4)
# fit log(Start) using a smoothing spline with 5 df.
y ~ Age + s(log(Start), df=5)
```

---

scatter.smooth   Scatterplot with a Smooth Curve   **scatter.smooth**

```
scatter.smooth(x, y, span = 2/3, degree = 1, family = c("symmetric",
    "gaussian"), evaluation = 50, ...)
```

ARGUMENTS

x: abscissas of points on scatterplot.

y: ordinates of points on scatterplot.

span: smoothing parameter.

degree: overall degree of locally-fitted polynomial. 1 is locally-linear fitting and 2 is locally-quadratic fitting.

family: the values are `"gaussian"` or `"symmetric"`. In the first case, local-fitting methods are used. In the second case, the default, local fitting is used together with a robustness feature that guards against distortion by outliers.

evaluation: number of values at which the loess curve is evaluated.

This function makes a scatterplot and adds a smooth curve using the loess fitting method. The fit is evaluated at `evaluation` equally spaced points from `min(x)` to `max(x)` and then graphed by connecting the successive plotting locations by line segments. This function replaces the use of the function `lowess()` for scatterplot smoothing.

Graphical parameters (see `par()`) may also be supplied as arguments to this function.

For an example of the output of `scatter.smooth()`, see Figure 8.5 on page 326.

SEE ALSO

loess, panel.smooth

EXAMPLE

```
scatter.smooth(x, y, span = 1, degree = 1, xlab = "predictor")
```

---

select.tree	Select Subtrees of a Tree Object	**select.tree**

```
select.tree(tree, nodes, drop = T)
```

ARGUMENTS

tree: fitted model object of class `tree`. This is assumed to be the result of some function that produces an object with the same named components as that returned by the `tree()` function.

nodes: an integer vector containing indices (node numbers) of all subtrees to be selected. If missing, user selects nodes as described below.

drop: if `TRUE`, and only one subtree was specified or selected, returns a `tree` object, not a list.

VALUE

returns a list (sometimes called a *stand*) of subtrees rooted at `nodes`. For noninteractive use, `select.tree(obj, nodes)` is longhand for the left-square-bracket method for trees `obj[nodes]`.

GRAPHICAL INTERACTION

A dendrogram of `tree` is expected to be visible on the graphics device, and a graphics input device (e.g., a mouse) is required. Clicking (the selection button) on a node selects the subtree rooted at that node. This process may be repeated any number of times. Warnings result from selecting the root or leaf nodes. Clicking the exit button will stop the selection process and return the resulting *stand* of `tree` objects.

See the documentation for the specific graphics device for details on graphical input techniques.

For an example of the output of `select.tree()`, see Figure 9.12 on page 398.

shrink.tree	Optimal Recursive Shrinking of Tree Objects	**shrink.tree**

```
shrink.tree(tree, k, newdata)
```

ARGUMENTS

  tree: fitted model object of class **tree**. This is assumed to be the result of some function that produces an object with the same named components as that returned by the **tree()** function.

  k: shrinkage parameter (0<k<1) defining either a specific shrunken version of **tree** or, if **length(k)>1**, the sequence of shrunken trees obtained by optimal shrinking for each value of **k**. By default the sequence **(1:10)/(20:11)** (roughly .05 to .91) is used.

  newdata: a data frame containing the values at which predictions are required. The sequence of optimally shrunken trees is evaluated on **newdata**. If missing, the data used to grow the tree are used. If supplied, the sequence is evaluated on **newdata**, The response as well as the predictors referred to in the right side of the formula in **tree** must be present by name in **newdata**. These data are dropped down each shrunken tree in the sequence and deviances calculated by comparing the supplied response to the prediction.

VALUE

  if **k** is supplied and is a scalar, a tree object is returned that has the same topology as the supplied **tree** but new values for the **yval/yprob**, **dev**, and n components of **tree$frame**. Otherwise, an object of class **tree.sequence** is returned. A **plot()** method exists for objects of this class. It displays the value of the deviance for each shrunken tree in the sequence. An additional axis displays the values of the shrinkage parameter for each tree. The object contains the following components:

  size: number of (effective) terminal nodes in each tree in the optimal shrinkage sequence.

  deviance: total deviance of each tree in the optimal shrinkage sequence.

  k: the value of the shrinkage parameter of each tree in the sequence.

  For an example of the output of **shrink.tree()**, see Figure 9.12 on page 398.

| **smooth.spline** | Fit a Smoothing Spline. | **smooth.spline** |

```
smooth.spline(x, y, w, df, spar, cv, all.knots, df.offset, penalty)
```

ARGUMENTS

x: values of the predictor variable. There should be at least ten distinct x values.

y: response variable, of the same length as x.

x and y can be supplied in a variety of different forms, along the lines of the function plot(); e.g., a list with components x and y, a two-column matrix, or simply a single vector, taken to be a time series.

w: optional vector of weights for weighted smoothing, of the same length as x and y. If measurements at different values of x have different variances, w should be inversely proportional to the variances.

df: one can supply the *degrees of freedom = trace(S)* rather than a smoothing parameter. Here S is the implicit smoother matrix. If both df and spar are supplied, spar is used unless it is 0, in which case df is used.

spar: the usual smoothing parameter for smoothing splines, which is the coefficient of the integrated second squared derivative penalty function. If spar is 0 or missing and df is missing, cross-validation is used to automatically select spar. If a value of spar greater than zero is supplied, it is used as the smoothing parameter.

cv: the ordinary or generalized cross validation score (CV or GCV) is computed according to whether cv is TRUE or FALSE. Default is FALSE.

all.knots: if FALSE, a suitable fine grid of knots is chosen, usually less in number than the number of unique values of x. If TRUE, the unique values of x are used as knots.

df.offset: allows an offset to be added to the df term used in the calculation of the GCV criterion: df=tr(S) + df.offset. Default is 0.

penalty: allows the df quantity used in GCV to be charged a cost = penalty per degree of freedom.

The last two arguments are experimental and typically will not be used. If used, the GCV criterion is RSS/(n - (penalty*(trace(S)-1) + df.offset +1)).

VALUE

an object of class `smooth.spline` is returned, consisting of the fitted smoothing spline evaluated at the supplied data, some fitting criteria and constants, and a structure that contains the essential information for computing the spline and its derivatives for any values of x.

x: ordered distinct x values

y: smoothing spline fits corresponding to x

w: weights used in the fit. This has the same length as x, and in the case of ties, will consist of the accumulated weights at.each unique value of x.

yin: y-values used at the unique x values (weighted averages of input y)

lev: leverage values, which are the diagonal elements of the smoother matrix S.

cv.crit: cross validation score (either GCV or CV)

pen.crit: penalized criterion

df: degrees of freedom of the fit estimated by the sum of lev. If df was supplied as the smoothing parameter, then the prescribed and resultant values of df should match within 0.1 percent of the supplied df

spar: smoothing parameter used in the fit (useful if df was used to specify the amount of smoothing)

fit: list containing details of the fits (knot locations, coefficients, etc.) to be used by `predict.smooth.spline()`.

call: the call that produced the fit

SEE ALSO

`predict.smooth.spline, print.smooth.spline`

COMMENTS

A cubic B-spline is fit with care taken to insure that the algorithm runs linear in the number of data points. For small data vectors (n<50), a knot is placed at every distinct data point, and the regression is fit by penalized least squares. For larger data sets the number of knots is chosen judiciously in order to keep the computation time manageable (if `all.knots=F`). The penalty spar can be chosen automatically by cross-validation (if `spar=0`), can be supplied explicitly, or supplied implicitly via the more intuitive df number.

The FORTRAN code supporting `smooth.spline()` was supplied by Finbarr O'Sullivan, whose original function `bart()` is known to some S users. Subsequent modifications allowing for user supplied `df` and other small changes were made by Trevor Hastie.

EXAMPLE

```
# smoothing spline fit and approximate 95% "confidence" intervals
    fit <- smooth.spline(x, y)                    # smooth.spline fit
    res <- (y - fit$y)/(1-fit$lev)                # jackknife residuals
    sigma <- sqrt(var(res))                       # estimated sd
upper <- fit$y + 2.0*sigma*sqrt(fit$lev)   # upper 95% conf. band
lower <- fit$y - 2.0*sigma*sqrt(fit$lev)   # lower 95% conf. band
matplot(fit$x, cbind(upper, fit$y, lower), type="plp", pch=".")
```

snip.tree	Snip Subtrees of a Tree Object	**snip.tree**

```
snip.tree(tree, nodes)
```

ARGUMENTS

tree: fitted model object of class `tree`. This is assumed to be the result of some function that produces an object with the same named components as that returned by the `tree()` function.

nodes: an integer vector containing indices (node numbers) of all subtrees to be snipped off. If missing, user selects branches to snip off as described below.

VALUE

a `tree` object containing the nodes that remain after specified or selected subtrees have been snipped off. For noninteractive use, `snip.tree(obj, nodes)` is longhand for the left-square-bracket method for trees `obj[-nodes]`.

GRAPHICAL INTERACTION

A dendrogram of `tree` is expected to be visible on the graphics device, and a graphics input device (e.g., a mouse) is required. Clicking (the selection button) on a node displays the total tree deviance and what the total tree deviance would be if the subtree rooted at the node were removed. Clicking a second time on the same node snips that subtree off and visually *erases* the subtree. This process may be repeated any number of times. Warnings result from selecting the root or leaf nodes. Clicking the exit button will stop the snipping process and return the resulting tree object.

See the documentation for the specific graphics device for details on graphical input techniques.

For an example of the output of `snip.tree()`, see Figure 9.11 on page 397.

---

split.screen   Split the Graphics Display into Multiple Screens   **split.screen**

```
split.screen(figs, screen, erase = T)
screen(n, new = T )
erase.screen(n, eject = F)
close.screen(n, all = F)
prompt.screen(delta = 1/8, draw = T)
```

ARGUMENTS

figs: a multiple-figure vector like `mfrow` or `mfcol`—e.g, `c(n,m)`—or an `N` by 4 matrix where `N` is the number of screens and each row specifies the position of a screen on the display. The first two columns specify the left and right coordinates (in the `0, 1, 0, 1` plane) for the screens, and the third and fourth columns the bottom and top coordinates. Use `prompt.screen()` to create this matrix interactively.

screen: screen to split up. By default it carves up the current screen, which initially is the entire display (denoted as screen 0).

erase: should the screen being split be cleared? Default is TRUE.

n: screen to activate for `screen()`, screen to erase for `erase.screen()`, or the vector of screens to close for `close.screen()` (the active screen cannot be closed). For `screen()` and `erase.screen()` the default is the currently active screen.

new: should screen be cleared? Default is TRUE.

eject: should current page be ejected? On noninteractive devices like PostScript printers, this will advance to a new page. On exit, the first valid screen is left active.

all: should all screens be closed? Set `all = TRUE` to exit the split-screen mode.

delta: spacing, in inches, used to *snap* or line up adjacent screens; if corners of a newly created screen are less than `delta` inches from a neighboring screen, screen borders are made to coincide.

draw: should display be cleared and screen borders be drawn? Default is TRUE.

VALUE

these functions provide a means of dividing up the graphics display into multiple screens or figures of various sizes. The split-screen mode is an alternative to the usual multiple-figure mode obtained by `par(mfrow)`, and it is useful for interacting with individual figures and for producing non-matrix arrangements; however, the two modes are not compatible—i.e., you cannot mix calls to `split.screen()` and `par(mfrow)`.

Graphics input and output is directed to and from screens by calling `screen()`. As part of the initialization, `split.screen(figs)` activates the first screen in `figs`.

SIDE EFFECTS

If any screen in `figs` is smaller than half of the horizontal or vertical dimensions of the graphics display, the character expansion parameter `cex` is set to 0.5 in all screens in `figs`. Outer margins are not defined in split-screen mode, thus the graphical parameters `oma` and `omi` should not be used.

For an example of the output of `split.screen()`, see Figure 9.14 on page 403.

EXAMPLE

```
split.screen(c(2,1)) # split display into to screens
[1] 1 2
split.screen(c(1,3), screen = 2) # now split the bottom half into 3
[1] 3 4 5
screen(3) # prepare screen 3 for output
plot(x, y)
close.screen(all = T) # exit split-screen mode
```

step	Build a Model in a Stepwise Fashion	**step**

```
step(object, scope, scale, direction, trace = T, keep, steps)
```

ARGUMENTS

object: a `glm` or `gam` object, or an object that inherits from either of these. This is used as the initial model in the stepwise search.

scope: defines the range of models examined in the stepwise search. This argument has a different form for `step.gam()` and `step.glm()`; see their detailed documentation.

scale: an optional argument used in the definition of the AIC statistic for selecting the models. By default, the scaled chi-squared statistic for the initial model is used, but if forward selection is to be performed, this is not necessarily a sound choice.

direction: the mode of stepwise search, can be one of `"both"`, `"backward"`, or `"forward"`, with a default of `"both"`. If the `scope=` argument is missing, the default for `direction` is `"backward"`.

trace: if `TRUE`, information is printed during the running of `step()`. This is a good choice in general, since `step()` can take some time for large models.

keep: a filter function whose input is a fitted `glm` or `gam` object and the associated AIC statistic, and whose output is arbitrary. Typically `keep()` will select a subset of the components of the object and return them. The default is not to keep anything.

steps: the maximum number of steps to be considered. The default is 1000 (essentially as many as required). It is typically used to stop the process early.

VALUE

the stepwise-selected model is returned, with up to two additional components. There is an `"anova"` component corresponding to the steps taken in the search, as well as a `"keep"` component if the `keep=` argument was supplied in the call.

A series of models is generated sequentially, where each model differs from its neighbors by a single term. The `step()` methods differ in the way they construct this sequence: both in the way the set of candidates are generated for each step, and in the way the candidates are evaluated for selection.

SEE ALSO

    step.glm, step.gam,glm, gam, drop1, add1

EXAMPLE

    step(glm.object)
    step(glm.object, list(upper = ~.^2, lower = ~ Age)
    step(gam.object, scope=list(
        "Age" = ~ 1 + Age + log(Age),
        "BP" = ~ 1 + BP + poly(BP, 2) + s(BP),
        "Chol" = ~ s(Chol, df = 4) + s(Chol, df = 7)
)

summary	Summarize an Object	**summary**

    summary(object, ...)

ARGUMENTS

object: any object, including a fitted model object of various kinds, a data frame, or a factor.

VALUE

a summary object is returned—usually a list-like object whose elements describe the contents of the argument to summary(). For example, the method for lm objects produces an object of class "summary.lm" with components "residuals", "correlation", "cov.unscaled", "r.squared", and more. There is a print() method corresponding to each "summary." class, so typing summary(object) will not save the summary, but rather produce a nicely formatted table of a selection of the components in the summary object. Simpler summary methods may get away without a special class; e.g., the summary for factors is the value of a call to table().

This is a generic function. Functions with names beginning in "summary." will be methods for this function.

EXAMPLE

    > summary(stackfit)
    Call: lm(formula = stack.loss ~ stack.x)
    Residuals:
            Min   1Q   Median   3Q   Max
     -7.237713 -1.763111 -0.455093 2.430138 5.697774
    Coefficients:
                    Value Std. Error t value

```
       (Int.) -39.9197    11.8960 -3.3557
    Air Flow    0.7156     0.1349  5.3066
  Water Temp    1.2953     0.3680  3.5196
  Acid Conc.   -0.1521     0.1563 -0.9733
Residual standard error: 3.243364 on 17 degrees of freedom
Multiple R-Squared: 0.8346228
Correlation of Coefficients:
            (Int.) Air Flow Water Temp
    Air Flow  0.1793
  Water Temp -0.1489 -0.7356
  Acid Conc. -0.9016 -0.3389    0.0002
```

| **terms.object** | Class of Objects for Terms in a Model | **terms.object** |

An object representing the information about a structural model, as specified by a formula. This object drives all the model-fitting and is returned as a component of the fitted object by the fitting functions lm(), etc. The object itself is of mode "expression", with one element for each of the terms, after expansion and possible simplification, and not including the intercept. The object also has a number of attributes:

formula: the formula, with abbreviations expanded.

variables: a vector of mode "expression", whose elements include those in the terms.object itself, as well as the response variable.

factors: a matrix of variables by terms, showing which variables occur in which terms. The entries are 0 if the variable does not occur, 1 if it appears and should be coded by contrasts, and 2 if it appears and should be coded by dummy variables for all levels.

order: the order of the terms (1 for main effects, 2 for two-way interactions, etc.)

term.labels: a character version of the terms expression, just to save converting later on.

intercept: 1 if there is an intercept, 0 otherwise.

response: the variable number corresponding to the response (the expression on the left side of the formula).

specials: if any of the specials actually appeared in the formula, this is a list with one component for each of the specials found. The value of this component says which of the terms was one of the specials. (See the "Error" component in the example below.)

EXAMPLE

```
> terms(Y~T*Co+Error(blocks), specials="Error")
expression(T, Co, Error(blocks), T:Co)
attr(, "formula"):
Y ~ T * Co + Error(blocks)
attr(, "factors"):
                T Co Error(blocks) T:Co
           Y 0  0              0    0
           T 1  0              0    1
          Co 0  1              0    1
Error(blocks) 0  0              1    0
attr(, "order"):
[1] 1 1 1 2
attr(, "variables"):
expression(Y, T, Co, Error(blocks))
attr(, "term.labels"):
[1] "T"            "Co"           "Error(blocks)" "T:Co"
attr(, "intercept"):
[1] 1
attr(, "response"):
[1] 1
attr(, "specials"):
attr(, "specials")$Error:
[1] 3
attr(, "class"):
[1] "terms"
```

---

trace   Trace Calls to Functions   **trace**

```
trace(what, tracer=, exit=, at=)
untrace(what)
```

ARGUMENTS

what: either the name of a function, or a character vector giving the names of several functions: trace(what) adds tracing to the functions and untrace(what)
removes it. Called with no arguments, untrace() removes tracing from all
functions.

tracer: a function, the name of a function, or a logical value. tracer will be called
on entry to the function what. If TRUE, a standard tracing function is
used, which prints the call to the traced function. Omitting all arguments
except what is equivalent to tracer=TRUE.

exit: trace the function on exit, either instead of or in addition to trace on entry. If supplied, this can either be TRUE, in which case a standard tracer is used, or it can be a function to use for tracing.

at: a number or character string describing a location inside the body of the traced function at which the tracing is to be inserted. The function body must consist of a braced expression; the call to the tracing function is inserted just before the expression identified by at.

Tracing is accomplished by putting a revised version of the functions into the session database; untrace removes those versions. There is no permanent effect on the functions themselves. Each call to trace for a given function replaces any previous tracing for that function.

WARNING

Because tracing inserts a modified version of the traced function in the session database, it is dangerous to attempt to edit the traced function while tracing is in effect. Use untrace before attempting to modify a traced function.

SEE ALSO

trace.on, sys.trace, std.trace, std.xtrace, browser

SIDE EFFECTS

Two session datasets are created or updated: .Tracelist is a character vector giving the names of the functions currently being traced. .Traceon is a logical value telling whether tracing is on or off.

EXAMPLE

```
trace(stem, exit=T)  # display when stem exits
trace(eval) # display all explicit calls to eval
trace(lm, exit=browser) #insert interactive browser on exit
trace(c("sin","cos"),browser)
```

tree	Fit a Tree-Based Model	**tree**

```
tree(formula, data, weights, subset, na.action,
     control, model = F, x = F, y = F, ...)
```

ARGUMENTS

formula: a formula expression as for other regression models, of the form `response ~ predictors`. See `lm` and `formula` for details.

data: an optional `data.frame` in which to interpret the variables named in the `formula`, `subset`, and `weights` arguments.

weights: optional prior observation weights; can have zeros, although the `subset` argument is preferred for deleting observations.

subset: an expression to be used to subset the rows of the `model.frame` prior to fitting.

na.action: a missing data filter function, which gets applied to the `model.frame`.

control: a list of iteration and algorithmic constants. See `tree.control` for their names and default values. These can also be set as arguments to `tree()` itself.

model: if `TRUE`, the `model.frame` is returned. If this argument is itself a `model.frame`, then the `formula` and `data` arguments are ignored, and `model` is used to define the model.

x: if `TRUE`, the `model.matrix` is returned.

y: if `TRUE`, the response variable is returned.

...: additional arguments for the fitting routines such as `tree.control()`; typically not used.

VALUE

an object of class `tree` is returned. The model is fit using *binary recursive partitioning*, whereby the data are successively *split* along coordinate axes of the predictor variables so that at any *node*, the split which maximally distinguishes the response variable in the left and the right branches is selected. Splitting continues until nodes are pure or data are too sparse; terminal nodes are called *leaves*, while the initial node is called the *root*. If the response variable is a factor, the tree is called a *classification tree*; otherwise, it is called a *regression tree*. Can be examined by `print()`, `summary()`, and `plot()`. Contents can be extracted using `predict()`, `residuals()`, `deviance()`, and `formula()`. Can be modified using `update()`. Other generic functions that have methods for `tree` objects are `text()`, `identify()`, `inspect()`, and `[.tree()`. Use `tree.object` for further details.

EXAMPLE

```
# fit regression tree to all variables
z.solder <- tree(Count ~ ., data = solder)
# fit classification tree to data in kyphosis data frame
z.kyphosis <- tree(kyphosis)
```

tree.object	Tree-Based Model Object	**tree.object**

This class of objects is returned from the `tree()` function to represent a classification or regression tree. Objects of this class have methods for the summary functions `print()`, `summary()`, and `plot()`; for the extractor functions `predict()`, `residuals()`, and `deviance()`; and for miscellaneous diagnostic functions `[.()`, `inspect()`, `identify()`, `text()`, and `labels()`.

COMPONENTS

The following components must be included in a legitimate `tree` object. Of these, only the `where` component has the same length as the data used to fit the tree object.

frame: data frame with one row for each node in the tree. The `row.names` of `frame` contain the (unique) node numbers that follow a binary ordering indexed by node depth. Elements of `frame` include `var`, the variable used in the split at each node (leaf nodes are denoted by the string `<leaf>`); `n`, the size of each node; `dev`, the deviance of each node; `yval`, the fitted value of the response at each node; `splits`, a two-column matrix of left and right split labels for each node. Classification trees have an additional element `yprob`, the matrix containing the class probabilities for each node.

where: vector, the same length as the number of observations in the root node, containing the row number of `frame` corresponding to the leaf node that each observation falls into.

terms: an object of mode `expression` and class `term` summarizing the formula. Used by various methods, but typically not of direct relevance to users.

call: an image of the call that produced the object, but with the arguments all named and with the actual formula included as the formula argument. To reevaluate the call, say `update(tree)`.

---

update	Update a Fitted Model Object	**update**

```
update(object, formula, ..., evaluate=T, class)
```

ARGUMENTS

object: any object with a component named `call`, which is the expression used to create itself.

formula: a modeling formula, such as y $\sim$ a + b. A single dot "." on either side of the $\sim$ gets replaced by the left or right side of the formula in `object`. The dot on the left can be omitted. By default, it refits `object` using the same formula as in `object`.

....: any other arguments that are appropriate for the particular call. These must all be named, and may be abbreviated, in the same manner they could be as arguments to the fitting function itself. Arguments in the previous fit; that is, in `object$call`, can be removed by putting nothing on the right side of the `=`. For example, the argument `x=`, in a call to `update()` causes the x argument, if present in `object$call`, to be removed.

evaluate: if `TRUE` (the default), the new call is evaluated; otherwise, the call is returned as an unevaluated expression.

class: the fitting class to be used for the new object; that is, the basic fitting function, such as `lm()`, `aov()`, `glm()`, etc. This argument allows the model to be switched from one kind to another, assuming the formula and other arguments make sense for the new model. Although suggestive, `class` is a slight misnomer since the object may already inherit from this new class.

VALUE

either a new updated object, or else an unevaluated expression for creating such an object.

This is a generic function. Functions with names beginning in `"update."` will be methods for this function.

EXAMPLE

```
# refit, unchanged
update(glmob)
# refit, adding term
update(glmob, ~ . + Age)
# transform response to log scale; drop intercept
update(lmob, log(.) ~ . -1)
# use all the 2nd order interactions of previous fit
update(lmob,~ .^2)
#remove the model argument, supply a subset
update(gamob, mod=, subset = Age>25)
```

References

Aitkin, M.A., Anderson, D.A., Francis, B.J., and Hinde, J.P. (1989) *Statistical Modeling in* GLIM. Oxford University Press, Oxford.

Akaike, H. (1973) Information Theory and an Extension of the Maximum Likelihood Principle, in *Second International Symposium on Information Theory* (eds. B.N. Petrov and F. Csàki). Akademia Kiadó, Budapest, 267–281.

Baker, R.J. and Nelder, J.A. (1978) *The* GLIM *System, Release 3, Generalized Linear Interactive Modeling.* Numerical Algorithms Group, Oxford.

Bates, D.M. and Lindstrom, M.J. (1986) Nonlinear Least-squares with Conditionally Linear Parameters. *Proceedings, Statistical Computing Section,* American Statistical Association.

Bates, D.M. and Watts, D.G. (1988) *Nonlinear Regression Analysis and its Applications.* J. Wiley & Sons, New York.

Baxter, L.A., Coutts, S.M., and Ross, G.A.F. (1980) Applications of Linear Models in Motor Insurance. *Proceedings of the 21st International Congress of Actuaries,* Zurich, 11–29.

Becker, R.A., Chambers, J.M., and Wilks, A.R. (1988) *The New S Language.* Wadsworth, Pacific Grove, California.

Belsley, D.A., Kuh, E., and Welsch, R.E. (1980) *Regression Diagnostics.* J. Wiley & Sons, New York.

Bentley, J.L. (1975) Multidimensional Binary Search Trees Used for Associative Searching. *Comm. Assoc. Comp. Mach.* **18**, 509–517.

Box, G.E., Hunter, W.G., and Hunter, S. (1978) *Statistics for Experimenters.* J. Wiley & Sons, New York.

Breiman, L. and Friedman, J.H. (1985) Estimating Optimal Transformations for Multiple Regression and Correlation (with discussion). *J. Am. Statist. Assoc.* **80**, 580–619.

Breiman, L., Friedman, J.H., Olshen, R., and Stone, C.J. (1984) *Classification and Regression Trees.* Wadsworth International Group, Belmont, California.

Brinkman, N.D. (1981) Ethanol Fuel—A Single-cylinder Engine Study of Efficiency and Exhaust Emissions. *SAE Transactions* **90**, No. 810345, 1410–1424.

Buja, A., Hastie, T., and Tibshirani, R. (1989) Linear Smoothers and Additive Models (with discussion). *Annals of Statist.* **17**, 453–555.

Buta, R. (1987) The Structure and Dynamics of Ringed Galaxies, III: Surface Photometry and Kinematics of the Ringed Nonbarred Spiral NGC7531. *The Astrophysical J. Supplement Ser.* **64**, 1–37.

Carroll, R.J. and Ruppert, D. (1988) *Transformations and Weighting in Regression.* J. Wiley & Sons, New York.

Cavendish, J.C. (1975) Local Mesh Refinement Using Rectangular Blended Finite Elements. *J. Comp. Physics* **19**, 211–228.

Chambers, J.M. (1977) *Computational Methods for Data Analysis.* J. Wiley & Sons, New York.

Chambers, J.M., Cleveland, W.S., Kleiner, B., and Tukey, P.A. (1983) *Graphical Methods for Data Analysis.* Wadsworth, Pacific Grove, California.

Chou, P.A. (1988) *Applications of Information Theory to Pattern Recognition and the Design of Decision Trees and Trellises.* Ph.D. Thesis, Computer Science Department, Stanford University.

Chou, P.A., Lookabough, T., and Gray, R.M. (1989) Optimal Pruning with Applications to Tree-structured Source Coding and Modeling. *IEEE Trans. Inf. Theory* **35**, 299–315.

Ciampi, A., Chang, C-H., Hogg, S., and McKinney, S. (1987) Recursive Partitioning: a Versatile Method for Exploratory Data Analysis in Biostatistics, in *Biostatistics* (eds. I.B. MacNeil and G.J. Umphrey). D. Reidel Publishing, New York.

Cleveland, W.S. (1979) Robust Locally-weighted Regression and Smoothing Scatterplots. *J. Am. Statist. Assoc.* **74**, 829–836.

Cleveland, W.S. *The Elements of Graphing Data*, 2d edition (forthcoming). Wadsworth, Pacific Grove, California.

Cleveland, W.S. and Devlin, S.J. (1988) Locally-weighted Regression: An Approach to Regression Analysis by Local Fitting. *J. Am. Statist. Assoc.* **83**, 596–610.

Cleveland, W.S. and Grosse, E. (1991) Computational Methods for Local Regression. *Statistics and Computing*, **1**.

Cleveland, W.S. and Grosse, E. (forthcoming) *Fitting Curves and Surfaces to Data.* Wadsworth, Pacific Grove, California.

Collomb, G. (1981) Estimation Non-parametrique de la Regression: Revue Bibliographique. *International Statistical Review* **49**, 75–93.

Comizzoli, R. B., Landwehr, J. M., and Sinclair, J. D. (1990) Robust Materials and Processes: Key to Reliability. *AT&T Technical Journal* **69** (No. 6), 113–128.

Cook, R.D. and Weisberg, S. (1982) *Residuals and Influence in Regression.* Chapman and Hall, New York.

Cox, D.R. and Hinkley, D.V. (1974) *Theoretical Statistics.* Chapman and Hall, London.

Dennis, J.E., Jr. and Schnabel, R.B. (1983) *Numerical Methods for Unconstrained Optimization and Nonlinear Equations.* Prentice-Hall, Englewood Cliffs, New Jersey.

Dongarra, J.J., Bunch, J.R., Moler, C.B., and Stewart, G.W. (1979) LINPACK *User's Guide* SIAM, Philadelphia.

Draper, N.D. and Smith, H. (1981) *Applied Regression Analysis*, 2d edition. J. Wiley & Sons, New York.

Federer, W.T. (1955) *Experimental Design Theory and Applications.* The Macmillan Company, New York.

Fisher, W.D. (1958) On Grouping for Maximum Homogeneity. *J. Am. Statist. Assoc.* **53**, 789–98.

Friedman, J.H. and Stuetzle, W. (1981) Projection Pursuit Regression. *J. Am. Statist. Assoc.* **76**, 817–823.

Gay, D.M. (1983) Algorithm 611: Subroutines for Unconstrained Minimization Using a Model/Trust-Region Approach. *ACM Trans. Math. Software* **9**, 503–524.

Gelernter, D. and Jagannathan, S. (1990) *Programming Linguistics.* The MIT Press, Cambridge, Massachusetts.

Golub, G.H. and Pereyra, V. (1973) The Differentiation of Pseudo-inverses and Non-linear Least-squares Problems Whose Variables Separate. *Journal of SIAM* **10**, 413–432.

Golub, G.H. and van Loan, C.F. (1989) *Matrix Computations*, 2d edition. Johns Hopkins University Press, Baltimore.

Hastie, T. and Pregibon, D. (1990) Shrinking Trees. *AT&T Bell Laboratories Tech. Report.*

Hastie, T. and Tibshirani, R. (1990) *Generalized Additive Models.* Chapman and Hall, London.

Healy, M.J.R. (1988) GLIM: *An Introduction.* Clarendon Press, Oxford.

Heiberger, R.M. (1989) *Computation for the Analysis of Designed Experiments.* J. Wiley & Sons, New York.

Hicks, C.R. (1973) *Fundamental Concepts in the Design of Experiments.* Holt, Rinehart and Wilson, New York.

Kalbfleisch, J.G. (1979) *Probability and Statistical Inference: Vol. 2: Statistical Inference.* Springer-Verlag, New York.

Kass, G.V. (1980), An Exploratory Technique for Investigating Large Quantities of Categorical Data. *Applied Statistics* **29**, 119–127.

Keene, S.E. (1989) *Object-Oriented Programming in COMMON LISP: A Programmer's Guide to CLOS.* Addison-Wesley, Reading, Massachusetts.

Lambert, D. (1991) Zero Inflated Poisson Regression, with an Application to Defects in Manufacturing. *Technometrics (forthcoming).*

Lawson, C.L. and Hansen, R.J. (1974) *Solving Least-Squares Problems.* Prentice-Hall, Englewood Cliffs, New Jersey.

Lehmann, E. L. (1986) *Testing Statistical Hypotheses.* J. Wiley & Sons, New York.

Macauley, F.R. (1931) *The Smoothing of Time Series.* New York: National Bureau of Economic Research.

Mallows, C.L. (1973) Some Comments on C_p. *Technometrics* **15**, 661–667.

McCullagh, P. and Nelder, J.A. (1989) *Generalized Linear Models*, 2d edition. Chapman and Hall, London.

McLain, D.H. (1974) Drawing Contours from Arbitrary Data Points. *Computer J.* **17**, 318–324.

Morgan, J.N. and Messenger, R.C. (1973) *THAID—A Sequential Analysis Program for the Analysis of Nominal Scale Dependent Variables.* Survey Research Center, Institute for Social Research, University of Michigan.

Phadke, M.S., Kackar, R.N., Speeney, D.V., and Grieco, M.J. (1983) Off-line Quality Control in Integrated Circuit Fabrication Using Experimental Design. *Bell System Technical Journal* **62**, 1273–309.

Pregibon, D. (1982) Resistant Fits for Some Commonly Used Logistic Models with Medical Applications. *Biometrics* **38**, 485–498.

Reade, C. (1989) *Elements of Functional Programming.* Addison Wesley, Wokingham, England.

Searle, S.R. (1971) *Linear Models.* J. Wiley & Sons, New York.

Seber, G. A. F. and C. J. Wild (1989) *Nonlinear Regression.* John Wiley & Sons, New York.

Sonquist, J.N. and Morgan, J.N. (1964) *The Detection of Interaction Effects.* Monograph 35, Survey Research Center, Institute for Social Research, University of Michigan.

Stone, C.J. (1977) Consistent Nonparametric Regression. *Ann. of Stat.* **5**, 595–620.

Stroustrup, B. (1986) *The C++ Programming Language.* Addison-Wesley, Reading, Massachusetts.

Tierney, L. (1990) *LISP-STAT: An Object-Oriented Environment for Statistical Computing and Dynamic Graphics.* J. Wiley & Sons, New York.

Wahba, G. (1990) *Spline Functions for Observational Data.* CBMS-NSF Regional Conference series. SIAM, Philadelphia.

Watson, G.S. (1964) Smooth Regression Analysis. *Sankhya* A **26**, 359–372.

Weisberg, S. (1980) *Applied Linear Regression.* J. Wiley & Sons, New York.

Whittaker, E. (1923) On a New Method of Graduation. *Proceedings of the Edinburgh Mathematics Society* **41**, 63–75.

Wilkinson, G.N. and Rogers, C.E. (1973) Symbolic Description of Factorial Models for Analysis of Variance. *Applied Statistics* **22**, 392–399.

Index

The index is arranged according to topics as well as distinctive words and phrases. Most chapters have at least two topic headings; e.g., *Linear models* and *Linear model functions*. The former lists some key topics in the linear models chapter and augments the table-of-contents information; the latter lists all those functions commonly used in fitting and examining that class of models.

A complete list of individual functions referenced in the book can be found under the index heading *Functions*. Page numbers in **bold** font refer to the detailed documentation in Appendix B. Documentation for all functions can be obtained online using ?; see the *Documentation* entry below.